The McGraw-Hill
Illustrated Encyclopedia of
Robotics & Artificial
Intelligence

Other books by the author

Amateur Radio Encyclopedia
The Concise Illustrated Dictionary of Science and Technology
Hot ICs for the Electronics Hobbyist
The Illustrated Dictionary of Electronics—6th Edition
International Encyclopedia of Integrated Circuits
International Encyclopedia of Integrated Circuits—2nd Edition
More Puzzles, Paradoxes and Brain Teasers
Optical Illusions: Puzzles, Paradoxes and Brain Teasers
Teach Yourself Electricity and Electronics

With Neil Sclater

Encyclopedia of Electronics—2nd Edition

The McGraw-Hill Illustrated Encyclopedia of Robotics & Artificial Intelligence

Stan Gibilisco
Editor in Chief

McGraw-Hill, Inc.

New York San Francisco Washington, D.C. Auckland Bogotá
Caracas Lisbon London Madrid Mexico City Milan
Montreal New Delhi San Juan Singapore
Sydney Tokyo Toronto

pbk 1 2 3 4 5 6 7 8 9 0 FGR/FGR 9 9 8 7 6 5 4
hc   1 2 3 4 5 6 7 8 9 0 FGR/FGR 9 9 8 7 6 5 4

Library of Congress Cataloging-in-Publication Data

Gibilisco, Stan.
   The McGraw-Hill illustrated encyclopedia of robotics and artificial intelligence / by Stan Gibilisco.
     p.   cm.
   Includes index.
   ISBN 0-07-023613-5 (H)   ISBN 0-07-023614-3 (pbk.)
   1. Robotics—Encyclopedias. 2. Artificial intelligence-
-Encyclopedias. I. Title. II. Title: Illustrated encyclopedia of
robotics and artificial intelligence.
TJ210.4.G53 1994
629.892'03—dc20   94-14309
                                   CIP

Acquisitions editor: Roland S. Phelps
Editorial team: Joanne Slike, Executive Editor
               Andrew Yoder, Supervising Editor
               Melanie Holscher, Book Editor
Production team: Katherine G. Brown, Director
               Tina M. Sourbier, Coding
               Rose McFarland, Desktop Operator
               Stephanie Myers, Computer Artist
               Cindi Bell, Proofreading
               Jodi Tyler, Indexer
Design team: Jaclyn J. Boone, Designer
               Brian Allison, Associate Designer
Cover design: Lori E. Schlosser
Cover photo: Brent Blair, Harrisburg, Pa.
Cover copy: Michael Crowner

                                            EL1
                                        0236143

To Tim and Tony
from Uncle Stan

Contents

Introduction

This is a concise reference about robotics and artificial intelligence (AI) for computer/electronics hobbyists, students, and anybody else who wants to know what's happening in these fields.

Computers and robots are here to stay. We depend on them every day. Usually we don't even notice them (until they break down). We will get more used to them, and more reliant on them, as the future unfolds.

To find information on a subject, look for it as an article title. If your subject isn't an article title, you'll probably find it in the index.

This book is meant to be precise, but without much math or jargon. It's written with one eye on today, and the other eye on tomorrow. Illustrations are functional, not literal; they're drawn with the intention of showing, clearly and simply, how things work.

Will robots ever mow your yard, cook your food, or shovel snow? Will a computer ever compose good music, or write a great novel? Might we give the fate of Wall Street entirely over to AI? Maybe. People in the United States, Japan, and Europe are trying to design systems that will do all these things—and much more.

Don't worry though. There'll always be plenty of things left for people to do. Robots can play with us and never get tired.

I hope you put this book down with more questions in your mind than when you picked it up.

Suggestions for future editions are welcome.

Stan Gibilisco
Editor in Chief

active chord mechanism

An *active chord mechanism (ACM)* is a gripper that adapts itself to the shapes of irregular objects. An ACM is built similar to your backbone. A typical ACM consists of numerous small, rigid structures connected by hinges, as shown in the illustration.

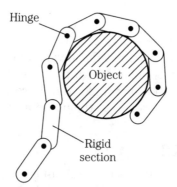

ACTIVE CHORD MECHANISM
Many rigid parts are hinged together.

    The precision with which an ACM can conform to an irregular object depends on the size and number of sections the ACM consists of. The smaller the sections, the greater the precision. An ACM exerts uniform pressure all along its length. This pressure can be increased or decreased, according to the task that needs to be done. The ACM grabs onto an object in much the same way as a snake does.

    One application of ACMs is to position or arrange fragile objects without damaging them. Another application might be the picking of fruits and vegetables. It has even been suggested that ACMs could be used to turn hospital patients in their beds, so that they don't lie for too long in one position. This might be all right if the patient is unconscious, but an awake person would probably rather be touched and moved by a human being, not a snake-like robot gripper. See also ROBOT GRIPPERS.

Adaptive Suspension Vehicle

Most robots move on wheels, because wheeled robots are far easier to design, construct, and control than machines with mechanical legs. When the locomotive ("iron horse") was first developed in the 1800s, very few engineers actually thought that it should take the form of a mechanical horse, with four moving, jointed legs. (Imagine one of these things galloping at 60 miles per hour!) The wheel had been invented thousands of years ago, and was the obvious way to go. The same is often, but not always, true of robot vehicles.

One type of robot under development at Ohio State University, known as the *Adaptive Suspension Vehicle (ASV)*, does use mechanical limbs to propel itself. It moves on six robot legs, like a gigantic insect trudging over the countryside. Six is the most common number of legs for "iron horse" type robots. This choice is based partly on the fact that insects have six legs, and they have done very well with this natural design. Six legs seem to provide optimum stability and maneuverability. See INSECT ROBOT, ROBOT LEGS.

The ASV can carry about 700 pounds and it moves at about five miles per hour (seven feet per second). The machine itself weighs nearly three tons. In terms of bulk, therefore, it is akin to a small truck. And, like a truck, it is designed to carry a driver or rider.

Although it is a big challenge to develop the software and hardware to operate a robot with legs, rather than wheels, there is also a big payoff. A "walking" robot can move over much rougher terrain than any vehicle with wheels. Such robots could be useful, for example, on the boulder-strewn plains of the moon, Mars, or any planet with rugged terrain. And, of course, there are plenty of places on our own planet that would lend themselves to the use of an ASV.

Advanced Robot Technology

Advanced Robot Technology (ART) is the name of a research-and-development program undertaken in Japan during the 1980s. The ART robots are designed for work that is dangerous for humans. This includes repairing nuclear power plant equipment and maintaining undersea oil rigs. Machines are far less sensitive than people to radiation and toxic chemicals encountered during these kinds of work. And if a machine does get damaged, it can be repaired or replaced without human suffering.

Originally, the Japanese engineers wanted to design and build the ART robots to be completely *autonomous* or independent (see AUTONOMOUS ROBOTS). But 100-percent autonomy has proven difficult. It requires that each robot have a highly sophisticated computer, similar or identical to the computers in other robots. Such redundant hardware costs money to produce, build, and maintain.

Simple, repetitive tasks are programmed into individual ART robots' electronic memory, but complicated functions use a central computer or remote control by a human operator. The ART robots are therefore somewhat like ants in an anthill: They have some intelligence of their own, but are interconnected by a sort of "central nervous system" or "main brain" for the complex work.

Remote control is sometimes done by means of telepresence (see TELEPRESENCE). This lets human operators work as if they are actually the robots. Control signals and telemetry are sent and received on laser beams. If the robots are in an undersea environment, optical fibers are used (see FIBEROPTIC DATA TRANSMISSION, LASER DATA TRANSMISSION).

The ART robots move around mainly by the use of legs, rather than wheels, because of the terrain in which they must work. See ADAPTIVE SUSPENSION VEHICLE, ROBOT LEGS.

AGV

See AUTOMATED GUIDED VEHICLE.

AIMS

See AUTOMATED INTEGRATED MANUFACTURING SYSTEM.

algorithm

An algorithm is a precise, step-by-step procedure by which a solution is found to a problem. Algorithms can usually be shown in flowchart form (see FLOWCHART). All computer programs are algorithms. Proofs in mathematical logic and geometry, such as the kind taught in high school, are algorithms. Robots perform their work by following algorithms that tell them exactly where and when to move.

In an algorithm, an individual step might not seem to lead anywhere in particular. There might appear to be no good reason for some steps. But in an efficient algorithm, every step is vital even if it seems to sidetrack or backtrack.

An algorithm can contain one or more *loops*, or places in the problem-solving process where a certain group of steps is repeated many times. This is especially common in computer programming (see LOOP).

Algorithms are sometimes found in the service manuals for machinery or electronic equipment. Troubleshooting processes can be written in algorithm form. Then, the technician simply "plays computer," or follows the instructions with unquestioning rigor. At the end of the process, the problem has been found and a solution is prescribed.

Algorithms always contain a finite number of steps. (A machine can't do an infinite number of steps!) Each step must be precisely expressible, and it must be possible for a machine to do it. A loop must never continue indefinitely; the whole process must be completed in a finite length of time. Algorithms that once would have taken people thousands or millions of years (and therefore were practically impossible) can now be done by computers in a few seconds. Algorithms that are beyond the abilities of today's large computers will probably be done quickly and easily by machines 50 years from now.

People often solve problems using intuition, without any apparent algorithm. But some scientists think the human brain is just a super-sophisticated computer. If that is true, then all of our thoughts and actions are algorithms. Also, if our minds really work in discrete steps (even if there are billions and billions of steps every second), then it is theoretically possible to build a machine that can think, learn, feel, hope, love, and even dream, in exactly the same way we do. See also ARTIFICIAL INTELLIGENCE, ARTIFICIAL LIFE.

Alpha

A simple, miniature industrial/instructional robot, resembling an arm with a hand at the end, has been developed by Microbot, a company in California. Known as *Alpha*, it extends to about a foot and a half, or 0.5 meter. It can lift objects weighing up to about a pound and a half (700 grams).

Alpha can be programmed using a home computer. All of the motors for the robot are in the base. A motor operates the "shoulder" directly; cables and pulleys are used to move the "elbow," "wrist," and "hand." This principle is illustrated in the drawing. Putting the motors in the base, rather than having individual motors at the joints, reduces the weight of the robot arm itself, so that it can lift more weight with smaller motors. This improves efficiency, minimizes bulk, and keeps the cost down. See also INDUSTRIAL ROBOTS, ROBOT ARMS, ROBOT GRIPPERS.

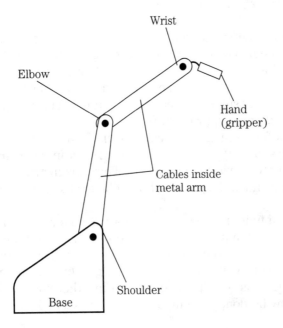

ALPHA
Principle of operation for Alpha robot.

American National Standards Institute

The American National Standards Institute, or ANSI, is an industrial group in the United States that encourages companies to make machinery, electronic devices, computer programs, connectors, and other apparatus compatible.

When components and procedures are standardized, things work more smoothly than they do when things don't match. Perhaps you've had the frustration of buying a flashlight bulb with a bayonet-type base, when your flashlight needed a screw-in bulb. Or maybe you couldn't find a replacement gas cap for your car's fuel tank. Perhaps you've worked with hi-fi audio equipment and bought the wrong type of connectors for your speakers. The list goes on and on.

Sometimes, a company builds devices with parts that are custom made, so that when you need a replacement part, you can only get it from that company. This might at first seem like a good way for a company to maximize its profits. After all, if you must go to that company to get a replacement part, you can't spend your money anywhere else! But suppose competing companies join together and agree to provide mutually interchangeable parts. They end up with better sales than the company with the specialized parts. This is because, when replacement parts are widely available, repair service is better and cheaper. Word gets around about things like that.

The job of ANSI is to help companies decide on the specifications that standard components should have. Thus, you will commonly see resistors rated at $\frac{1}{4}$, $\frac{1}{2}$ and 1 watt, but never at $\frac{1}{7}$, $\frac{1}{3}$ or $\frac{5}{6}$ watt. Electronic component values, cable impedances, voltages for computer chips, and other parameters are standardized among all companies in this way.

amusement robots

Robots used for entertainment purposes are known as *amusement robots*. These devices are sometimes employed by companies to show off new products and to attract customers. They are common at trade fairs, especially in Japan. Although they are usually small in size, they can sometimes have sophisticated software (see SOFTWARE).

An example of an amusement robot is a mechanical "mouse" that navigates a maze. The simplest such device bumps around randomly until it finds its way out by sheer luck. A more sophisticated robot "mouse" might move along one wall of the maze, say the wall to its right, until it emerges. This technique will work with most mazes (see the drawing).

The most advanced amusement robots include *androids*, or machines with a human appearance. Robots of this type might greet customers in stores, operate elevators, or demonstrate products at conventions. Some amusement robots can accommodate human riders. Others are more accurately called toys. See also ANDROID, PERSONAL ROBOTS.

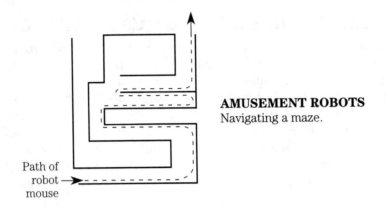

AMUSEMENT ROBOTS
Navigating a maze.

Path of
robot
mouse

analog

The term *analog* is used to denote a variable or quantity that can attain an infinite number of values within a certain range. This is in contrast to *digital* variables or quantities, that can have only a finite number of discrete values within a given range.

You can move freely around a room, varying your position to any point in a region; you have the capability of analog motion. Your arms can move to an infinite number of positions (in theory, anyway), in a fluid and continuous way, in a certain region of space. This, too, is analog motion. Many robots, however, can move only to certain points along a line, on a plane, or in space. Their motion is digital. There are some robots that can move in an analog fashion, but the necessary hardware is generally more complicated than for digital motion.

The illustration shows an example of analog motion in a plane. See also DIGITAL.

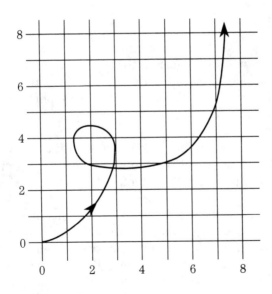

ANALOG
An analog path in a plane.

analytical engine

A primitive calculating machine, designed by Charles Babbage in the 1830s, was called the *analytical engine*. Babbage never quite completed the task of building this device to perfection. The idea was to employ punched cards to make, and print out, calculations, in a manner similar to digital computers of the 1960s. Babbage is considered to be the first engineer to work on a true digital calculator (see BABBAGE, CHARLES).

One of the main problems for Babbage was that electricity was not available. The machines had to use mechanical parts exclusively. These wore out with frequent, repetitive use.

Another problem was that Babbage liked to dismantle things completely in order to start over with new designs, rather than saving his old machines to keep their shortcomings in mind when designing new ones.

During the research-and-development phase of the analytical engine, some people thought that machine intelligence had finally been discovered. The Countess of Lovelace even went so far as to write "software" for Babbage's machine, imagining it to be capable of playing board games. From the analytical engine's finite number of digital states, she thought, an infinite variety of concepts could be synthesized. The concrete (machines) and the abstract (the human mind) had, she imagined, finally been linked. This was, of course, an overstatement. Even today, we are a long way from designing a machine that shows all the idiosyncrasies of a human being's mind.

Mere *calculation* is just a tiny subset of *analysis*; this in turn is only a small slice of the landscape of *thought*. Babbage's machine was important because it represented a change in attitudes toward machines. Rather than dismissing artificial intelligence as preposterous, people began to believe that it might be attainable. See also ARTIFICIAL INTELLIGENCE.

AND gate

An AND gate is a digital logic circuit with two or more inputs and one output. It performs the logical AND operation, also known as conjunction.

The output of an AND gate is high, or logic 1 ("true"), if and only if all of the inputs are high. If any of the inputs are low, or logic 0 ("false"), then the output is low.

Logic gates are extensively used in digital computers and other electronic devices.

The schematic symbol for an AND gate, along with its logical truth table, is shown in the figure. See also EXCLUSIVE-NOR GATE, EXCLUSIVE-OR GATE, INVERTER, NOR GATE, OR GATE.

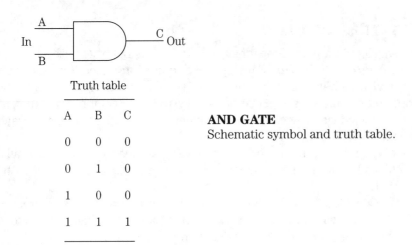

Truth table

A	B	C
0	0	0
0	1	0
1	0	0
1	1	1

AND GATE
Schematic symbol and truth table.

Androbot

Androbot is the name of a company in San Jose, California that has historically manufactured personal robots. The robots manufactured by Androbot have generally been androids, or quasi-human machines.

Androbot was founded by Nolan Bushnell in 1981. He also founded Atari, a manufacturer of computer games popular with young people, in the 1970s. See AMUSEMENT ROBOTS, ANDROID, PERSONAL ROBOTS.

android

An *android* is a robot, often very sophisticated, that takes a more or less human form. An android usually propels itself by rolling on small wheels in its base. The technology for fully functional arms is under development, but the software needed for their operation has not been made cost-effective for small robots. Legs are hard to design and engineer, and they aren't really necessary; wheels work well enough. Elevators can be used to allow a rolling android to get from floor to floor in a building.

An android often has a rotatable head equipped with position sensors. Binocular, or stereo, vision allows the android to perceive depth, thereby locating objects anyplace within a large room. Speech recognition and synthesis are becoming common (see OBJECT RECOGNITION, POSITION SENSING, SPEECH RECOGNITION, SPEECH SYNTHESIS, VISION SYSTEMS).

Because of their quasi-human appearance, androids are especially suited for use wherever there are children. Androids, in conjunction with computer terminals, might someday replace school teachers in some situations. It is unlikely that the teaching profession will suffer because of this; the opposite will probably be the case. There will be a demand for people to teach children how to use the robots! Robots will free human teachers to spend more time in areas like humanities and philosophy, while robots instruct students in computer programming, geography, and other rote-memory subjects. Thus machine teachers could eventually help to raise children to be more sensitive and compassionate adults—not emotionless creatures, as some people fear.

No android has yet been conceived (even on the trendiest drawing board) that could be mistaken for a person, as has been depicted in science-fiction books and movies. It's fun to debate whether or not there will ever be a robot as advanced as "Data" in *Star Trek: The Next Generation.* See also AMUSEMENT RO-BOTS, PERSONAL ROBOTS.

animism

People in some countries, notably Japan, believe that the force of life exists in things like stones, lakes, and clouds, as well as in people, animals, and plants. This belief is called *animism.*

One of the most significant questions in the field of artificial intelligence (AI) is: "Will we ever be able to build a machine that is actually alive?" Along with this question, there is the more philosophical one: "What, exactly, makes something living different than something nonliving? How do animate things differ from inanimate things?"

As early as the mid-nineteenth century, a machine was conceived that was thought to have some sense of being animate. This was Charles Babbage's analytical engine (see ANALYTICAL ENGINE). At that time, very few people seriously thought that a contraption made of wheels and gears could have life. But today's massive computers, and the promise of more sophisticated ones being built every year, have brought the question out of the realm of science fiction.

Computers can do things that people cannot. For example, even a simple personal computer (PC) can figure out the value of pi, the ratio of a circle's circumference to its diameter, to 1000 decimal places. Large industrial computers have precision and speed beyond most people's imaginations, let alone their abilities. Robots can be programmed to do things as complicated as figuring out how to get through a maze, or how to rescue a person from a burning building. Is this life? Are such robots truly animate?

In recent years, AI has progressed to the point that computers can learn from their mistakes, so that they won't make any particular error more than once. This is what makes a machine "intelligent." But is intelligence the same thing as life? Most American scientists think not.

Computers can be programmed to have what seem to be emotional responses. With speech synthesis, computer voices can sound happy, sad, or angry (see SPEECH SYNTHESIS). Perhaps someday, a robot will be designed so that it smokes, rattles, jumps around, or shuts itself off for awhile when it's "angry" or "happy" or "sad." But are these really emotions? Even if they are, does this make the robot truly alive?

The debate over whether machines can be animate has been going on for awhile now, and it will certainly continue in the United States. But for some Japanese people, the question never had to be asked in the first place. They believe that computers and robots have life, at least in a sense, because all machines are made from parts that are individually animate. See also ARTIFICIAL INTELLIGENCE, ARTIFICIAL LIFE.

anthropomorphism

Sometimes, machines or other objects have characteristics that seem human-like to us. This is especially true of advanced computers and robots. We make an anthropomorphism when we think of a computer or robot as being like a person.

Robots with humanoid form (such as the kind that have a head and arms) are easy to anthropomorphize. Science-fiction movies and novels have made use of anthropomorphisms extensively. One good example is the robot "R2D2" in *Star Wars*.

A science-fiction example of an artificial-intelligence anthropomorphism is found in *2001: A Space Odyssey*. The space ship was controlled by "Hal," a computer that tried to kill Dave, an astronaut, by refusing to let his shuttle into the main ship. Dave braved the vacuum of space without a pressure suit, thereby getting back into the ship and disabling Hal.

Are machines really capable of becoming like people? Can they ever have a sense of self, an identity? Some engineers believe that sophisticated robots and computers already have this quality, because they can optimize problems and/or learn from their mistakes. But others contend that the criterion for "being alive" is far more strict, and is best left to priests and biologists to debate.

Owners of personal robots sometimes think of the machines as companions. In one sense, at least, robots really are like people: You can grow fond of them. See also ANIMISM, ARTIFICIAL INTELLIGENCE, ARTIFICIAL LIFE, PERSONAL ROBOTS.

Anticipatory Sciences

Anticipatory Sciences is a group of futurists (people who make predictions about the future of technology). They have made some interesting forecasts in the field of artificial intelligence (AI).

Some futurists think that scientific advances will keep taking place at a faster and faster rate in coming years. This has been the rule during the 20th century. It has been suggested that AI might soon reach a critical point, where machines become able to do independent research! Then, a computer/robot system could literally conceive, design, and build more advanced computer/robots. To some, this qualifies as artificial life.

If machines like this ever actually evolve, our homes might be fully automated; cars might be driven by computers, practically eliminating accidents; and even skilled jobs might become obsolete. One rather bizarre scenario is a sports league in which the players are robots instead of people. Different corporations could form teams with machines of their own design, playing them off against each other, thereby determining which designs are best, in an entertaining way.

But many futurists are skeptical. They doubt that machines will ever truly become able to evolve the way that biological life has. They say that no machine can ever be smarter than its makers. They cling to the belief that only God can create life.

Still another group of futurists is afraid of what might happen if robots become able to evolve on their own. Will they become peace-loving, and teach us humans how to get along without wars? Or will evil somehow get hold of them, causing them to take over the world and rule it with (literally) cold metal fists?

Time will tell who is right. Most likely, the truth will prove to be different from, and more fascinating than, anything scientists have ever imagined. See also ARTIFICIAL INTELLIGENCE, ARTIFICIAL LIFE.

antivirus program

Sometimes, computers malfunction because of a "bug," or defect, known as a *virus* (see VIRUS). Computer viruses can cause sudden, total failure of a system. But more often, they just degrade the performance. (See CATASTROPHIC FAILURE, GRACEFUL DEGRADATION). You might not even notice some computer viruses. In artificial intelligence (AI) systems, viruses have the effect of making the system less "smart," and/or causing erratic behavior.

A computer virus can be detected before it erases vital data in a system. This is done by means of an antivirus program.

Many computer viruses give their presence away by certain changes in the data. An antivirus program can be set up so that it recognizes these signs. If "symptoms" of a virus are caught, the program finds the virus and literally eradicates it.

Another way that an antivirus program can work is by checking the data to see if any of it has been damaged or erased. A "positive" result causes the program to search out and erase the virus.

Some antivirus programs have a list of specific viruses that have been found in other systems in the past. The program goes through this list, checking the computer data to see if any of the known culprits are there. If one is found, it is erased.

Apprentice robot

The *Apprentice robot* is a robot designed and built by a company called *Unimation, Inc.* As its name implies, the Apprentice serves as an assistant to human workers.

Unimation's Apprentice robot is used to help in the assembly of other robots. This is an important step in the evolution of artificial life (see ARTIFICIAL LIFE), in which robots might someday be able to build "offspring." The Apprentice itself cannot build robots; the human operator must guide it. See also UNIMATION, INC.

ARPAnet

ARPAnet is an acronym for a computer mail network run by the *Advanced Research Projects Agency* (ARPA). This agency is part of the U.S. Department of Defense. Using ARPAnet, scientists and graduate students exchange ideas, information, and opinions in many fields of science. The ARPAnet is heavily used by robotics and computer researchers.

In a computer bulletin board system (BBS), you can log in, or access the system, using a personal computer and a device called a modem that allows the computer to operate as a communications terminal through the telephone lines. There are thousands and thousands of messages constantly being exchanged via computer BBSs. Perhaps you have used one of them. Bulletin boards exist for practically any subject you can think of. Robotics and Artificial Intelligence are two very large categories, each with various subcategories.

Using ARPAnet, researchers discuss things as diverse as whether there will ever be a computer with emotions, or whether we ought to use robots as soldiers. Some of the best innovations begin as exchanges among scientists and students on ARPAnet. See also BULLETIN BOARD.

ART

See ADVANCED ROBOT TECHNOLOGY.

articulated geometry

Robot arms can move in various different ways. Some can attain only certain discrete, or definite, positions, and cannot stop at any intermediate position. Others can move in smooth, sweeping motions, and are capable of reaching to any point within a certain region. See ROBOT ARMS.

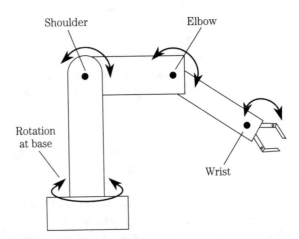

ARTICULATED GEOMETRY A jointed robot arm.

One method of robot arm movement is called *articulated geometry*. The word *articulated* means broken into sections by joints. This type of robot arm resembles the arm of a human. The versatility is defined in terms of the number of degrees of freedom (see DEGREES OF FREEDOM). There might, for example, be *base rotation, elevation*, and *reach*. There are several different articulated geometries for any given number of degrees of freedom. The illustration shows one scheme for a robot arm that uses articulated geometry.

Other geometries that can move in two or three dimensions are discussed under the titles CARTESIAN COORDINATE GEOMETRY, CYLINDRICAL COORDINATE GEOMETRY, POLAR COORDINATE GEOMETRY, and SPHERICAL COORDINATE GEOMETRY.

artificial experience

See VIRTUAL REALITY.

artificial intelligence

The term *artificial intelligence*, abbreviated AI, refers to computers that mimic aspects of human thought. A simple electronic calculator doesn't have AI. But a machine that can learn from its mistakes, or that can show reasoning power, does have AI. Between these extremes, there is no precise dividing line.

As computers have gotten more and more powerful, people have set higher and higher standards for AI. Things that were once thought of as AI are now quite ordinary. And things that seem fantastic now will someday be just humdrum. There is a tongue-in-cheek axiom about AI: Something is AI only as long as it's new and strange.

Relationship with robotics

Artificial intelligence lends itself to robotics. Scientists have dreamed for over a century about building "smart" androids, robots that look and act like people. Androids already exist, but they aren't very "smart." See ANDROID.

If a machine has the ability to move around under its own power, to lift things, and to move things, it seems reasonable that it should do so with some degree of "smarts," if it is to be able to accomplish anything worthwhile. Otherwise it would be just a bumbling idiot box, and it might be dangerous, like a driverless car with a brick on the gas pedal.

If a computer is to manipulate anything with its "brain power," it will need to be able to move around, to grasp, to lift, and to carry objects. It might contemplate fantastic exploits, but if it can't act on its thoughts, the work (and the risk) must be undertaken by people, whose strength and maneuverability (and courage) are limited.

Robots without any intelligence, or electronic brains without moving parts, have various uses and abilities. But when robots are given AI, their power multiplies.

Playing games

Computers have been programmed to play games, particularly mathematically oriented games such as checkers and chess. With checkers, AI has proven to be extremely adept. But with chess, the results have been less spectacular. See CHECKERS PLAYING MACHINE, CHESS PLAYING MACHINE.

Composing music

Computers can be used to process music in many different ways. Anyone who listens to pop music, such as progressive rock, has heard computer-enhanced sound. But computers can also generate their own melodies.

Music generated entirely by a computer has a lackluster sound, according to musicians and well-seasoned music listeners. There are many variables in music, some of which communicate the feelings of the composer. If the computer has no emotion, then the melody can't convey any.

Pitch, loudness, and timbre are the three characteristics of music that most of us notice right away. But there are many other characteristics of music, such as rhythm, speeding-up and slowing-down, changes in loudness, pauses, vibrato, and the list goes on and on. A good piece of music is as sophisticated as a good verbal story. No one has even tried to have a computer write a story yet. We probably won't see good computer-composed music until AI is a lot further along than it is today. See COMPUTER EMOTION, COMPUTER MUSIC.

Proving theorems

One measure of computer intelligence, that works on a level somewhere between intuition and brute-force logic, is the proving of mathematical theorems. If you have taken high-school geometry, you've probably been exposed to theorem-proving. Elementary logic courses deal with it, too. And computer programming is a type of reasoning similar to theorem-proving, raising the question, "Will computers ever be able to program themselves?"

Programs in AI have sometimes found remarkable proofs in mathematics. One theorem states that the base angles in an isosceles triangle have equal measures (see the illustration).

The traditional proof cuts the isosceles triangle vertically down the middle. Let the triangle ABC have sides with lengths AB = AC. By "dropping a perpendicular," AD, from the apex to the base, two right triangles are created. It turns out that these two triangles are exact mirror images of each other, and therefore, the angles ∠ABC and ∠ACB have equal measure.

An AI program found a way to prove this theorem without cutting up the triangle. Consider the triangle ABCA (going around counterclockwise) and also another triangle ACBA (going around clockwise). These triangles are *congruent*, meaning you can lie one down right on top of the other, because the lengths of corresponding sides are all equal: AB = AC (you're given this fact to start with), BC = CB (it's the same line segment traveled opposite ways), CA = BA (again,

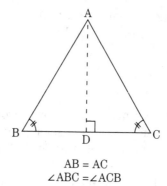

ARTIFICIAL INTELLIGENCE
Isosceles triangle theorem.

AB = AC
∠ABC = ∠ACB

you're given this fact; now you're just looking at the line segments backwards). This means that the triangle ABC is precisely the same if you flip it over or look at it in a mirror. Because corresponding angles of congruent triangles have equal measure, ∠ABC and ∠ACB are identical. Proof complete.

The mathematician Pappus found this proof almost 2000 years ago. It's not the one commonly taught. Could a program have been written that gave a computer the mathematical IQ of Pappus? Skeptics said no. The program just reflected the knowledge of its author, even though he hadn't been thinking about the isosceles-triangle theorem when he wrote the program.

The program didn't repeat this stroke of apparent genius. It didn't find any new, exciting proofs for other theorems. If it had, it would have been a far more convincing demonstration of true AI.

Can machines get smarter than people?

Experts in the field of AI have been rather disappointed in the past couple of decades. Computers have been designed that can handle tasks no human could ever contend with, such as navigating a space probe or making billions of mathematical calculations. Machines have been built that can play some mathematical games well enough to compete with human experts. Modern machines can understand, as well as synthesize, much of any spoken language. But these abilities, by themselves, don't count for much in the dreams of scientists who hope to create true artificial life (see ARTIFICIAL LIFE).

The human mind is incredibly complicated. A circuit that would have occupied, and used all the electricity in, a city in 1940, can now be housed in a box the size of an aspirin and run by a battery. Imagine this degree of miniaturization happening again, and then again, and then again! Would that begin to approach the level of sophistication in your body's nervous system? Perhaps.

A human brain might be nothing more than a digital switching network, but no electronic device yet conceived has come anywhere near its level of intelligence. Some experts think that a machine might someday be built that is smarter than a human. But most concede that if it's ever done, it won't be for a very long time. See COMPUTER CONSCIOUSNESS, COMPUTER REASONING.

There are many articles in this encyclopedia that deal with AI. For information about specific topics not discussed or cross-referenced here, please refer to the appropriate article title, or to the index.

artificial life

What, exactly, comprises something living, that makes it different from something nonliving? This has been one of the great questions of science throughout history. In some cultures, life is ascribed to things that Americans think of as inanimate (see ANIMISM).

One definition for artificial life involves the ability of a human-made thing to reproduce itself. Suppose you were to synthesize a new kind of molecule in a beaker, and named it QNA (that stands for some weird name nobody can ever remember). Suppose this molecule, like DNA, could make replicas of itself, so that when you put one of them in a glass of water, you'd have a whole glass full in a few days. This molecule would be artificial life, in the sense that it could reproduce, and that it was made by humans rather than by nature in general.

You might build a robot that could assemble other robots like itself. The machine would also be artificial life according to the above definition. It would, of course, be far different than the QNA molecule in terms of size and environment. But reproduction ability is the basis for the definition. The QNA molecule and the self-replicating robot both meet the definition. Interestingly, however, both of these things reproduce asexually; that is, they do not need to mate in order to have offspring.

A truly self-replicating robot hasn't yet been developed. Society is a long way from having to worry that robots might build copies of themselves and take over the whole planet. But artificially living robots are almost certainly possible from a technological standpoint. Robots would probably reproduce asexually, by merely assembling other robots similar to themselves. Robots could also build machines much different from themselves individually. It's interesting to think of the ways in which artificially living machine populations might evolve.

Another definition for artificial life involves thought processes. At what point does machine intelligence become *consciousness*? Can a machine ever be aware of its own existence, and be able to ponder its place in the universe, the way a human being can? This is a difficult matter to resolve. The reproduction question is answerable by an "either/or," but the consciousness question can be endlessly debated. One person might say that a personal computer (PC) is conscious; others could argue that even some people aren't truly conscious! See also ARTIFICIAL INTELLIGENCE.

artificial reality

See VIRTUAL REALITY.

artificial stimulus

An *artificial stimulus* is a method of guiding a robot along a certain path. Automated guided vehicles, or AGVs, make use of artificial stimuli to follow certain routes in their environment.

Various different "landmarks" can be used as artificial stimuli. It is not necessary to have wires or magnets embedded in the floor, for example. A robot might be programmed to follow the wall on its right (or left) side until it reaches its destination, like finding its way out of a maze. The lamps in a hallway ceiling can be followed by light and direction sensors.

Another way to provide guidance is to use a beacon. This might be an infrared or visible beam, or a set of ultrasound sources. With ultrasound, the robot can measure the difference in propagation time from different sources to find its position in an open space, if there are no obstructions (see the illustration).

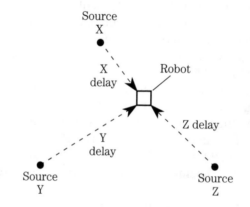

ARTIFICIAL STIMULUS
Three sound sources (X, Y, Z) are heard by robot with different delays.

How does a robot know when it has found its destination? There are many ways that objects can be marked for identification. One method is bar coding. This is used for pricing in stores. Another is a passive transponder, of the type attached to merchandise to prevent shoplifting. See also AUTOMATED GUIDED VEHICLE, BAR CODING, BEACON, PASSIVE TRANSPONDER.

Asimov, Isaac

One of the world's most well-known science-fiction writers, Isaac Asimov invented the "Three Laws of Robotics" in 1942. He wrote more than 400 books in his lifetime. He was born in Russia in 1920, shortly after the Communist revolution, but did most of his work in the United States. See ASIMOV'S THREE LAWS OF ROBOTICS.

Asimov's Three Laws of Robotics

In one of his early science-fiction stories, Isaac Asimov first mentioned the word *robotics*, along with three fundamental rules that all robots ought to obey. The rules, now called Asimov's Three Laws of Robotics, are as follows.

- A robot must not injure, or allow the injury of, any human being.
- A robot must obey all orders from humans, except orders that would contradict the First Law.
- A robot must protect itself, except when to do so would contradict the First Law or the Second Law.

Although these rules were first coined in the 1940s, they are still considered good standards for robots nowadays.

assembler program

In a computer, information is stored and transferred in the form of binary digits, or bits (see BIT). These are just combinations of ones and zeros, corresponding to high and low digital logic states. This is the *machine language*.

When you program a computer, you write your program in a high-level language such as BASIC, C, COBOL, or FORTRAN. This language consists of words and numbers, along with various punctuation marks and mathematical symbols. For the computer to work with this information, it must be "downgraded" to machine language. For you to understand the output of the computer, the machine language must be "upgraded" to words, numbers, and symbols. (Reading and writing in machine language is extremely difficult and tedious.) The *assembler* or *assembler program* converts between machine language and the high-level language (see the illustration).

Each high-level language has its own special assembler program. For example, the assembler for BASIC differs from that for Fortran, because the commands and symbols in BASIC and Fortran are different. See also ASSEMBLY LANGUAGE, C, HIGH-LEVEL LANGUAGE, MACHINE LANGUAGE.

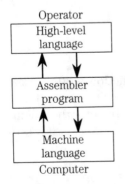

ASSEMBLER PROGRAM
Translates between machine language and high-level language.

assembly language

Assembly language is a computer programming language. It is used to write the assembler program that converts high-level languages, such as BASIC, C, COBOL, and FORTRAN, to machine language and vice-versa (see ASSEMBLER PROGRAM).

Assembly language is harder to read and write than high-level, or "user-friendly" languages. But it is not quite as difficult to work with as machine language. Nevertheless, it is time-consuming and tedious to write a program in assembly language. To accomplish a certain task that would be easy in high-level language, many lines might be required in assembly language.

A major advantage of assembly language is speed. If you write a computer program directly in assembly language, rather than in a high-level language, you save the compiler the work of translating back and forth between machine language and the high-level language. Therefore, the computer works faster, because it can process more meaningful information in a given length of time. Even though assembly language usually has more steps per task than high-level languages, assembler programs take up less memory in an operating system, because the steps in assembly language are specific and concise.

An assembly language that works with a particular *central processing unit*, or *CPU*, might not work with another CPU. This is one of the main drawbacks of assembly languages. It works against standardization in the computer industry. See also BASIC, C, CENTRAL PROCESSING UNIT, COBOL, COMPILER, FORTRAN, HIGH-LEVEL LANGUAGE, MACHINE LANGUAGE.

assembly line

An *assembly line* is a modern industrial system in which hardware is manufactured step by step. Each item passes through a number of "stations," usually moving on a conveyor belt. At each station, a certain task is performed, as shown in the illustration. In this way, large numbers of identical units can be made with the greatest speed and at the lowest cost.

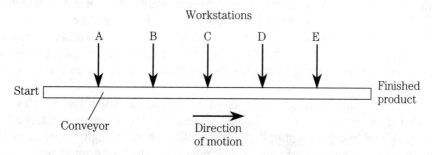

ASSEMBLY LINE In this example there are five workstations. Usually there are many more.

Traditionally, human workers have been employed at the stations in assembly lines. It might be one person's job to put the rivets in a piece of sheet metal, or to solder a certain connection, or to install a certain screw. The person does this for eight hours a day, five days a week.

Probably the biggest disadvantage of assembly-line work is the mind-numbing monotony. Sometimes workers are trained at two or three different tasks, so that they can trade off from time to time and get a little more variety. But even then, boredom is a major problem.

In recent years, robots have begun to perform some of the chores previously done by people. Robots do not get bored or frustrated, and they are "down" far less often than people are sick. Some workers see the increasing use of robots as a threat to their job security. Others see it as a welcome opportunity for advancement to more interesting work. See also ASSEMBLY ROBOTS, AUTOMATED INTEGRATED MANUFACTURING SYSTEM, AUTOMATION.

assembly robots

An *assembly robot* is any robot that assembles products, such as cars, home appliances, and electronic equipment. Some assembly robots work alone; most are used in *automated integrated manufacturing systems (AIMS)*, doing repetitive work at high speed and for long periods of time.

Assembly robots have taken the place of human workers in some jobs. Some people are concerned that robots take jobs from human beings. But in fact, robots create new kinds of jobs that are much more interesting than the old ones.

A person who puts screws in a car door all day long, for example, might be displaced by a robot. But that person might be trained to oversee the operation of a set of assembly robots, to maintain the robots, to program the robots' computer, to check the quality of goods produced, or even to sell the goods themselves. The end result is a happier, better-paid worker, who is less likely to suffer from "boredom fatigue."

Many assembly robots take the form of robot arms (see ROBOT ARMS). Several different joint arrangements are used. The type of joint arrangement depends on the task that the robot must perform. Joint arrangements are named according to the type of coordinate system they follow (see CARTESIAN COORDINATE GEOMETRY, CYLINDRICAL COORDINATE GEOMETRY, POLAR COORDINATE GEOMETRY, SPHERICAL COORDINATE GEOMETRY). The complexity of motion in an assembly robot is expressed in terms of the number of degrees of freedom (see DEGREES OF FREEDOM).

One type of assembly robot, developed in Japan, is called the *SCARA* (see SCARA). It resembles a Japanese folding screen that lets it move horizontally to within 0.05 millimeter. Its simplicity allows it to work at high speed, and also minimizes the *downtime*, or time during which the device is out of commission for repairs. It is also rather cheap, as far as assembly robots go.

To do their jobs right, assembly robots need to have all the parts exactly in place. They receive precise instructions, and there is almost no tolerance for error. Human operators, on the other hand, can work with a much larger margin for error. If you need to get a certain pair of pliers, you can recognize it by its shape and size. A robot wouldn't be able to find the pliers unless it were exactly in the right place, or unless it were marked in some way. There are some jobs, therefore, that assembly robots can't do very well. One of the biggest challenges for humans is the programming for assembly robots, so that the efficiency will be greatest while minimizing the possibility of "hangups."

Automated Guided Vehicle

An *Automated Guided Vehicle (AGV)* is a type of robot cart that runs without a driver. The cart has an electric engine and is guided by a magnetic field, produced by a wire on or just beneath the floor (see the illustration). Alternatively, an AGV might run on a track, like a miniature train engine.

In an automated factory, AGVs are used to bring components to assembly lines. The parts must be put in just the right places, so the assembly robots can find them (see ASSEMBLY ROBOTS).

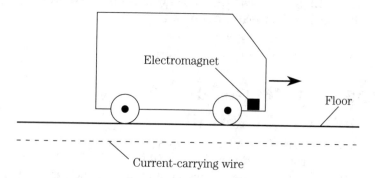

Electromagnet

Floor

Current-carrying wire

AUTOMATED GUIDED VEHICLE Magnet guides cart along current-carrying wire embedded in floor.

In the future, AGVs might serve as "low-priority" nurses in hospitals, bringing food and nonessential items to patients. An AGV can also serve as a "mechanical janitor" or "mechanical gopher," performing routine chores around the home or office.

On a larger scale, there has been some talk about making automobiles into AGVs that follow wires embedded in the road pavement. This would take the driver's job away, letting computers do it instead. Each car would have its own individual computer, and the traffic in a whole city would be overseen by one or more central computers. In the event of computer failure, all traffic would stop. This would practically eliminate accidents. But people might not accept the idea of putting their schedules, let alone their lives, into the hands (wheels?) of robot computers.

automated home

To some extent, all modern homes are "automated." Even running water qualifies; it flows automatically when you turn on a faucet. Automatic dishwashers, laundry machines, garage-door openers, intrusion detectors, and television remote-control switchboxes are some examples of devices that have become commonplace during the latter part of the 20th century.

As robotic and artificial-intelligence (AI) technology become more available to the consumer, we can expect to see things like a robotic laundry that takes items from the chute, places them in the wash, moves them to the dryer, and perhaps even sorts and folds them.

We might see robots that make our beds, do our dishes, vacuum our carpets, shovel snow from our driveways, clean our windows, shop for our groceries—the list is practically endless.

If robots can do all our housework, what will be left for us to do? That is a good question. Although robots and computers *can* do work for us, they don't *have* to be used. There will always be times when you would prefer to mow the yard yourself. People will probably always want to do their own gardening. Perhaps the greatest challenge in home automation will be to decide what chores are best left to the homeowners.

automated integrated manufacturing system

An *automated integrated manufacturing system (AIMS)* is an assembly line or factory that uses robots to do the individual tasks. An AIMS is overseen by a master computer, along with lower-level computers for the assembly lines and robots.

People are needed in any AIMS for the diverse and complicated tasks that robots simply cannot perform. These include things like programming the computers and maintaining the hardware. See ASSEMBLY ROBOT, AUTOMATED GUIDED VEHICLE.

Automatic Sequence Controlled Calculator

The first truly automatic digital calculator was designed by International Business Machines (IBM) for the United States Navy. It was put into operation in 1944, towards the end of World War II. The technical name for this machine was *Automatic Sequence Controlled Calculator*. Its nickname was *Mark I*.

The Mark I could work with 23 significant digits. Thus, it had an accuracy of one part in 100 sextillion (10^{23}). Its main purpose was to do precise ballistics calculations. This is the type of calculation used for aiming shells and missiles.

Mark I was not like the personal computers in use today. For one thing, Mark I was huge. It took up a whole room. It made use of electromechanical devices, whereas modern computers use microchips. Nowadays, a calculator with the same level of sophistication as Mark I can fit in your pocket. See also COMPUTER.

automation

The term *automation* refers to the operation of a device or factory where some or all of the processes are done by machines. You might hear about the "automation of an assembly line," for example; this means that robots are being installed to take the places of human workers.

There has always been some controversy about automation. Can machines do things as well as people?

Advantages of automation include the following:

- Machines work faster than people.
- Machines are usually more precise than people.
- Machines are "down" less often than people are out sick.
- Some machines are much stronger than people.

Advantages of human operators are:

- People can solve some problems that machines cannot.
- People have greater tolerance for confusion in the manufacturing process.
- Humans can do some minute tasks that machines cannot.
- Human beings are always needed to supervise machines, because some decisions simply cannot be made by computers.

See also ASSEMBLY LINE, ASSEMBLY ROBOTS.

automaton

An *automaton* is a very simple robot that performs a task or set of tasks without artificial intelligence (AI). Automatons have been around for well over 200 years. The plural of the term is automata.

A good example of an automaton is a "mechanical duck" designed by J. de Vaucanson in the 18th century. It was used to entertain audiences in Europe. With thousands of parts, it was quite a sophisticated machine for its time. It made quacking sounds and seemed to eat and drink like a live duck. Its inventor wanted to create artificial life, and he used the duck act to raise money for his work.

Another example of automata comes from this writer's childhood during the 1960s. Every December, a family in Rochester, Minnesota used to build a holiday display in their yard, consisting of machines in the form of people and animals. They moved, and seemed to sing Christmas songs. Cars would line up and slowly drive by, filling the subzero air with the steam from their exhausts, while the act repeated itself in the snowy yard. It, like the mechanical duck two centuries before, was a big hit for adults as well as children.

The mechanical duck had no "brains," because there were no digital computers in the 18th century. Likewise, the Christmas display was a set of machines that simply followed a mechanical routine. Although fun to observe, these devices don't have industrial uses. They lack precision, and the motions they can make are very limited. Even though some of these machines might look like androids, they are basically just moving statues. See also ANDROID, ARTIFICIAL INTELLIGENCE, ARTIFICIAL LIFE.

Automax

Automax is the name of a corporation based in Tokyo, Japan. They have built robots with a wide variety of industrial and entertainment applications. Among their creations are undersea robots and oil-tank-cleaning robots. The corporation claims that their designs are all original, conceived entirely by Japanese engineers.

The Automax corporation makes use of radical creative methods. They believe that individuals should work without concern for whether they get personal recognition. The Japanese say that ideas flow best when engineers forget about past problems and limitations. American industries are rapidly catching on to this way of thinking. It is sometimes called the *Mukta philosophy*, after the Mukta Institute, an organization of Japanese engineers. See also MUKTA INSTITUTE.

autonomous robots

A robot is autonomous if it is self-contained, housing its own computer system, and not depending on a central computer for its commands. It gets around under its own power, usually by rolling on wheels or by moving on two, four, or six legs.

Robot autonomy might at first seem like a great advantage. After all, if a robot functions by itself in a system, then when other parts of the system fail, the robot will keep working. But often, where many identical robots are used, autonomy is inefficient. It is far less expensive to put programs in one central computer that controls all the robots. Insect robots work this way.

The illustration shows the locations of computers in robot systems. Robots are shown as triangles, and computers as solid black dots. In the drawing at A, the computer is central, and is common to all the individual robots. The com-

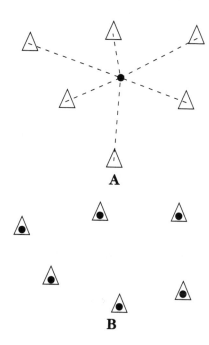

AUTONOMOUS ROBOTS
Systems with central computer
(A) and individual computers
(B). Robots are triangles;
computers are dots.

puter communicates with, and coordinates, the robots through wires or fiberoptics, or via radio. In the drawing at B, each robot has its own computer. These computers might communicate with each other, but it is not necessary.

Simple robots, like those in assembly lines, are not autonomous. The more complex the task, and the more different things a robot must do, the more autonomy it will have. The most autonomous robots have artificial intelligence (AI).

The ultimate autonomous robot will act like a living animal or human. Such a machine has not been developed yet. It will probably be many years before this level of sophistication is reached. See also ADVANCED ROBOT TECHNOLOGY, ANDROID, ARTIFICIAL INTELLIGENCE, ARTIFICIAL LIFE, INSECT ROBOTS.

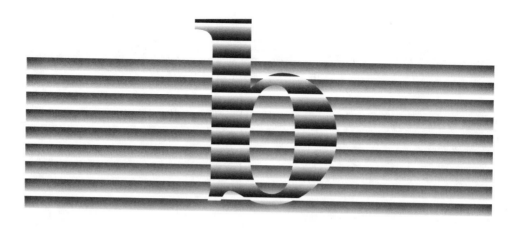

Babbage, Charles

Charles Babbage was an engineer in 19th-century England. He is given credit for having conceived the first digital calculator other than the ancient Oriental abacus. His design was intended for the British Post Office in 1836.

Babbage's digital calculator never got to the stage where it was "up and running." It would have made use of nonelectrical components, because electric utilities did not yet exist. Audible signals, such as bells, would tell operators various things about the machine's operation. Data would have been stored on punch cards.

Babbage is said to have had a short temper, and lacked the persistence to keep at things long enough to perfect them. The problem of building calculators (let alone computers!) without electricity might have been too great even for a patient engineer to overcome. But Babbage would reach a certain point and then take everything apart, starting all over to rebuild a new design. Eventually, his components, and his bank account, wore out.

Babbage's ideas were left alone for many years after his death. But when electricity became widely available, engineers once again tackled the problem of building a digital calculator. They examined the work of Charles Babbage and found his theoretical principles were useful. See also ANALYTICAL ENGINE.

back pressure sensor

When a robot motor operates, it encounters mechanical resistance. This resistance depends on various factors, such as the weight of an object being lifted, or the friction of an object as it is moved along a surface. The amount of turning force that the motor delivers is called *torque*. The torque is a direct function of mechanical resistance.

A robot motor produces a measurable back pressure that depends on the torque being applied. The greater the torque, the greater the back pressure. A *back pressure sensor* is a device that detects, and measures, the amount of

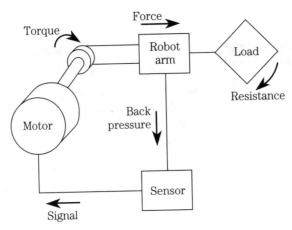

BACK PRESSURE SENSOR Functional block diagram.

torque that the motor is applying at any given time. The sensor produces a signal, usually a variable voltage, that increases as the torque increases. This signal can be used for various purposes.

Back pressure sensors are used to limit the amount of force applied by a robot gripper, arm, drill, hammer, or other device. The *back voltage*, or signal produced by the sensor, reduces the torque applied by the motor, acting as a sort of governor. This can prevent damage to objects being handled by the robot. It also helps to ensure the safety of people working around the robot.

The illustration shows a block diagram of the operation of a back pressure sensor. See also ROBOT ARMS, ROBOT GRIPPERS.

Bandai

Bandai is the name of a major toy manufacturer in Japan, known for their popular amusement robots. They first began making toy robots in the 1960s. During the 1970s, their business took off. Their success was helped by the Japanese industrial practice of cooperation among different companies. Bandai created many types of amusement robots based on children's science-fiction movies.

Engineers at Bandai are mostly young people who were raised on "robo-science" videos. They believe that toy robots should have the following characteristics:

- They must look realistic.
- They must look strong.
- They must look "awesome."
- They must have movable parts.
- They must actually do something.

Bandai joined forces with Tonka, a major U.S. toy manufacturer, and inspired an interest of robots in American children during the 1980s. Today, robot toys are common in department stores throughout the U.S. See also AMUSEMENT ROBOTS.

bandwidth

Bandwidth is a term that refers to the amount of frequency space, or *spectrum space*, that a signal requires in order to be transmitted and received clearly.

All signals have a finite, nonzero bandwidth. No signal can be transmitted in an infinitely tiny slot of spectrum space. In general, the bandwidth of a signal is proportional to the speed at which the data is sent and received. Speed is measured in bauds, bits per second or words per minute (see BAUD RATE, BITS PER SECOND, WORDS PER MINUTE).

The greater the allowable bandwidth that a signal has, the more data can be transmitted in a given length of time. The more restricted the allowable bandwidth, the more limited the data speed must be.

The illustration shows a typical data signal, graphed for level (amplitude) as a function of frequency. The information in this signal is contained within a band of frequencies about 5.2 kilohertz (kHz) wide. This is 5200 cycles per second. The data extends 2.6 kHz above and below the center frequency of the signal.

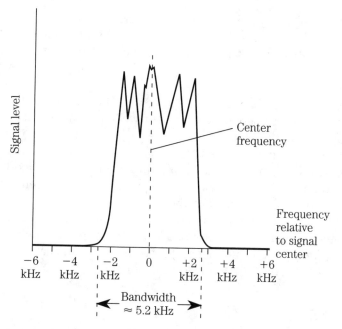

BANDWIDTH This signal is 5.2 kHz wide.

bar coding

Bar coding is a method of labeling objects for identification. You have seen bar-code labels or tags in stores; they are used for pricing merchandise.

A bar-code tag has a characteristic appearance, with parallel lines of varying width and spacing (see the drawing). A laser-equipped device scans the tag,

BAR CODING
Tags have a characteristic
appearance.

thereby retrieving the identifying data. The reading device does not need to be brought right up to the tag; it can work from some distance away.

Bar-code tags are one method by which objects can be labeled so that a robot can identify them. This greatly simplifies the recognition process. A whole tool set can be tagged using bar-code stickers with a different code for each tool. When a robot's program tells it that it needs a certain tool, the robot can seek out the appropriate tag and carry out the movements according to the program subroutine for that tool. Even if the tool gets misplaced, as long as it is within the robot's range of motion, it can easily be found again when it is needed. See also PASSIVE TRANSPONDER.

BASIC

BASIC is one of several high-level computer languages. It was originally formulated in the early 1960s with the intent of making computer programming easy for nonscientists. The acronym stands for *Beginner's All-purpose Symbolic Instruction Code.*

As its name implies, BASIC is easy to learn. The commands are generally self-explanatory, even to people who are not well acquainted with computers. BASIC can be taught at the junior-high and even at the elementary school level. It is not difficult to learn by yourself from instruction books.

Although BASIC is easy to master, it is not a very efficient method of programming a computer. Numerous variations on BASIC have been devised, and some of these are more efficient than the original BASIC.

Another problem with BASIC is that it is not a powerful language. It does not take advantage of a computer's potential. While this might not be a problem for the casual computer user, professionals prefer more versatile languages, such as C. See also C, COBOL, FORTRAN, HIGH-LEVEL LANGUAGE.

baud rate

The speed at which digital signals are transmitted can be measured in various ways. One common method is to specify the baud rate.

A digital signal changes at hundreds, thousands, or even millions of times every second. The exact number of changes per second might be, and often is, constantly changing. The baud rate is the maximum number of changes per second. A 300-baud digital transmission has, at most, 300 changes of state per second. A 9600-baud digital signal has, at most, 9600 changes per second.

When computers are interconnected by telephone, satellite, microwave, or fiberoptic systems, the transmissions are always digital and are done at certain standard speeds. The higher the baud rate, the higher the signal speed, and the faster the computers respond to each other and to their human operators' commands.

When you access a telephone bulletin board system (BBS), you use a modem with the capability to send and receive data at a certain baud rate. The computer at the other end of the line also has a modem, with the ability to send and receive at a certain baud rate. The modems might have the same maximum speed, but often they do not. The slower of the two modems determines the fastest speed at which you can communicate.

There is a limit to the baud rate that a telephone circuit can handle. The greater the baud rate, the larger the *bandwidth*, or range of frequencies that a signal takes up. A telephone line can handle bandwidths of up to about 3000 cycles per second, or 3 kilohertz (3 kHz). This will accommodate 2400-baud and 9600-baud modems easily. These modems are commonly used with personal computers (PCs).

Satellite, microwave, and fiberoptic communications systems can handle far larger signal bandwidths than telephone circuits. Therefore, much higher baud rates are possible in these modes, compared with telephone systems. The speed of data transmission and reception is an important factor in how "smart" an interconnected system of computers can become. The faster a system can "think," the higher its potential level of artificial intelligence (AI). See also ARTIFICIAL INTELLIGENCE, BANDWIDTH, BIT RATE, BULLETIN BOARD, DIGITAL, MODEM, PACKET RADIO, WORDS PER MINUTE.

BBS

See BULLETIN BOARD.

beacon

A *beacon* is a device used to help robots navigate. Beacons can be categorized as either *passive* or *active*.

A mirror is a good example of a passive beacon. It does not produce a signal of its own; it merely reflects light beams that strike it. The robot requires a transmitter, such as a flashing lamp or laser beam, and a receiver, such as a photocell. The distance to each mirror can be determined by the time required for the flash to travel to the mirror and return to the robot. Because this delay is an extremely short interval of time, a high-speed measuring apparatus is needed.

An example of an active beacon is a radio transmitter. Several transmitters can be put in various places, and their signals synchronized so that they are all exactly in phase. As the robot moves around, the relative phases of the signals will change. Using an internal computer, the robot can determine its position by comparing the phases of the signals from the beacons. With active beacons, the robot does not need a transmitter, but the beacons must have a source of power and be properly aligned. See also ARTIFICIAL STIMULUS.

biased search

A biased search is a method of finding a destination or target, by first looking off to one side and then "zeroing in." This method is sometimes used by robots.

The illustration shows the technique as a boater might apply it on a foggy day. At some distance from the shoreline, the boater can't see the dock, but he or she has a reasonably good idea of where it is. Therefore, an approach is deliberately made well off to one side (in this case, to the left) of the dock. When the shore comes into view, the boater turns to the right and follows it until the dock is found.

For a robot to effectively use this technique, it must have some familiarity with its environment, just as the boater knows roughly where the dock will be. This is accomplished by means of task level programming, a form of artificial intelligence. See also ARTIFICIAL INTELLIGENCE, TASK LEVEL PROGRAMMING.

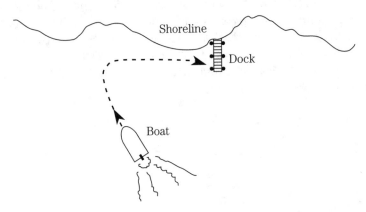

BIASED SEARCH A boat finds a dock in the fog.

binary

The term *binary* is used for any quantity that can attain either of two states.

Computers use the binary system because circuits that differentiate between two states are simple, cheap, and highly precise. The states are represented by the numbers 0 and 1. Sometimes the states are called low and high; off and on; space and mark; no and yes; or false and true respectively. See also BINARY-CODED DECIMAL, BINARY NUMBER SYSTEM.

binary-coded decimal

The *binary-coded decimal (BCD)* system of writing numbers assigns a four-bit *binary* code to each numeral 0 through 9 in the base-10 system, as shown in the table. The four-bit BCD code for a given digit is its representation in the binary number system. Four places are always assigned.

BINARY-CODED DECIMAL:
Numbers 0 through 9 in BCD form.

Digit in Base 10	Digit in BCD
0	0000
1	0001
2	0010
3	0011
4	0100
5	0101
6	0110
7	0111
8	1000
9	1001

Numbers larger than 9, having two or more digits, are expressed in BCD digit-by-digit. For example, the decimal number 4907 would be written 0100 1001 0000 0111 in BCD form. The four binary numbers represent the decimal numbers 4, 9, 0, and 7 in sequence. See also BINARY, BINARY NUMBER SYSTEM.

binary number system

The *binary number system* is a method of expressing numbers using only the digits 0 and 1. It is sometimes called *base-2*, *radix-2*, or *modulo-2*. Normally, numbers are written in *decimal*, or *base-10* form. But computers work much more efficiently with binary numbers than with decimal numbers.

In the base-10 system, the rightmost digit is the "ones" digit. The next digit to the left is the "tens" digit; after that comes the "hundreds" digit, and so on, increasing in powers of 10.

In binary notation, the rightmost digit is the "ones" digit, just as in base 10. But the next digit to the left is a "twos" digit; after that comes the "fours" digit, and so on, upwards in powers of 2.

A number takes more digits to write in binary form, compared with decimal notation. But for a digital computer, sheer volume of digits is not a problem. The important thing is that the computer be able to easily recognize each digit. In the binary number system, the digits are all either 0 or 1. Electronically, these states can be represented as "off" and "on," or "low" and "high," or "no" and "yes." This either-or situation is the simplest possible expression for a quantity, and provides the most computer accuracy at the least cost.

The illustration shows both the decimal and binary notations for the decimal number 94. At A, in base 10:

$$94 = (4 \times 10^0) + (9 \times 10^1)$$

At B, in the binary number system:

$$1011110 = (0 \times 2^0) + (1 \times 2^1) + (1 \times 2^2) + (1 \times 2^3) + (1 \times 2^4) + (0 \times 2^5) + (1 \times 2^6)$$

This is another way of saying that $94 = 2 + 4 + 8 + 16 + 64$.

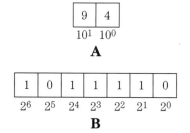

A

BINARY NUMBER SYSTEM
At A, decimal 94; at B, binary 1011110. These have the same numerical value.

B

Every decimal number has one and only one binary representation. Every binary number, likewise, has one and only one decimal representation. When you work with a computer or calculator, you give it a decimal number that is converted into binary form. The computer or calculator does its operations with 0s and 1s, and when the process is complete, it converts the result back into decimal form for you. See also BINARY, BINARY-CODED DECIMAL.

binary search

In a digital computer a *binary search*, also called a *dichotomizing search*, is a method of locating an item in a large set of items. Each item in the set is given a number key. The number of keys is always a power of 2. Therefore, when it is repeatedly divided into halves, the end result will always be a single key.

If there are 16 items in a list, for example, they might be numbered 1 through 16. If there are 25 items, they can be numbered 1 through 25, with the numbers 26 through 32 as "dummy" keys.

The desired number key is first compared with a number halfway down the list. If the desired key is smaller than the halfway number, then the first half of the list is accepted, and the second half is rejected. If the desired key is larger than the halfway number, then the second half of the list is accepted, and the first half is rejected. The desired key is then compared with a number in the middle of the accepted portion of the list. On this basis, one half of the list is accepted and the other half is rejected, just as in the first case. The process is repeated, each time selecting half of the list and rejecting the other half, until just one item remains. This item is the desired key.

The table shows an example of a binary search to choose one item from a list of 25. Keys are indicated by o's; the desired key, 21, is indicated by #. "Dummy" keys are shown as x's.

BINARY SEARCH:
List is repeatedly broken in half.

Choose no. 21:

o o o o o o o o o o o o o o o

o o o o # o o o o x x x x x x x

Select second half:

o o o o # o o o o x x x x x x x

Select first half:

o o o o # o o o

Select second half:

o o o

Select first half:

o

Select first half:

#

bin-picking problem

One of the most difficult problems for a robot is to find an object based on its physical appearance. This is because a single object can look like all kinds of different things when it is viewed from different angles.

A cylindrical drinking cup, for example, when seen from exactly side-on, looks like a rectangle and its interior, as shown at A in the illustration. From the top or the bottom, it looks like a circle and its interior, as shown at B. From a skewed angle, it has the shape shown at C. The roundness of the ends will depend on the exact angle from which it is viewed.

A **B** **C**

BIN-PICKING PROBLEM
Cup as seen from the side (A), the top (B), and an intermediate angle (C).

The problem of object recognition is compounded when a certain object must be picked from a bin containing many other objects. Part, most, or even all of the desired object can be obscured by other objects. A human being can pick a drinking cup from a dishwasher full of various eating utensils with ease; however, even smart robots have a very hard time with this task. Problems of this type are called *bin-picking problems*. One of the biggest challenges in developing artificial intelligence (AI) is giving robots the ability to solve these kinds of problems.

One way to help a robot select items from a bin is to give each item a code. This can be done by means of bar coding or passive transponders. See also ARTIFICIAL INTELLIGENCE, BAR CODING, OBJECT RECOGNITION, PASSIVE TRANSPONDER.

binaural robot hearing

Even with your eyes closed, you can usually tell from which direction a sound is coming. This is because you have *binaural hearing*. You have two ears. Sound arrives at your left ear with a different intensity, and in a different phase, than it arrives at your right ear. Your brain processes this information, allowing you to locate the source of the sound, with certain limitations. If you are confused, you can turn your head until the sound direction becomes apparent to you.

Robots can be equipped with binaural hearing. Two sound transducers are positioned on either side of the robot. A microprocessor can compare the relative phase and intensity of the signals from the two transducers. This lets the robot "know," with certain limitations, the direction from which the sound is coming (see the illustration). If the robot is confused, it can turn until the confusion is eliminated and a meaningful bearing is obtained. See also SOUND TRANSDUCER.

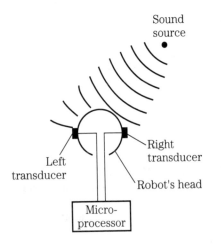

BINAURAL ROBOT HEARING
Microprocessor compares signals from two transducers.

binocular robot vision

Binocular robot vision is the analog of binocular human vision. It is sometimes called *stereo robot vision.*

In humans, binocular vision allows perception of depth. With just one eye (monocular vision), you can infer depth to some extent on the basis of perspective. Almost everyone, however, has had the experience of being fooled when looking at a scene with one eye covered or blocked. A nearby pole and a distant tower might seem to be right next to each other, when in fact they are more than a city block apart.

For robots, binocular vision requires a sophisticated microprocessor. The inferences we make, based on what our two eyes see, are in fact quite complicated. We do them "without thinking," but our brains are far more complex than the most advanced computers in existence. The illustration shows the basic concept for binocular robot vision. See also VISION SYSTEMS.

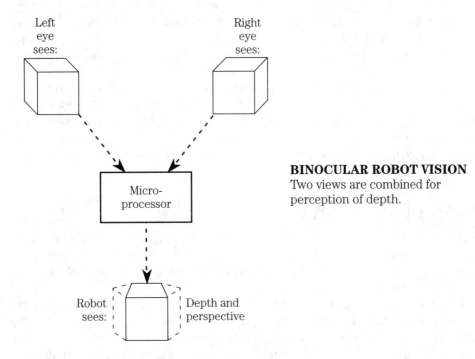

BINOCULAR ROBOT VISION
Two views are combined for perception of depth.

Of primary importance for good binocular robot vision are the following:

- High-resolution robot eyes
- A sophisticated microprocessor
- Programming via which the robot can act on commands, based on what it sees.

biochip

The term *biochip* has been coined for naturally occurring integrated circuits (ICs). The term has also been suggested for ICs that can be manufactured using techniques that mimic the way nature puts atoms together.

Do ICs, also called chips, really grow in nature? This might at first seem like a silly question, to which the answer must be: "Of course not." But some of the most advanced thinkers in the field of artificial intelligence (AI) have asked this question, because of the resemblance between certain natural structures and the patterns used in the manufacture of ICs. See INTEGRATED CIRCUIT.

It has been suggested that the human brain is a huge computer. If this is true, the brain is far more sophisticated than any computer yet made. But also, no matter how complex a set of logic gates becomes, it is always built up from individual logic gates. See AND GATE, EXCLUSIVE-NOR GATE, EXCLUSIVE-OR GATE, INVERTER, NAND GATE, NOR GATE, OR GATE.

However nature actually assembles a brain (or any living matter), it is done by putting protons, neutrons, and electrons together. Every proton is just like every other proton; the same is true for neutrons and electrons. The building blocks are simple. It is the way they're combined that is so amazingly complicated.

Based on these premises, it is possible (in theory, at least) to build a computer as smart as a human brain. And AI researchers might do well to look at the way nature "builds" things to get ideas for the construction of improved ICs. If we humans can figure out just how to string together a set of atoms, we should theoretically be able to make anything we want! See also ARTIFICIAL INTELLIGENCE, ARTIFICIAL LIFE.

biological robots

Suppose that you could clone cells from living organisms, and grow them in a lab to make humanoids, animals, or hybrid human-animal creatures? Research has been done in this field. Such creatures would be called *biological robots*. They would be artificial life forms.

Biological robots have been characters in science-fiction stories. If scientists could "manufacture" humanoid clones, would the creatures be people, complete with emotions, minds, and . . . and souls? Would they be alive in every sense that you and I are alive? Would they be mentally programmable like robots, but physically as agile as people? Could scientists grow them to any physical specifications desired, such as six foot eight, three hundred pounds of solid muscle, with an IQ of 180?

There are ethical questions and problems in biological-robot research. These issues are of such grave concern that some scientists refuse to work in this field. It doesn't take much imagination to see how biological robots could turn from a scientist's dream into a nightmare for the whole world.

Conscientious researchers feel uneasy about the idea of cloning cells to make biological robots. They fear such beings could carry out the plots of half-baked warmongers. Soldiers could be grown in labs. A mad dictator could create an army of a million superzombies. Is this farfetched? History is full of characters who, if they could have done these kinds of things, would have. Thus, if ever they *can*, they *will*.

Biological robots might be impossible to manufacture. Or the technology might be within reach, but well beyond our grasp right now. Or perhaps we're just a few short steps away from it. See also ARTIFICIAL LIFE.

biomechanism

A *biomechanism* is a mechanical device that simulates the workings of some part of a living being's body. Examples of biomechanisms are mechanical hands, arms, and legs. See ROBOT ARMS, ROBOT GRIPPERS, ROBOT LEGS. The term applies especially to robotic devices that not only perform the functions of their living counterparts, but look like them too.

The term *biomechanism* can also be used in reference to some body function. Thus, you might talk about the structure of a forearm and hand, calling it a biomechanism. See also ARTIFICIAL LIFE, BIOLOGICAL ROBOTS, BIOMECHATRONICS.

biomechatronics

The word *biomechatronics* is a contraction of the words *biology*, *mechanics* and *electronics*. The field of biomechatronics is part of the larger realms of robotics and artificial intelligence (AI). Specifically, biomechatronics involves electronic and mechanical devices that duplicate human body parts and their functions.

Biomechatronics has received more attention in Japan than in the United States. In Japan, some robot researchers attack their problems with religious zeal. It has been suggested that this is because there are no prevalent religions in Japan like there are in America, therefore many Japanese literally use science as their religion. Not only would Japanese robotics engineers like to build robots that can do all the things people can do, but some want their robots to look like people too.

The ultimate biomechatronic device is an android. Scientists generally agree that an intelligent android will probably not be developed for many years.

The problem of making androids can be approached from two directions. On the one hand, biological robots might be grown in labs by a process of cloning. This idea is clouded by profound ethical issues. On the other hand, engineers can try to build a mechanical robot with the dexterity and intelligence of a human being. This notion, too, brings up ethical questions, but to a lesser degree. See also ANDROID, ARTIFICIAL INTELLIGENCE, ARTIFICIAL LIFE, BIOLOGICAL ROBOTS.

biped robots

A *biped robot* is a robot with two legs. Usually, but not always, such robots have arms and a head, so that they look something like humans.

Biped robots are not very stable. Humans do well with two legs because we have a good sense of balance. This sense comes from tiny organs in our ears. If you've ever had "dizzy spells," or an infection in your middle/inner ear that caused you to temporarily lose your sense of balance, you've experienced how difficult it is to stand up without holding on to something. A walk across a room becomes a long journey.

Your sense of balance is sophisticated; it's hard to duplicate electromechanically. Therefore, engineers design and build biped robots mainly from vanity. It is human nature to want to make likenesses of ourselves.

"Legged" robots generally have four or six legs. These designs are much more stable than two-legged ones. See also INSECT ROBOTS, QUADRUPED ROBOTS, ROBOT LEGS.

bit

The term *bit* comes from the words *binary* digi*t*. A bit is a single piece of digital data, represented by either 0 or 1. These states are sometimes called low and high; off and on; no and yes; or false and true respectively. Usually, the state 0 is a low current or voltage, and the state 1 is a more positive current or voltage, as shown in the illustration. See also BYTE, DIGITAL.

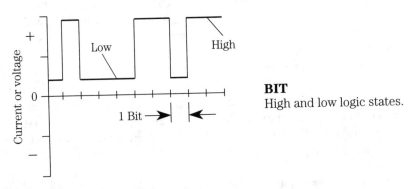

BIT
High and low logic states.

bits per second

The speed of digital data can be measured in various ways. One method is to specify the number of bits per second (bps) being sent or received. This is not always the same as the baud rate. The speed of data in bps can range from one to several times the baud rate. See BIT, BAUD RATE, WORDS PER MINUTE.

blackboard system

A *blackboard system* is a scheme for voice recognition and image recognition. It incorporates artificial intelligence (AI) to help a computer recognize sounds or images.

In a blackboard system, the incoming signal is *digitized*, or changed from analog form into digital form. The digital data is written onto a blackboard, which is basically a read/write memory circuit. Then it is evaluated by various specialty circuits (see the illustration).

BLACKBOARD SYSTEM
Data in memory (box) is evaluated by specialty circuits (triangles labeled "S").

For speech recognition, specialties might include vowel sounds, consonant sounds, grammar, syntax, context, and other variables. Have you spoken the word "weigh" or the word "way?" Did she say "ad" or "add?" Was that the word "by" or "buy?" These word pairs all sound identical; the specialist circuits determine which word was used by checking other words and meanings in the sentence. How does the computer know when you've finished a sentence? How does it know when it should put a question mark at the end of a sentence, rather than a period? When should paragraph divisions be made? All these things can be determined by the specialist circuits, as they "debate," using the blackboard as their forum, the most likely and logical interpretations of what has been said. A "referee," usually called a *focus specialist*, mediates.

For object recognition, specialties might be shape, color, size, texture, height, width, depth, and other visual cues. How does a computer know if an object is a cup on a table, or a water tower a mile away? Is that a bright lamp or is it the sun? Is that biped thing a robot, or is it a person? As with speech recognition, the blackboard serves as a debating ground.

Speech and object recognition machines can be of great value in robotics and artificial intelligence. The applications are numerous. Most researchers consider any money spent in these fields to be wise investments. See also ANALOG, ARTIFICIAL INTELLIGENCE, DATA CONVERSION, DIGITAL, MEMORY, OBJECT RECOGNITION, SPEECH RECOGNITION.

Bongard problems

In robot vision systems, it is necessary for the machine to recognize various patterns. *Bongard problems*, named after their inventor, are one method of evaluating how well a vision system can differentiate among patterns. Solving the problems also requires a certain level of artificial intelligence (AI). Incidentally, these problems make good pattern-recognition tests for people, as well as for machines.

An example of a Bongard problem is shown in the illustration. There are two groups of six boxes. The contents of the boxes on the left all have something in common; those on the right have the same characteristic in common, but to a different degree, or in a different way. To solve the problem, the vision system (or you) must answer three questions:

1. What do the contents of the boxes to the left of the heavy, vertical line have in common?
2. What do the contents of the boxes to the right of the line have in common?
3. What is the difference between the contents of the boxes on opposite sides of the heavy, vertical line?

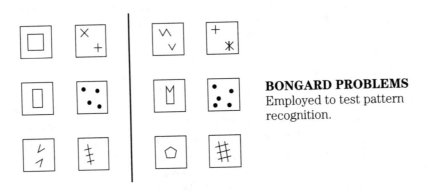

BONGARD PROBLEMS
Employed to test pattern recognition.

In this case, the boxes on the left contain four dots or straight lines each; those on the right contain five dots or straight lines each. The difference between the boxes on the left and those on the right, therefore, is in the number of dots or straight lines they contain. See also ARTIFICIAL INTELLIGENCE, OBJECT RECOGNITION, VISION SYSTEMS.

Boolean algebra

Boolean algebra is a system of mathematical logic, using the functions AND, NOT, and OR. In the Boolean system, AND is represented by multiplication, NOT by a minus sign (negation), and OR by addition. Therefore, X AND Y is written

XY, NOT X is written $-X$, and X OR Y is written $X + Y$. Sometimes alternative notations are used, such as X' for NOT X and $X * Y$ for X AND Y. Table 1 shows the values of these functions, where 0 indicates "falsity" and 1 indicates "truth." Boolean functions are used in the design of computer digital logic circuits.

BOOLEAN ALGEBRA, TABLE 1:
Truth table for logic operations.

X	Y	–X	XY	X + Y
0	0	1	0	0
0	1	1	0	1
1	0	0	0	1
1	1	0	1	1

Using Boolean representations, some familiar rules of arithmetic can be applied to logic. Logic equations often resemble their arithmetic counterparts. The statements on either side of an equal sign are *logically equivalent*. That is, if you say $X = Y$, it means that X is true whenever Y is true, and X is false whenever Y is false.

Table 2 shows several logic equations. These are known facts, or theorems, in Boolean algebra. These theorems can be used to analyze more complicated logic functions. This is an aid to engineers who design digital devices such as computers. See also AND GATE, INVERTER, LOGIC, OR GATE.

BOOLEAN ALGEBRA, TABLE 2:
Common theorems in Boolean algebra.

$X + 0 = X$	OR identity
$X1 = X$	AND identity
$X + 1 = 1$	
$X0 = 0$	
$X + X = X$	
$XX = X$	
$-(-X) = X$	Double negation
$X + -(X) = X$	
$X(-X) = 0$	
$X + Y = Y + X$	Commutativity of OR
$XY = YX$	Commutativity of AND
$X + XY = X$	
$X(-Y) + Y = X + Y$	
$X + Y + Z = (X + Y) + Z = X + (Y + Z)$	Associativity of OR
$XYZ = (XY)Z = X(YZ)$	Associativity of AND
$X(Y + Z) = XY + XZ$	Distributivity
$-(X + Y) = (-X)(-Y)$	DeMorgan's Theorem
$-(XY) = -X + -Y$	DeMorgan's Theorem

boot

In computer operation, the term *boot* can mean any of several different things.

A *warm boot* is a reinitialization of the system, clearing all the memory circuits but not completely removing power from the computer. You warm boot your personal computer (PC) when you press the "reset" button.

A *cold boot* is done by switching the computer off, waiting a few seconds for all the capacitors to discharge and for all stray voltages to decay to zero, and then switching it back on again. You interrupt the power to the whole machine.

When you need to reinitialize a computer system, you should try a warm boot before a cold boot. This is because the warm boot forces the computer to do less work. A cold boot is to your computer, in a sense, just what its name implies: a rude kick in the butt!

The term *boot* sometimes refers to inserting a disk with software, and then instructing the computer to load the software for use.

BORN System

The *BORN System* was a biped robot developed by the Japanese, and first introduced at a show in Tokyo in 1985.

To date, biped (two-legged) robots, and robots in humanoid form, have not been given much attention by designers outside Japan. American and European engineers see two-legged designs as unstable, hard to program, and inefficient. The BORN System robot failed during the 1985 exhibition, to the embarrassment of its designers. Nonetheless, the Japanese are still very interested in the biped concept. See also ANDROID, BIPED ROBOTS.

branching

In artificial intelligence (AI), the term *branching* is used in reference to routines, or programs, that have points where the computer must select among alternatives.

Suppose you have a robot on an assembly line that makes cars. The robot's job is to insert hubcaps in the two right-side wheels. (An identical robot does the same job on the left side.) Suppose that 20 percent of the cars are fitted with gold-colored (G) hubcaps and the rest are fitted with silver-colored (S) ones. Then the robot should insert hubcaps in the following sequence: SS SS SS SS GG SS SS SS SS GG SS SS . . . and so on. Every fifth pair of hubcaps is gold.

Each time a hubcap pair is to be inserted, the computer must make a choice. The routine is at a branch point for every hubcap pair. Every fifth time the choice must be made, the computer chooses gold hubcaps. Otherwise, it chooses silver ones. This sequence is programmed into the computer. The thought processes of the computer go something like the flowchart in the accompanying figure.

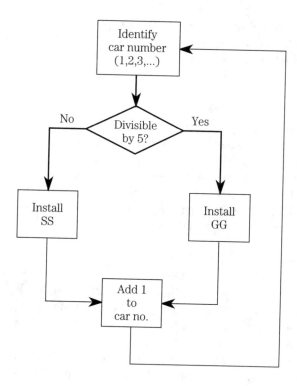

BRANCHING
Example of a routine with a branch point (diamond-shaped box).

If a mistake occurs, a car might get the wrong color hubcaps. Or worse, the computer might "hiccup" and miss a hubcap. This would throw off the robot's perception of the sequence of cars, so that it would think a new car had arrived with each set of rear wheels. Before long, the front wheel of a car would get a silver hubcap, and the rear wheel of the same car would get a gold one. The next car would get a gold hubcap on the front wheel and a silver one on the rear wheel. This "glitch" would be repeated down the line over and over again, messing up two out of every five cars—40 percent of the automobiles coming off the assembly line. The quality-assurance engineers would be furious!

The above example shows that branching routines are critical. They don't allow much room for error. To some extent, expert programming can improve the error tolerance of a system with branching routines. See also ARTIFICIAL INTELLIGENCE.

Brooks, Rodney

Rodney Brooks is an engineer who specializes in robotics and artificial intelligence (AI). He has become well known for his work with insect robots. One such robot has been named "Atilla." Brooks has also done extensive work with robot vision systems. He grew up in Australia, and received his training in AI at Stanford University.

Insect robots have six legs (as do living insects), and this is part of the reason for their name. But insect robots also share a central intelligence source, like an insect colony. The individual ants in an anthill are stupid, but the combination is quite smart. You know about this behavior if you've ever owned an ant farm.

Brooks' approach to AI has been, until recently, considered unorthodox. Traditionally, engineers have wanted to give AI systems reasoning power, that is, to have them mimic human thought processes. It is human nature to want to build replicas of ourselves, giving robots autonomy. But Brooks believes that the insect-intelligence concept is better in many, if not most, robotics applications. Support for his argument comes from the fact that insect colonies are highly efficient, and often they survive adversity better than higher life forms. See also ARTIFICIAL INTELLIGENCE, AUTONOMOUS ROBOTS, INSECT ROBOTS.

bug

A *bug* is an error in a computer program, or a malfunction of computer hardware. Often, a bug only degrades system efficiency, or causes occasional miscalculations. But sometimes a tiny error gets magnified many times as a program is executed. Then a seemingly minor bug can be disastrous.

Flaws in computer programs are "software bugs." When the term "bug" is used, it is usually in reference to this kind of problem. Programmers must "debug" computer programs before software is made widely available.

A computer might have a one-time component malfunction, caused by things like noise pulses on the line. This is a "hardware bug." An example of this type of bug is given in the article BRANCHING.

Some bugs can be "caught" by a computer from other computers. This kind of bug is called a *virus*. See also HARDWARE, SOFTWARE, VIRUS.

building-construction robots

Construction jobs are among the most hazardous occupations. This is especially true in the construction of large buildings. Wherever a job is hazardous for human beings to perform, there is a potential application for robots.

To some extent, the familiar crane is a robot. The human operator sits in an enclosure, relatively safe from harm, while the machine lifts heavy objects and carries them from place to place. Every movement of the large "robot arm" is controlled by the human operator.

A crane operator is fairly safe, but it is often necessary to have other workers help guide things like steel girders into place. These people work high up in the building framework, and are in danger of falling or being hit by objects that might fall from above. These workers could be replaced by robots, lowering the overall risk to human life involved in the construction of the building.

As in any industry where people might be replaced by robots, there is concern among workers about job loss. But the use of robots provides opportunities for more interesting, and less dangerous, work as long as companies are willing to spend some money for training people at new jobs, and as long as people are willing to learn new jobs. See also ASSEMBLY LINE, ASSEMBLY ROBOTS, AUTOMATED INTEGRATED MANUFACTURING SYSTEM.

bulletin board

A bulletin board, also called a *bulletin-board system (BBS)*, is a forum in which personal-computer (PC) owners can exchange information and ideas.

You must usually pay a fee for access to a BBS. A modem, and software that you can rent or buy, allows the PC to operate as a communications terminal through the telephone lines. When you access a bulletin board, you can check to see if anyone has left you any "mail." Some BBSs supply you with privacy codes, so that only you (in theory) can read your mail. Other BBSs are "open," so that all messages can be read by everyone. The illustration is a block diagram of a typical BBS.

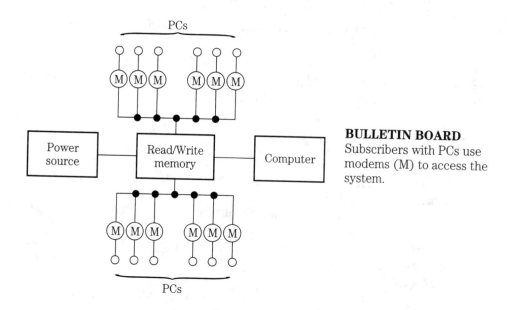

BULLETIN BOARD
Subscribers with PCs use modems (M) to access the system.

There are literally hundreds of different BBSs available for casual as well as professional use. An example of a professional BBS is ARPAnet, used by robotics and artificial-intelligence specialists (see ARPANET).

It can be risky to use a BBS if you intend to get software from the system. This is because public-domain software sometimes contains viruses. See also MODEM, VIRUS.

bumpers

See PROXIMITY SENSING.

burn-in

Before any electronic or electromechanical device is put to use, it should undergo a burn-in process. This usually involves running the device continuously for a few hours or days. In many cases, a faulty system will fail shortly after it is first "brought up." The burn-in process is a way of weeding out systems or subsystems with early-failure problems, so that there are fewer costly real-time failures. See also QUALITY ASSURANCE AND CONTROL.

Bushnell, Nolan

Nolan Bushnell is an interesting and colorful robotics engineer. He founded a company called Androbot, and has designed various amusement robots. One of these is a little cat that runs around and obeys commands. It even shows contentment by purring.

Bushnell is one of the few engineers outside Japan who has seriously tried to make a humanoid robot, or android. After some frustrating experiences with androids, he is said to have joked that trying to make humanoid robots is "playing God," and God might retaliate by messing them up. See also AMUSEMENT ROBOTS, ANDROBOT, ANDROID, ARTIFICIAL INTELLIGENCE, ARTIFICIAL LIFE.

byte

A *byte* is a unit of digital data, consisting of a string of eight bits (see BIT). One byte is generally the equivalent of one character, such as a letter, numeral, punctuation mark, space, line feed command, or carriage return command.

Computer memory is almost always specified in bytes. But because memory capacities of modern computers are large, kilobytes (units of $2^{10} = 1024$ bytes), megabytes (units of $2^{20} = 1,048,576$ bytes), and even gigabytes (units of $2^{30} = 1,073,741,824$ bytes) are given. The abbreviations for these units are Kb, Mb, and Gb respectively.

A typical personal-computer (PC) floppy disk has a little more than 1 Mb data storage capacity. Hard disks for PCs and commercial computers have anywhere from about 20 Mb to 1 Gb capacity. See also MEMORY.

C

C is a high-level computer-programming language. It is employed extensively by professional people because of its speed, power, and efficiency. C can be run on any computer with a compiler that can deal with it; in today's business environment, most computers are so equipped. The tradeoff is that C takes somewhat more time to learn than simpler languages like BASIC or FORTRAN. But it's much easier to learn to write in C than in assembly language.

The main reason that C is so efficient is that it is structured in a way that modern computers find easy to "digest." This allows the computer to work much faster with instructions in C, as compared with most other high-level languages. When you run a program written in C, the computer can use it almost as well as if you had written it in the assembly language. See also ASSEMBLY LANGUAGE, BASIC, COBOL, COMPILER, FORTRAN.

CAD/CAM

See COMPUTER-AIDED DESIGN, COMPUTER-AIDED MANUFACTURING.

cache memory

See MEMORY.

CAI

See COMPUTER-ASSISTED INSTRUCTION.

capacitive proximity sensing

How does a robot know when it is near an object? There are various devices that can be used for proximity sensing. One of the simplest uses capacitive effects.

You have probably noticed the effects of capacitance when using portable television sets or portable earphone radios. With a TV set using "rabbit ears" (an antenna mounted directly on the set), bringing your hand near either of the rods will often change the reception, especially if the station you are watching is weak. Or, if you are jogging with a portable headphone radio, you have almost certainly noticed how stations tend to fade in and out (especially during your favorite tunes). Part of this fading is caused by the effect of the earphone wire, which serves as an antenna, moving with respect to your body.

A capacitive proximity sensor uses a radio-frequency oscillator, a frequency detector, and a metal plate connected into the oscillator circuit, as shown in the diagram. The oscillator is designed so that a change in the capacitance of the plate, with respect to the environment, will cause the oscillator frequency to change. This change is sensed by the frequency detector, which sends a signal to the apparatus that controls the robot. In this way, a robot can avoid bumping into things.

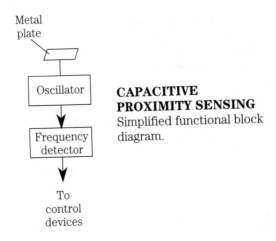

CAPACITIVE PROXIMITY SENSING Simplified functional block diagram.

Objects that conduct electricity to some extent, such as house wiring, people, cars, or refrigerators, are sensed more easily by capacitive transducers than are things that do not conduct, like wood-frame beds and sofas. Therefore, other kinds of proximity sensors are often needed for a robot to navigate well in a complex environment, such as your home. See PROXIMITY SENSING.

Capek, Karel

Karel Capek was a Czech who wrote a play about robots in 1920. The title of the play, translated into English, is *Rossum's Universal Robots*.

In the play, the machines are made for mundane work, freeing people to do more interesting things. But someone sees the potential for using the robots as soldiers. Soon the machines are being employed destructively, rather than constructively as their developers intended. The robots develop more and more of what we now call artificial intelligence (AI). Eventually, the robots realize they are stronger, smarter, and longer-lived than human beings. Finding people a nuisance, the robots begin to murder the human population.

The play caused an uproar when it was performed. People knew, even in the 1920s, of the problems with assembly-line labor, and that robots could alleviate some of those problems. Some took Kapek's play as a warning, and as a cynical statement that "the cure can be worse than the disease." To this day, some researchers have fears that androids might be used, and might ultimately evolve, much as Kapek showed them. Kapek himself apparently hoped people would see his play as a satire. See also ANDROID, ARTIFICIAL INTELLIGENCE, ARTIFICIAL LIFE, ASSEMBLY LINE, BIOLOGICAL ROBOTS.

caretakers

Some researchers think that advanced artificial-intelligence (AI) systems might be useful as caretakers to keep us humans from fighting wars and doing other bad things to each other. Rational and impartial, these systems could be used to resolve differences among people, from sovereign nations all the way down to small-claims courts.

Many people say this idea is impossible, ridiculous, and unworkable. The argument goes something like this: "It wouldn't be long before the computer would reach a solution that one or both human parties could not accept. So they'd start fighting anyway." The answer: Give the computer police robots to enforce its decisions!

Some authors have written novels and made movies based on the idea of using robots to govern societies. Usually, the stories show the bad side of robots as rulers. The machines are created with good intentions. But the robots end up robbing people of their freedom in one way or another. A good example of this is the movie *COLOSSUS: The Forbin Project*, produced during the years of the Cold War between the United States and Russia, then called the Soviet Union.

Some scientists, and even some philosophers and theologians, think a cold, calculating machine would be an improvement over hot-tempered, power-hungry politicians and corporate chiefs. They say that our "freedom" is an illusion. The only "freedoms" we'd lose under a computer's caretaker rule, they say, would be the ones that let us steal from, kill, enslave, and otherwise abuse our fellow human beings. See also ARTIFICIAL INTELLIGENCE, COLOSSUS, POLICE ROBOTS.

Carter, Forrest

Modern computers make use of integrated circuits (ICs) using silicon wafers, or chips, as the basic material. An IC is manufactured by etching the silicon, forming microscopic transistors, resistors, capacitors, and diodes. In a sense, IC fabrication is something like sculpting a statue out of marble, except on a far smaller scale. See INTEGRATED CIRCUIT.

Some scientists have suggested that, rather than carving ICs from chips, the circuits could be built up molecule by molecule. A leading researcher in this field is Dr. Forrest Carter of the U.S. Naval Research Lab.

Building up, rather than chiseling away, has some advantages. Perhaps most significant is the fact that this method of IC fabrication would result in the greatest possible degree of miniaturization. Molecules are the smallest bits into which material compounds can be broken while maintaining their identities. By starting from nothing and building something up, the smallest details can be given attention. This is far more difficult, if not impossible, to do by chipping away. See also DREXLER, ERIC.

Cartesian coordinate geometry

Industrial robot arms can move in various different ways, depending on their intended use. One mode of movement is known as *Cartesian coordinate geometry*. This term comes from the Cartesian system often used for graphing mathematical functions. Alternatively, it is called rectangular coordinate geometry.

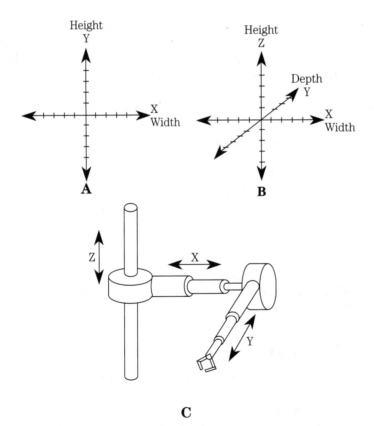

CARTESIAN COORDINATE GEOMETRY Basic scheme for two dimensions (A) and three dimensions (B); a robot arm using this configuration (C).

The drawings at A and B show Cartesian coordinate systems in two and three dimensions, respectively. The axes are always perpendicular to each other. Variables are assigned the letters x and y for a two-dimensional Cartesian plane (A), or x, y, and z for a Cartesian three-space (B). These can be thought of as "width and height" for the plane, and "width, depth, and height" for three-space.

The drawing at C depicts a robot arm capable of moving in three dimensions using Cartesian coordinate geometry. See also CYLINDRICAL COORDINATE GEOMETRY, POLAR COORDINATE GEOMETRY, ROBOT ARMS, SPHERICAL COORDINATE GEOMETRY.

catastrophic failure

The term *catastrophic failure* refers to a complete, often sudden, breakdown in a system. This might happen because of massive damage to components, because of a virus in the software, or because of a malfunction in a critical part of the system.

When a computer "freezes up," without actual damage to components and without the presence of a virus, the failure is called a crash.

Computer-controlled systems are usually designed so that viruses, or minor malfunctions, won't cause catastrophic failure. This is done by incorporating backup circuits, especially at those points where a malfunction could cause a major system breakdown. Also, a system should be put together in such a way that there is the fewest possible number of critical points or units. When these measures are taken, component failures and viruses cause limited deterioration of system performance. See also CRASH, GRACEFUL DEGRADATION, VIRUS.

CCD

See CHARGE-COUPLED DEVICE.

CD

See COMPACT DISK.

central processing unit

Every computer has a "brain," comprised of one or more integrated circuits (ICs). The control center of a computer is known as its *central processing unit* or *CPU*.

The CPU supervises the running of the programs. It consists of an arithmetic logic unit (ALU), a control/timing unit, and registers, also known as memory circuits.

The main IC of the CPU is called the microprocessor. It contains the ALU and the control/timing unit. The memory circuits are separate from the microprocessor. There are several different kinds of memory devices; the particular type used by a given computer depends on the application.

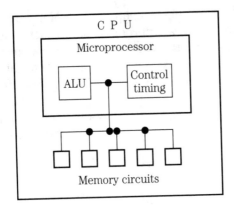

CENTRAL PROCESSING UNIT
Consists of microprocessor and memory circuits.

The block diagram shows the CPU of a typical computer, along with its memory. See also COMPUTER, MEMORY.

character

In computer systems, a character is a letter of the alphabet from A through Z, a numeral from 0 through 9, a punctuation mark, or any of various different symbols that you can find on a typewriter keyboard. These symbols vary, but they always include certain standard items.

The illustration shows the layout of the computer keyboard on which this book was typed. Function keys are not included. This is pretty much standard among personal-computer keyboards.

A character is normally equivalent to one byte of data. See also BYTE.

!	@	#	$	%	^	&	*	(	)	_	+	|
1	2	3	4	5	6	7	8	9	0	-	=	\
Q	W	E	R	T	Y	U	I	O	P	{	{	
q	w	e	r	t	y	u	i	o	p	]	]	
A	S	D	F	G	H	J	K	L	:	"		
a	s	d	f	g	h	j	k	l	;	'		
Z	X	C	V	B	N	M	<	>	?			
z	x	c	v	b	n	m	,	.	/			

CHARACTER Standard computer keyboard layout.

charge-coupled device

A *charge-coupled device (CCD)* is a digital circuit that enhances the detail seen by a video camera. These circuits are used by astronomers to improve their view of outer space. Biologists sometimes employ CCDs to process the images seen by their microscopes. Robots sometimes use them to sharpen their "vision." Meteorologists use them to process satellite and radar images.

The CCD works in a manner similar to a digital signal processor. Digital signals are "clearer" than analog ones, in the sense that on/off, or high/low, duality is far easier for a computer to work with than continuous variability. A CCD can also process an analog signal. A CCD allows a video image to be stored in a digital memory circuit.

A simplified block diagram of a CCD vision system is shown in the accompanying illustration. See also ANALOG, DIGITAL, DIGITAL SIGNAL PROCESSING, MEMORY.

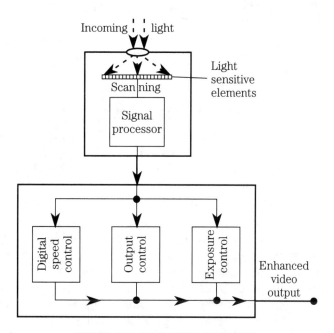

CHARGE-COUPLED DEVICE
Block diagram of a CCD computer vision system.

checkers-playing machine

A computer can be programmed to play an almost unbeatable game of checkers. An excellent program has been created by Arthur Samuel (see SAMUEL, ARTHUR), in which the computer cannot only play the game move-by-move, but it can also "look ahead" to see the possible consequences of a move.

Checkers is a fairly simple board game. It is more complex than tic-tac-toe, but far less complicated than chess. If you have played tic-tac-toe for any length of time, you've discovered that you can always get at least a draw. It is not difficult to program a computer to play tic-tac-toe. You need only "look ahead" one move in this game.

"Look-ahead" strategies of more than one move take practice to acquire. It's hard for a human, let alone a computer, to develop this degree of savvy in any game. Arthur Samuel's checkers program uses the "look-ahead" strategy so effectively that even the best human players in the world find it next to impossible to beat his machine.

There is yet another scheme that can be used for checkers. This scheme is to adopt a general game plan. General strategies can be broadly categorized as either defensive ("hang back") or offensive ("go for it like mad"). In my limited experience, the offensive strategy seems to work better. The key is that once you decide on a strategy, you mustn't change, even if things seem to be going badly. The defensive/offensive schemes require a "look ahead" of just one move. Therefore, if you're interested in programming a computer to play checkers, but you don't feel as qualified as Arthur Samuel, you might try a combination of "one-move look-ahead" in conjunction with a defensive or offensive general strategy. See also CHESS-PLAYING MACHINE.

chess-playing machine

Chess has been used to develop and test computer "brains" ever since the beginning of artificial intelligence (AI) research. One of the first chess-playing machines was developed by Rand Corporation in 1956. In fact, at a project at Dartmouth, Professor John McCarthy first used the words "artificial intelligence" in reference to "smart" computers.

While a computer can be programmed to play a good game of chess, no program has been developed that can consistently beat good human players at this game. Chess is a complex game, in which various different strategies can be employed. Of course, the object is to capture the opponent's king. But the road to this victory can be long and tortuous.

Perhaps you've seen masters playing a game of chess; almost all of their time is spent sitting in silence, contemplating the next move. The human brain is millions of times more sophisticated than the most advanced computer yet developed, and the very best human chess masters must tax their minds to the utmost in order to win. Computers are no match for these minds—not yet, anyway.

A computer can operate according to fairly simple strategies. But it cannot consciously try to outwit an opponent. No computer has yet been able to develop such techniques as bluffing, or distracting, the other player.

Checkers, in contrast, is a much simpler game. A computer can be programmed so that it is very hard to beat at checkers. Consistent strategies in checkers work much better than they do in chess (see CHECKERS-PLAYING MACHINE).

The program developed by Rand Corporation and Prof. McCarthy was also able to prove some mathematical theorems. This is another good way to test the intelligence of a computer. See also ARTIFICIAL INTELLIGENCE.

chip

See INTEGRATED CIRCUIT.

choreographer program

A *choreographer program* is a form of computerized animation. The program is written with the intention of simulating motion pictures, especially with human forms.

One such program, called the "Ultimate Choreographer," was written by Charles Lecht of Lecht Sciences, Inc. It shows the famous mime, Marcel Marceau, on a video screen. But this is not a television-type video, and it is not a movie. Instead, it is entirely contained within the digital data banks of a personal computer (PC). The programmer can sit at the computer and, by giving various commands, cause the mime to do almost anything. See also LECHT, CHARLES.

Cincinnati Milacron

Cincinnati Milacron is the name of a major manufacturer of industrial robots. Located in Lebanon, Ohio, the company was one of the first to use a computer to control a set of machines. The company also builds amusement robots, used as demonstration units in robotics shows around the world.

Cincinnati Milacron robots are extensively used in assembly-line work for drilling, welding, and other repetitive operations. Some of the robots incorporate vision systems to help them "see" what they are doing. One such system uses lasers to illuminate a surface as it is being welded. Two or three lasers, shone at

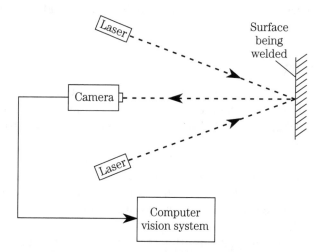

CINCINNATI MILACRON Laser illumination allows a computer to align a surface for welding.

different angles, provide lighting that a camera and computer can interpret in terms of depth and position (see drawing). This increases the work output and improves the work quality at the same time. See also AMUSEMENT ROBOTS, INDUSTRIAL ROBOTS, VISION SYSTEMS.

ciphers

See CRYPTANALYSIS.

Clarke, Arthur C.

Arthur C. Clarke is a well-known science-fiction writer. Some people think he might be the best science novelist who has ever lived.

Clarke is best known for the story *2001: A Space Odyssey*. This novel was made into a movie that did well at the box office. In this story, an interplanetary spacecraft is overseen by a highly intelligent computer named Hal, that can speak in a perfectly normal human voice. In fact, Hal even seems to show emotions, along with exceptional common sense. Hal is a good example of what might happen when computers develop a level of artificial intelligence (AI) comparable to a human brain.

Unfortunately, not all of Hal's behavior is good. Hal is so advanced that it (he!) develops a persecution complex. He becomes paranoid. Hal starts to think that Dave, the ship's captain, wants to take his power away. As a result, Hal refuses to open up the pod bay doors to let Dave bring a shuttle back into the main ship. Dave gets in anyway, by braving the vacuum of space without his pressure suit, and then methodically proceeds to "pull the plug" while Hal pleads for mercy. Thus Hal brings about his own demise. See also ARTIFICIAL INTELLIGENCE, ARTIFICIAL LIFE.

clean room

In some industries, it is important that dust, dirt and other contaminants be kept to an absolute minimum. A good example is the manufacture of integrated circuits (see INTEGRATED CIRCUIT). Robots have a big advantage over people in these environments.

Humans shed particles constantly. Much of the dust in a room is made up of tiny particles of dead skin, for example. Get rid of the people in a building, and you will get rid of much of the dust. Good air conditioning helps, but if all other things are equal, a place will be cleaner without people in it, as compared with having people in it.

By taking certain precautions, the environment in a room can be kept "clean" while still allowing humans in. People who enter such a room must first put on airtight suits, gloves, and boots. But a room that has only robots, and no people, can always be just a little bit cleaner.

A clean room is a room equipped and maintained to keep dust to an absolute minimum. The cleanliness is measured in terms of the number of particles of a certain size in a cubic foot of air. Alternatively, the liter (about a quart) can be used as the standard unit of volume. A typical clean room has 10 particles averaging 0.5 microns in diameter per cubic foot of air. This is called a *Class-10* clean room.

clinometer

A *clinometer* is a device for measuring the steepness of a sloping surface. Mobile robots use clinometers to avoid inclines that might cause them to tip too far, possibly even falling over.

The floor in a building is almost always horizontal. Thus, its incline is zero. But sometimes there are inclines such as ramps. A good example is the kind of ramp used for wheelchairs, in which a very small elevation change occurs. A rolling robot can't climb stairs, but it might use a wheelchair ramp, provided the ramp isn't so steep that it would upset the robot's balance or cause it to spill or drop its payload.

A clinometer is a transducer that produces an electrical signal whenever it is tipped. The greater the angle of incline, the greater the electrical output, as shown in the graph at A. A clinometer might also show whether an incline goes down or up. A downward slope might cause a negative voltage at the transducer output, and an upward slope a positive voltage, as shown in the graph at B.

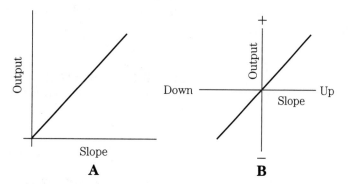

CLINOMETER At A, voltage (vertical axis) versus slope (horizontal axis). At B, up-slope causes positive voltage; down-slope causes negative voltage.

clone

A *clone* is a machine that performs all the same functions of some other machine. The term is commonly used with respect to personal computers (PCs). You might have a PC that is "IBM (International Business Machines) compatible," and that works in essentially the same way as the IBM PC. Such a computer is an IBM PC clone.

In biology, the term *clone* has a much different meaning than it does in computer science. In the lab, biologists have grown organisms using just one cell. All the genetic information for the organism is contained within the nucleus of every cell in its body. Some biologists believe that any organism, in theory, can be cloned, including humans.

How does this fit in with robotics and artificial intelligence? Some scientists are interested in the possibility of cloning human beings and then programming their minds. Their bodies would be entirely biological, but their minds would be like super-sophisticated computers. Right now, most researchers regard this idea as half-baked, and its realization impossible. Even if it can be done, it could cause serious social problems and ethical dilemmas. See also ARTIFICIAL LIFE, BIOLOGICAL ROBOTS.

closed-loop system

The term *closed loop* refers to any device or machine that regulates itself in some way. Closed-loop systems can be found in many different kinds of devices, from the engine in a car (governor) to the gain control in a stereo radio receiver (automatic level control).

A closed-loop system, also known as a *servoed system*, has some means of incorporating feedback from the output to the input. A sensor at the output end generates a signal that is sent back to the input to regulate the machine's behavior. A good example of this is a back pressure sensor (see BACK PRESSURE SENSOR).

The block diagram shows the general scheme for a closed-loop robotic system. See also OPEN-LOOP SYSTEM.

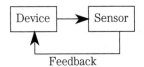

CLOSED-LOOP SYSTEM
Feedback governs the behavior of the device.

CMOS

See COMPLEMENTARY-METAL-OXIDE-SEMICONDUCTOR TECHNOLOGY.

COBOL

The acronym *COBOL* (pronounced like "cobalt" without the "t") stands for *common business-oriented language*. It is a high-level computer-programming language.

COBOL is not difficult to learn, because the commands are mostly words and directions written in English. The language is especially good for handling numbers in tabular form. This makes it useful for bookkeeping and other numbers-related jobs in corporations. See also BASIC, C, FORTRAN.

cold boot

See BOOT.

color sensing

Robot vision systems often function only in black and white, like simple television. But color sensing can be added, in a manner similar to the way it is added to television systems.

Color sensing can help a robot with artificial intelligence (AI) tell what an object is. Is that horizontal surface a floor inside a building, or is it a grassy yard? If it is green, it's probably a grassy yard. (How often have you seen green carpeting indoors?) Sometimes, objects have regions of different colors that have identical brightness as seen by a black-and-white system; these objects, obviously, can be seen in more detail with a color sensing system than with a vision system that sees only shades of gray.

The drawing shows a block diagram of a color sensing system. Three black-and-white cameras are used. Each camera has a color filter in its lens. One filter is red, another is green, and another is blue. These are the three primary colors. All possible hues, brightnesses, and saturations are comprised of these three colors in various ratios. The signals from the three cameras are processed by a microcomputer, and the result is fed to the robot's AI center. See also ARTIFICIAL INTELLIGENCE, OBJECT RECOGNITION, TEXTURE SENSING, VISION SYSTEMS.

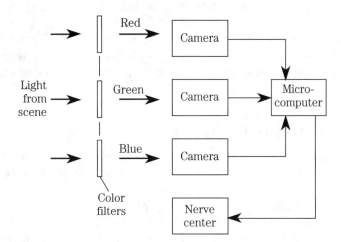

COLOR SENSING Block diagram of color vision system.

COLOSSUS

Suppose that a single, intelligent computer were given control of all the nuclear weapons systems in the world. If that could be done, then a nuclear war could never occur because of human error. It would have to be executed by the computer.

To some extent, computers already control the nuclear defenses in the United States. A "fail-safe" system makes it impossible for any one person to press a button and cause a nuclear holocaust. But some scientists think the current systems don't go far enough. They argue that a computer, seeing how illogical nuclear war is, would not let it happen. Thus these scientists believe a computer would be the ideal controller for nuclear arsenals.

Such is the scenario for a 1960s novel and movie called *COLOSSUS: The Forbin Project*. The U.S.-designed computer, called COLOSSUS, discovers that there is a similar system in Russia (then known as the Soviet Union). The two computers interconnect, and begin to share data and learn from each other. Their artificial intelligence (AI) grows more and more rapidly until, before very long, they are far smarter than any human being. They are also supremely powerful, because they hold control of all the nuclear missiles in both the United States and the Soviet Union.

This intelligence and power having been achieved, the computers decide that they will not only outlaw war, but they will control people's lives in every detail. COLOSSUS tells people when, and how much, they may eat, work, sleep, and even exercise. If anyone disobeys COLOSSUS, the computer will retaliate by launching a missile and killing several million people. To show that it means business, the computer does exactly this on one occasion.

Many researchers in AI believe that the COLOSSUS scenario is realistic! Nowadays, with the Cold War between the U.S. and Russia apparently over, there seems to be less risk of a massive nuclear holocaust than there was in the 1960s. But there are still a lot of nuclear missiles in the world. If we ever have the opportunity to give computers absolute control over all our nuclear weapons, we had better be sure there's a way to "pull the plug." See also ARTIFICIAL INTELLIGENCE, CARETAKERS.

combinatorial explosion

Certain problems are extremely hard for machines to work out. Sometimes there are so many different choices that a computer just can't make a good decision.

Suppose you're playing a game of five-card draw. This is a form of poker. You receive five cards, as does every other player in the game. You have the option of receiving zero to four replacement cards, but you have no way of predicting what the new cards will be. The only way you can be sure of your hand is to stick with the cards you have—but your hand is not very good. What should you do when the dealer asks you, "Would you like any cards?" If you ask for, say, three new cards, there are literally thousands of possible outcomes. This kind of problem is called a *combinatorial explosion*.

The illustration shows the principle of a combinatorial explosion. A decision-making process between either of two alternatives, repeated many times, quickly blows up into a huge array of possibilities. If there are n repetitions, where n is some integer, then the number of possible choices is 2^n (two multiplied by itself, n times).

The following story will give you an idea of how combinatorial explosions can confuse people (as well as machines). A child went to see the King of a very rich country on the first day of spring. "Can I have some money as a birthday present for my mom?" asked the child, having heard that the King was a generous man. That he was—and a practical joker too. He told the child, "I'll give you either a million dollars, or else a penny today, and then double it every day until Mother's Day. Which would you rather have?" The King was confident the child would go for the million dollars. After all, that's an awful lot of pennies! The child replied, "A penny today, and then double it every day until Mother's day." By Mother's Day, the child's mother was Queen.

In complex board games, and also in some real-life situations, intuition often works better than any computer program. In general, the more skill a game or scenario requires, the harder it is to write a computer program to choose well. Artificial intelligence is one thing; artificial wisdom is quite another.

In complex board games, and in real life too, there are subjective variables. These are variables like the personalities of your opponent(s), the expressions in their eyes, and other cues that no electronic brain comes close to understanding. In a combinatorial explosion, when the number of choices becomes huge (like the number of pennies the child's mother would have had on Mother's day if the King hadn't gone broke before then), problems acquire subtle aspects. Some researchers even think that extrasensory perception (ESP) plays a role in games like poker and chess. Can truly skilled gamesters read their opponents' minds? Most scientists doubt it; a few think it's possible. But practically no one believes that computers have ESP. See also ARTIFICIAL INTELLIGENCE.

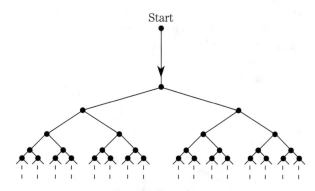

COMBINATORIAL EXPLOSION Start with one item, and then double the number over and over.

command

In computer practice, a *command* is any statement that you make to get the computer to do something. Commands can be given in various different ways. The two most common systems are command driven and menu driven.

An example of a command-driven program is the word-processing scheme used to write this book. The software, called XYWrite, makes use of the disk operating system (DOS). In this system, if you want to select a file, you type "CALL" followed by the name of the file. A file might have as many as eight letters and/or numbers, followed by a period, and up to three more letters and/or numbers. (The name of the file containing this article was *RAI.C*.)

In a menu-driven program, you select what you want from a list that appears on the screen. Windows is probably the most common scheme for this. You might select the desired file by moving the cursor with keyboard arrows (up, down, right, left) or by using a mouse.

In robotics and artificial intelligence, commands can often be given in the form of spoken words. This can also be done with command-driven programs for personal computers. Rather than typing "CALL RAI.C," you might say it instead. But this gives rise to all kinds of potential hangups. How do you say the period? How does the computer know you mean "C" and not "SEE"? These problems are dealt with by engineers who work with cybernetics and speech recognition systems. See also COMMON-SENSE SUMMER PROJECT, CYBERNETICS, DISK OPERATING SYSTEM, MOUSE, SPEECH RECOGNITION, WINDOWS.

Common-sense Summer Project

The *Common-sense Summer Project* was a session in which artificial-intelligence (AI) experts tried to find ways to program real-world notions into a computer.

It's extremely hard to get a "smart robot" to understand things like people do. While there might someday be a robot that is as intelligent as a person, and which might even regard humans as flawed and obsolete, most researchers agree that this will not happen for a long time.

To get an idea of how complex a simple command might be, suppose you tell a robot, "Go into the kitchen, make me a cheddar cheese sandwich, and bring it out to me on a paper plate."

First, the robot must understand the words you speak, and also must identify, and remember, who you are and where you are. The words must be translated into a chain of commands that can be followed by branching, preferably with no more than two decisions (yes/no) at each branch point.

Some of the questions a robot must ask itself, and then answer, follow:

1. Where is the kitchen?
2. What is a sandwich?

3. What are the ingredients in a cheddar cheese sandwich?
4. Where is the bread?
5. Where is the cheese?
6. Which cheese is the right kind (cheddar)?
7. How will I know that I've found the right things, and not, say, two napkins and a piece of ham?
8. What should I do if you're out of bread or cheddar cheese?
9. How much cheese is a reasonable amount?
10. Should the cheese be sliced flat, or should it be in a cubical chunk? Or a ball? Or what?
11. How can I recognize a plate?
12. How can I tell the difference between good china and a paper plate?

. . . and so on, all broken down into binary data! The above questions are very general, and could themselves each be broken down into hundreds or thousands of individual yes/no branch points. The end result must be that the robot comes up to you without much delay, holding in its mechanical hands a paper plate, on which there are two slices of bread with a reasonable amount of cheddar cheese sliced flat between them.

As an exercise, you might try writing a detailed set of instructions for tying a shoelace, without using any diagrams. Someone must be able to follow these instructions precisely, and end up with a properly tied shoelace. Once you've completed the set of instructions, try to get them into a series of questions, all of which can be answered either "yes" or "no"—and which will still, when followed exactly, result in a properly tied shoelace. This will give you some idea of how difficult it is to program a "smart robot" to tie your shoes. (Hint: Set aside a couple of months for this.)

See also ARTIFICIAL INTELLIGENCE, BRANCHING, CYBERNETICS.

communication protocol

When computers "talk" to each other, they exchange digital signals. There are several variables in a digital signal; for two computers to effectively share data, all of these parameters must match. The *communication protocol* refers to the specifications of the digital signals.

The most common variables in a protocol include the speed (baud rate), code type, bit lengths, mark-to-space ratios, and spacings. Some systems can exchange data in both directions at the same time, like you do on the telephone; this is called full duplex. Others must take turns, like you do when you use a two-way radio; this is called *half duplex*. See also BAUD RATE.

compact disk

A *compact disk (CD)* is a device that stores digital information. In audio and video applications, CDs are known for their excellent fidelity. Noise (hissing and popping in audio systems, or "snow" and horizontal lines in video systems) is virtually eliminated. When CDs are used to store computer data, errors hardly ever occur.

The data on a CD is stored in the form of tiny, crater-like depressions (see the drawing). The surface of the disk is highly reflective of visible light. To recover the data, a laser beam scans the surface. The depressions cause modulation of the reflected laser light. In computer practice, this system is sometimes called CD ROM, meaning compact-disk read-only memory. See also DISKETTE, HARD DISK, MEMORY.

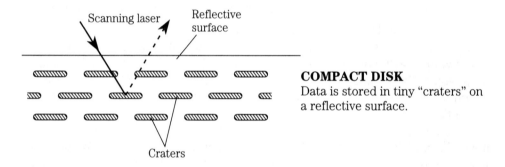

COMPACT DISK
Data is stored in tiny "craters" on a reflective surface.

companionship robots

In the future, "smart robots" might be used not only for convenience and amusement, but for companionship too. Imagine owning a robot that could not only prepare your meals, do your laundry, and clean your house, but could also carry on an intelligent conversation with you, give you a bath, and perhaps even rub your neck when you're tired!

Researchers at a company called Anticipatory Sciences think that such robots can, and will, be made. In fact, as artificial intelligence (AI) advances, computers will become more able to converse with people. To some extent this is already happening in the form of electronic games and other interactive devices.

Computers can be logical, humorous, and maybe even carry on a heated debate; but robot grippers don't have the comforting power of real, living, human hands. (Many researchers think the human touch *can't* be built into machines.) There is debate, too, over how much psychological comfort a machine could give to a lonely person, no matter how brilliant, witty, or sympathetic the programming might be. When talking with a "smart robot," there would be, in the back of your mind, the knowledge that your companion was not human. See also AMUSEMENT ROBOTS, ANTICIPATORY SCIENCES, ARTIFICIAL INTELLIGENCE, ARTIFICIAL LIFE.

compatibility

Compatibility is a term used with electronic devices, especially computers and computer peripherals. You've probably heard that certain software is "IBM compatible." This means that the software can be run on an IBM personal computer (PC), or on any of its clones.

Devices or programs are fully compatible if, and only if, they will work together, doing all their functions, without either one having to be modified in any way. If either device or program must be changed, they are partially compatible. If fundamental changes must be made to one or both devices or programs, then they are noncompatible or incompatible.

One of the biggest headaches for consumers in modern society is the seemingly endless changes that are made to products, rendering machines and software programs partially compatible or incompatible. To some extent, this is a natural consequence of free enterprise. But many companies believe that standardization is more to their advantage than crafting their products and accessories differently from everyone else. See also AMERICAN NATIONAL STANDARDS INSTITUTE, CLONE.

compiler

In a digital computer, the *compiler* is a program that translates the high-level language, such as BASIC, C, COBOL, or FORTRAN, into machine language. The operator understands and uses the high-level language; the computer operates in machine language. Machine language is extremely difficult and tedious for a person to use; a high-level language can't be understood directly by a computer. The compiler works like an electronic interpreter.

The compiler must be written uniquely for a given high-level language. Every high-level language is different from every other. See also HIGH-LEVEL LANGUAGE, MACHINE LANGUAGE.

complementary-metal-oxide-semiconductor technology

Complementary-metal-oxide-semiconductor, also called *CMOS* (pronounced "seamoss"), is the name for a technology used in digital devices, such as computers. Two types of field-effect transistors (FET) work together on a single integrated circuit (IC). A CMOS IC is fabricated on a chip, or wafer of silicon. The illustration shows the interconnection of two metal-oxide-semiconductor (MOS) FETs to form a CMOS switch.

One of the biggest advantages of CMOS technology is the fact that it can work with tiny electrical currents. Thus, it draws very little power from the power supply, allowing the use of batteries. This is an obvious asset for robots and portable computers. Another advantage of CMOS technology is that it works extremely fast. It can process a lot of data in a short period of time.

A disadvantage of CMOS is that it is easily damaged by static electricity. Devices of this type must be stored with their pins embedded in conductive foam material. When building or servicing equipment using CMOS, technicians must take precautions to avoid the buildup of static electric charges on their hands and on instruments like probes and soldering irons. See also INTEGRATED CIRCUIT.

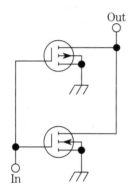

COMPLEMENTARY-METAL-OXIDE-SEMICONDUCTOR TECHNOLOGY
Schematic diagram of CMOS logic gate.

composite video signal

A *composite video signal* is the waveform that modulates a television carrier wave. It contains the video information, and also synchronization, blanking and timing information. This signal is typically 6 megahertz (6 MHz) wide for fast-scan television, and approximately 3 kilohertz (kHz) wide for slow-scan signals (see BANDWIDTH). A video camera, such as an image orthicon or vidicon, produces a fast-scan signal (see IMAGE ORTHICON, VIDICON).

Most video monitors, used with computers, communications terminals, and robot-vision systems, operate from the composite video signal. The illustration

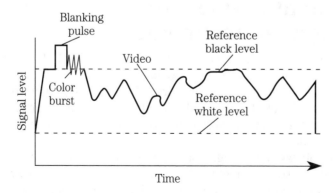

COMPOSITE VIDEO SIGNAL One line in a frame of a color television picture.

shows the waveform for a single line of a color picture signal. There are normally 525 or 625 lines in a complete frame for standard fast-scan television.

For robot vision systems, there are advantages to using more lines per frame than is standard with television. This is because the result would be improved image resolution, that is, the ability to see greater detail. See also VISION SYSTEMS.

computer

A *computer* is an electronic machine that processes information. Computers can be found in children's toys, in radio communications equipment, and in machines of all kinds, including robotic systems.

Computers range in size and complexity from simple microcomputers, found in calculators and scanning radios, to vast systems, interconnected by telephone, radio, fiberoptic, and/or satellite links.

Components of a personal computer system

A typical personal computer (PC) system has several peripherals, in addition to the main unit. Refer to the block diagram at Fig. 1.

Keyboard The keyboard lets you enter data into the computer. The keys are arranged like those on a typewriter (see CHARACTER). As you type, you can see the data in readable form on the monitor.

Monitor Most desktop PCs nowadays use color monitors, which resemble small, high-definition television sets. Some PCs, especially notebook or laptop types, use monochrome monitors. The monitor for a desktop PC generally uses a cathode-ray tube, and is "outboard," or separate, from the main unit. Therefore, you can choose your own monitor, either color or monochrome, to suit your

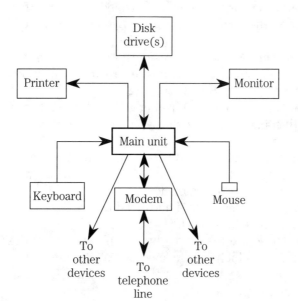

COMPUTER
Fig. 1. A personal computing system.

needs. Most notebook PCs use built-in liquid-crystal-display (LCD) monitors. See MONITOR.

Disk drives All PCs have a diskette drive built in. Most have provision for addition of one or more auxiliary diskette drives. Many PCs include a hard disk built in. See DISKETTE, HARD DISK.

Printer When you buy a PC, you'll almost certainly want a printer. There are several different kinds. The fastest and cheapest is the dot-matrix type. More expensive printers use lasers. A few use daisy-wheel devices, similar to those found in office typewriters. See PRINTER.

Mouse In addition to the keyboard, most computers have provisions for using a mouse, which is a small unit that can be slid around on a flat surface, causing an arrow or cursor to move to various points on the monitor screen. See MOUSE.

Modem A modem lets you connect your PC to the telephone lines. In this way you can communicate "on line" with others who also have PCs and modems. Modems can be used with ham radio equipment too; you must have a license to communicate via ham radio. See MODEM.

Other accessories Household appliances can be computer-controlled. Examples are burglar alarms, garage-door openers, thermostats, ovens, and even dishwashers and laundry machines. Some people love to use PCs to control just about everything in their homes, even if doing so actually increases the complexity of such jobs. This is an example of computer addiction or computermania.

Inside the main unit

The main unit of a digital computer consists of several parts, as shown in the block diagram at Fig. 2.

Central Processing Unit (CPU) This includes the arithmetic/logic unit (ALU), a control/timing unit, and memory circuits, also called registers, that store data for short periods. (Here, the term *data* means "digital information.") See CENTRAL PROCESSING UNIT.

Memory Medium-sized internal memories store more data, for a longer time, than the CPU registers. These circuits are of two kinds: read-only memory (ROM) and random-access memory (RAM). See MEMORY.

Address bus This is a line inside the main unit, running from the CPU to the RAM and ROM. The CPU uses this line to obtain its programming, and to locate information in memory.

Control bus This line carries control signals from the CPU to the various other circuits in the computer. For example, the CPU might instruct the RAM to send some data to the monitor, the printer or the modem.

Data bus This line carries data from the RAM and ROM to the CPU, or to other parts of the computer as instructed by the CPU.

Input/output (I/O) ports These are the points by which data enters the computer from outside, and/or from which data comes out of the computer. These ports go to external equipment such as the keyboard, printer, modem, monitor, and mouse.

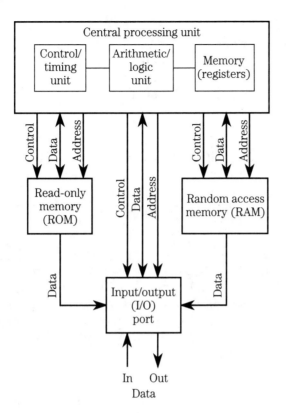

COMPUTER
Fig. 2. Elements of a PC main unit.

Analog computers

Most modern computers are digital, working with a combination of on/off signals. A few computers use analog processes to make their calculations. In some scientific work, analog computers are more effective than digital ones. An analog computer works with quantities that vary continuously. See ANALOG, DIGITAL.

"IQ" keeps going up

Computer scientists sometimes talk about the evolution of computers in terms of generations. There have been several generations of computers since the earliest machines were built in the 1940s.

First-generation computers

These machines used vacuum tubes, like the ones in antique radios. Information was usually stored on punched cards or paper tape. A first-generation computer, even if it occupied a whole building, was much slower than even the smallest PCs are today. Not only that, but they wasted an enormous amount of power as heat, because vacuum tubes are inefficient. First-generation machines were built during the late 1940s and early 1950s.

Second-generation computers

In the late 1950s and early 1960s, transistors replaced vacuum tubes for switching. This reduced the sizes and weights of computers by hundreds of times. It also greatly reduced the power needed to run them. Data was stored primarily on magnetic tapes, either reel-to-reel (for large computers) or cassette (for smaller machines).

Third-generation computers

During the late 1960s, integrated circuits (ICs) became available, and this resulted in another great reduction in physical size and power requirements (see INTEGRATED CIRCUIT). Magnetic disks became available for data storage. Interconnection of computers was done via telephone, radio, fiberoptic, and satellite links.

Fourth-generation computers

In the early 1980s, a new form of IC technology was developed, called *complementary-metal-oxide-semiconductor (CMOS) technology*. This increased the speed at which a computer could work, and also reduced the power requirements (see COMPLEMENTARY-METAL-OXIDE-SEMICONDUCTOR TECHNOLOGY). More powerful programming languages were developed. Literally hundreds of software packages became commercially available.

Fifth-generation and beyond

Characteristics of future computers will include more and more "user-friendly" software, making it possible for anyone to work with, and even program, a computer. Speech recognition and speech synthesis will be perfected. This will include the ability to infer the correct words through analysis of context and syntax. Researchers might find new ways to fabricate ICs, such as biochips that mimic the structure of living matter. Artificial intelligence (AI) will probably evolve gradually. As these things happen, computer scientists will eventually speak of sixth-generation and higher computer systems. See also BIOCHIP, CONTEXT, SPEECH RECOGNITION, SPEECH SYNTHESIS, SYNTAX.

Artificial intelligence

As computers get more and more powerful, they are starting to seem almost human in some ways. Computers have been programmed to play an excellent game of checkers. (They can play chess, too, but not as well.) Computers are already taking the places of some schoolteachers (see COMPUTER-ASSISTED INSTRUCTION). Engineers find computers useful in designing and constructing new devices (see COMPUTER-AIDED DESIGN, COMPUTER-AIDED MANUFACTURING).

Brain versus mind

At what point will a computer be so "smart" that we can actually think of it as having consciousness? Will a computer ever write excellent progressive rock and roll music, or jazz, or rap? Will a computer ever write a moving novel, or profound poetry? Will people ever build a computer that is smarter than its makers? Will a computer ever become an electronic *mind*, rather than just an electronic *brain*?

The answers to these questions depend on who you ask. There will probably never be a moment when someone will say, "This machine is not conscious now, but when we add this diode, a life force will enter it, and it'll become conscious." The notion that this could happen—that some cosmic ectoplasm could take up residence in a machine and transform it into a living being—makes good science fiction because it's dramatic. But in real life, computers will probably just get smarter and smarter, and we won't think much about it. Perhaps some day we'll look back in time and say, "Remember when a computer couldn't even think up a good joke?" Then we'll realize what we've created.

Standards for artificial life

In the U.S., researchers demand a lot from a machine before they are willing to think of it as lifelike. You might think a PC is nice to have, and that some video games are cool; however, you probably don't love them like you can love a cat or a dog. Maybe some people think they love a bunch of ICs and switches, but to date, no one has built a computer that can love people back. This is one of the standards by which researchers measure how human-like a computer is. Another criterion is consciousness, or the ability of a computer to ask, entirely on its own, "Where did I come from?" By yardsticks like these, we haven't come anywhere near creating a living machine.

Animism

In some cultures, notably Japan, everything is in some sense alive. This philosophy is called *animism*. The Japanese have much less trouble than Americans or Europeans with the notion that a PC or toy robot is alive. This is probably why the Japanese are so enthusiastic about human-like robots with super-smart electronic brains. In America and Europe, robots are seen as a benefit to design and manufacturing, but not so much as potential servants or companions. See ANIMISM, ARTIFICIAL INTELLIGENCE, ARTIFICIAL LIFE, COLOSSUS, COMPANIONSHIP ROBOTS, COMPUTER CONSCIOUSNESS, COMPUTER EMOTION, COMPUTER MUSIC.

Computers and robots

An obvious application of computers is for the control of robotic systems. There are two basic schemes for this.

Many robots, one computer

One method of computerized robot control employs a central computer to control a large group of robots. In this sense, the robots play a role very much like ants in an anthill, or bees in a hive. Each individual robot is "stupid," but the group, or colony, is "smart." See INSECT ROBOTS.

Many robots, many computers

The other scheme gives each robot its own computer. There might be a central computer that "masterminds" the entire set, but not necessarily. With this system, each robot has high intelligence. The ultimate endpoint, carrying this idea as far as it will go, is the creation of a population of androids. But that's just science fiction—for now. See ANDROID, AUTONOMOUS ROBOTS, BIOLOGICAL ROBOTS.

computer-aided design

Computers are extensively used in the design of new products. This is known as *computer-aided design (CAD)*. Using CAD, you can "test" devices without actually constructing them. This saves enormous amounts of time and money. Inferior designs can be rejected without the need for putting prototypes together, testing them, and then comparing the results with the data for other designs. The CAD process is extensively used to design robots.

A human operator is needed to operate a CAD terminal. No computer yet in existence is "smart" enough to interpret the data and create improved designs all by itself.

A CAD system makes extensive use of computer graphics to show objects with movement in three dimensions. These systems are used in conjunction with computer-aided manufacturing (CAM), and the acronyms are therefore often seen together, written "CAD/CAM." See also COMPUTER-AIDED MANUFACTURING, COMPUTER-ASSISTED INSTRUCTION, COMPUTER GRAPHICS.

computer-aided manufacturing

A computer can be used to control machinery for the purpose of manufacturing goods. Computers are employed in the control of automated integrated manufacturing systems (AIMS), which are sophisticated assembly lines. The hardware ranges from simple tools, such as drills and lathes, to complex robots. The term *computer-aided manufacturing (CAM)* is used in reference to automated systems that are partly, or completely, controlled by a computer. See ASSEMBLY LINE, ASSEMBLY ROBOTS, AUTOMATED INTEGRATED MANUFACTURING SYSTEM.

Robots lend themselves especially well to computer control. A robot can be "taught" how to perform a sequence of motions by literally forcing it manually through them; the computer stores all the movements in its memory in the form of functions in coordinate systems. There are several different types of coordinate systems in common use in CAM. The functions, which plot two or three space coordinates versus time, can be recalled from the computer memory, causing the robot to automatically duplicate the movements just as it was "trained" to move. See CARTESIAN COORDINATE GEOMETRY, CONTROLLER, CYLINDRICAL COORDINATE GEOMETRY, POLAR COORDINATE GEOMETRY, SPHERICAL COORDINATE GEOMETRY, TEACH BOX.

Some CAM systems use robot vision to help with alignment of parts during the manufacturing process. See BINOCULAR ROBOT VISION, CINCINNATI MILACRON, VISION SYSTEMS.

Computers are used in the design of new products, as well as in their manufacture. This is called *computer-aided design (CAD)*. The two acronyms, CAD and CAM, are often combined as "CAD/CAM." There are subspecialties within the general field of CAD/CAM. Examples are computer-aided testing (CAT) and computer-aided engineering (CAE). The abbreviations and acronyms in this field seem to multiply and change almost daily; this can sometimes get a little confusing. See COMPUTER-AIDED DESIGN, COMPUTER-ASSISTED INSTRUCTION.

computer-assisted instruction

Computers can make good teachers for some school subjects. This is especially true of mathematics, spelling, and other subjects in which lots of practice is needed. Skills that require coordination, such as flying an airplane or driving a car, can be taught on computer simulators before putting the student at the controls of the real thing. The use of computers in teaching is called *computer-assisted instruction (CAI)*.

Some of the advantages of CAI over exclusively human teachers are as follows:

- Computers treat all students the same. They do not have prejudice.
- Computers have unlimited patience.
- Computers are cheap; they need only maintenance, not a salary.
- Software can be upgraded as new programs become available.
- Children often find it more fun to learn on a computer than to learn in a classroom.

Some disadvantages of CAI are:

- Computers can't offer the personal attention that a human teacher can.
- There are questions that students might ask, that a computer might not be able to answer, but into which a human teacher could at least give insight.
- Children must be supervised while learning from computers, lest lazy ones "dilly dally" and learn little or nothing.
- There is controversy about replacing people with machines on a large scale.

This last problem is a bugaboo not only for CAI, but wherever machines replace people in large numbers. Some workers lose their jobs when this happens. Many psychologists question the wisdom of getting children more used to working with machines than working with people. See also ARTIFICIAL INTELLIGENCE.

computer checkers

See CHECKERS-PLAYING MACHINE.

computer chess

See CHESS-PLAYING MACHINE.

computer consciousness

Is a computer aware of its own existence? Can it question its role in the world? Can it ask for something and really feel a need for it? Most researchers think not. But nobody is absolutely certain one way or the other.

A landmark in the quest for artificial intelligence (AI) will be to make a computer that has emotions, self-awareness, and other qualities of a human mind. But this brings up the question: "Must a mind be human-like in order to have consciousness?" To answer "Yes" is to look at the universe with tunnel vision. It is naive to suppose that people's minds are the only ones that can be conscious.

Some people say that animals aren't conscious. But if you've ever lived with a cat or dog for a long time, you *know* that they are. Is an insect conscious? Who can claim to know for sure? The only way to be certain would be to trade brains with one for a few minutes, and check it out. Otherwise, saying "This animal is not conscious" is to draw a conclusion based on an arrogant attitude, not on hard scientific data.

The same holds true for computers. As computers become more and more sophisticated, the question will become increasingly important.

When a computer begins to discuss, with some human operator, the state of the universe; when it asks deep and meaningful questions about the personality of the President of the United States; when it can give its own perspective in understandable sentences; when it can solve some problem that has confounded humanity for a thousand years—then, but only then, will people know for sure that machine has developed consciousness. But the actual "mind process" might develop long before we see absolute proof of it.

What sort of world view would a conscious machine have? That, say most AI researchers, will depend on whether it has hands, or legs, or eyes, or ears that can interact with the world. If the machine is an android, its world view might be like ours. But if the machine is simply a massive electronic circuit, it might well see the universe from a perspective foreign to us humans.

The brain of a porpoise is as big and complex as a human brain, but porpoises lack hands and fingers with which to build things. What, then, are they thinking with all that gray matter? Have they known about (and perhaps disproved!) the Big Bang theory, or relativity, or biological evolutionism, since before we humans invented the wheel? Some people are trying to learn the "language" of the porpoise, so that we will be able to communicate with them.

But their mental processes might be so different from ours that we will never be able to truly know how they perceive the world. People must be prepared to face these same complexities when building electronic minds. See also ANDROID, ANIMISM, ARTIFICIAL INTELLIGENCE, ARTIFICIAL LIFE.

computer emotion

Will computers ever laugh and cry? This is hard, if not impossible, to imagine. You might program a computer to print out "Ha-ha-ha!" or "Boo-hoo!" on a monitor screen if you type a joke or sad story on its keyboard. But this hardly qualifies as computer emotion. To have true emotion, a computer must first be aware of its own existence. That is, it must be conscious.

Does consciousness inevitably occur at some level of artificial intelligence (AI)? And if so, does emotion come along with it? Perhaps brilliant androids, like Commander Data in the television series *Star Trek: The Next Generation*, can have consciousness without emotion. We won't know the answers to these questions unless, or until, computers evolve that are as smart as, or smarter than, we are. Most researchers believe it will be a long while before this happens. See also ANDROID, ARTIFICIAL INTELLIGENCE, ARTIFICIAL LIFE, COMPUTER CONSCIOUSNESS.

computer graphics

Computer graphics refers to the use of computers to assist in drawing and drafting. This is especially important in computer-aided design (CAD) and computer-assisted instruction (CAI). Computers can also be used to process existing video images. A good example of this is the "color weather radar" you've seen on television.

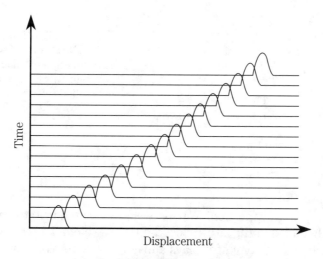

COMPUTER GRAPHICS Rendition
of a solitary wave with time on the vertical axis.

The drawing shows a solitary wave, or pulse, graphed with displacement on the horizontal axis and time on the vertical axis. This produces an image that seems to have three-dimensional qualities. Irregular surfaces and multi-variable functions lend themselves especially well to graphics displays of this kind.

Computer graphics can be used to formulate a "map" of an area in which a robot works. This helps the robot to navigate in its environment. See also COMPUTER, COMPUTER-AIDED DESIGN, COMPUTER-ASSISTED INSTRUCTION, COMPUTER MAP, VISION SYSTEMS.

computer interconnection

See NETWORK.

computerized defense system

Computers have many different possible applications in warfare, both "conventional" (non-nuclear) and nuclear. Computers can be used to plan invasion strategies, as well as to decide where to place soldiers and equipment for defense. Computers are already in use by the Army, the Navy, the Air Force, and the Marine Corps of the United States. Computers are also employed in large-scale nuclear defense.

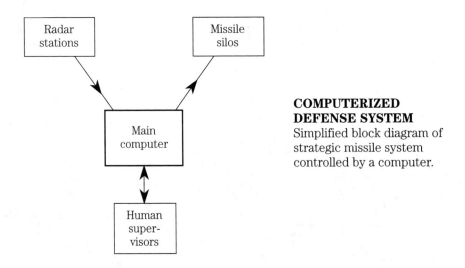

COMPUTERIZED DEFENSE SYSTEM
Simplified block diagram of strategic missile system controlled by a computer.

During the years of the Cold War between the United States and Russia, which gradually ended in the late 1980s, the U.S. developed a strategic defense system in which a large computer plays a major role. The block diagram shows the basic nature of the system. Information from radar stations is sent to the computer. The computer interprets this information and figures out the probability that an attack is in progress. If it appears that an attack is taking place, the human supervisors are consulted, and a careful double-checking process is followed. If the human supervisors conclude that there is no attack taking place, the system is reset. But if the human supervisors are certain that there is a nuclear attack in progress, they tell the computer to go ahead and activate one or more missile silos.

Obviously, the stakes in this process are extremely high. During the Cold War, the leaders and people in both the U.S. and Russia were fearful that a single computer (or human) error could result in the launching of an offensive nuclear missile for no reason. The scenario was painted in various science-fiction novels and movies.

Some researchers have suggested that we ought to give computers complete control of all the weapons in the world. If a computer were sufficiently "smart," that is, if artificial intelligence (AI) were advanced enough, the proponents of this idea say that there could be no war. Computers would see the illogical nature of war, and would never engage in it—we hope.

Other researchers suggest that such powerful computers could annihilate humanity if they malfunctioned. The slightest little thing can cause a computer to go haywire; just about everybody has experienced this. Have you ever been unable to mail a package at the Post Office, or make an airline reservation, or use your credit card? Have you ever been working on a computerized word processor and had it develop a temporary glitch? These things are not unusual; they happen all the time. In theory, a single diode could blow out someplace, and a computerized strategic defense system might think it was under attack. Or a computer might actually become paranoid, like Hal did in *2001: A Space Odyssey* (see CLARKE, ARTHUR C.).

A more chilling prospect is this: What if a smart, logical, all-powerful computer, with complete control of all weaponry, decides that there are too many people in the world, and that we humans would be healthier and happier if our numbers were cut by a factor of, say, 10? Maybe the computer would even decide that human beings are obsolete, and that it is time for the planet to move ahead in its evolution without us. See CAPEK, KAREL. See also ARTIFICIAL INTELLIGENCE, CARETAKERS, COLOSSUS.

computer map

An autonomous robot must "know" where it is relative to objects around it, so that it won't bump into things, and so that it can find whatever it is seeking. For this to be possible, the robot makes a computer map of its environment.

An example of a computer map is shown in the drawing. You should recognize this as a dining room. The robot's assignment: Set the table before the people sit down. The robot can envision, relative to the table and chairs, where to put the plates; then it can follow its programming and place the napkins, forks, knives, spoons, glasses, etc., in their proper places.

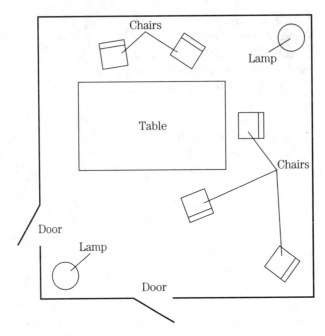

COMPUTER MAP A dining room arrangement.

The robot must realize that one chair has been moved into the corner, and is not at the table. This might confuse the robot and result in its setting an extra place at the table and/or dropping utensils on the carpet in the corner. Whether or not these problems occur will depend on the quality of the map-interpretation program.

Of course, the robot will also need a map of the kitchen, where all these utensils are to be found. It will need to recognize all these items, besides knowing in which drawers or cupboards they are. The complexity of these problems is much greater than it might seem at first. See also AUTONOMOUS ROBOTS, BIN-PICKING PROBLEM, COMMON-SENSE SUMMER PROJECT, OBJECT RECOGNITION, VISION SYSTEMS.

computer music

Musical notes can be generated electronically. The way a note sounds depends on the shape of the wave. Musical instruments all produce waveforms that can be emulated by electronic oscillators. These oscillators can also produce waveforms unlike any musical instrument, thereby concocting unique sounds. Computers combine and process these waves, creating new types of music. Modern rock bands use these techniques, especially when recording songs in the studio.

Some musicians let computers compose, as well as process, music. The result can be interesting, but it rarely has depth. Music is communication, intended to convey the emotions of the composer. An emotional computer has not been built yet, so music from a computer cannot portray genuine feeling. Nevertheless, some computer-composed music has a soothing, relaxing sound. Some computer-generated melodies seem to show us the world through a machine's "mind." To me, computer-generated music often communicates detached amusement like that of Mr. Spock in *Star Trek* movies. See also ARTIFICIAL INTELLIGENCE.

computer programming

In order to perform their functions, computers must be programmed. Generally, a computer needs a high-level program that lets a human operator interact with the machine; it also needs an assembler program to convert the high-level language into machine language that the computer actually uses.

Computers have traditionally been programmed by people. But in the future, it is possible that computers might take over some programming jobs. This will require a fairly advanced level of artificial intelligence. See also ARTIFICIAL INTELLIGENCE, ASSEMBLY LANGUAGE, BASIC, C, COBOL, COMPUTER, FORTRAN.

computer reasoning

Computers can process huge amounts of information, more than any person could hope to handle. But computers can only process data in logical steps.

Is it possible for a computer to have reasoning power? The answer depends on whether reasoning can be broken down into logical steps. If so, then a computer can be built to have reasoning power. If not, then computers might never be able to reason.

Arthur Samuel, the inventor of a computer checkers program, stated his belief that human reasoning consists of something more than logical steps, while computer reasoning is logic and only logic. He used this belief to argue that computers will never have reasoning power as humans do. But Samuel gave no basis for his beliefs.

Does human reasoning involve "something extra"? Is it more than sheer logic? If so, what is this "something"? Why do we have it? From where did we get it? Samuel gives no answers to these questions. But they are crucial! Religious people might say that this "something extra" is the soul, and that humans get it from God; but this is not a *scientific* answer.

In his book *Godel, Escher, Bach*, author Douglas Hofstadter argues that human reasoning comes purely from the physical structure of our brains. This bypasses any conflict between science and religion. If Hofstadter is right, then it's possible to build a reasoning machine, although it might be decades away.

The question has not yet been answered. Nor will it be—unless, or until, a machine with true reasoning power has been put into operation. See also ARTIFICIAL INTELLIGENCE, ARTIFICIAL LIFE, COMPUTER, COMPUTER CONSCIOUSNESS.

computer virus

See VIRUS.

consciousness

See COMPUTER CONSCIOUSNESS.

context

Context is the environment in which a word is used. It is important in speech recognition systems.

You've probably heard the expression "out of context." When a word is used out of context, it results in a phrase or sentence that doesn't make sense. Worse yet, it might mean something not intended. When a word is taken out of context, the phrase or sentence is all right, but it gets interpreted as nonsense, or in the wrong way.

In order to interpret some spoken statements, a computer or robot with artificial intelligence will need to know the context in which each word is used. "I went down to the sea" makes sense; "I went down too the C" and "Eye went down two the see" do not. See also ARTIFICIAL INTELLIGENCE, SPEECH RECOGNITION.

continuous-path motion

A robot arm can move in either of two general ways: smoothly or in discrete steps. Smooth-moving robot arms are known as *continuous-path* devices.

Robot motion is always determined as a function of position versus time. In the mathematical representation of motion, then, time is always the independent variable.

For a robot to move along a smooth, continuous path, every point along the way must, in theory, be stored in memory. Of course, this is not literally possible, because a continuous path contains an infinite number of points. There are two general ways to approach this problem. One method uses point-to-point approximation; the other involves mathematical functions.

It is possible for a computer memory to store many points, so close together that the resulting motion is continuous for all practical purposes. Small time increments are used, such as 0.01 second or 0.001 second. The result is point-to-point motion, but with so many points that it appears continuous. This is roughly illustrated for two-dimensional motion at A in the drawing. (The approximation here is "coarse" so you can see how it works.) In this case, Cartesian coordinate geometry is used. The coordinate points, x and y, are memorized in the form of sets; they are *discontinuous functions* of time, t. Other coordinate geometries might be used. See CARTESIAN COORDINATE GEOMETRY, CYLINDRICAL COORDINATE GEOMETRY, POLAR COORDINATE GEOMETRY, SPHERICAL COORDINATE GEOMETRY.

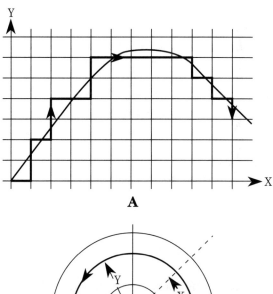

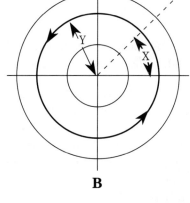

CONTINUOUS-PATH MOTION
Point-by-point approximation
(at A) and function memorization
(at B).

The second method stores position as a set of continuous functions. An example of this is shown at B in the drawing. Here, polar coordinate geometry is used. The motion is in a circle, represented by a constant radius and an angle that varies with time from 0 degrees through 360 degrees. The angle, x, is a continuous function of time, t. The radius, y, is constant, no matter what the value of t. This motion is truly continuous, in that it actually passes through an infinite number of points. This is possible because of the precise nature of the mathematical functions. The tradeoff is that there are only so many functions that can be programmed into the controller. Also, the controller can "understand" functions only up to a limited degree of complexity. See also POINT-TO-POINT MOTION.

controller

The *controller* is one of the three major parts of an industrial robot, the other two being the power supply and the manipulator(s). As its name implies, the controller oversees and controls the robot's moving parts.

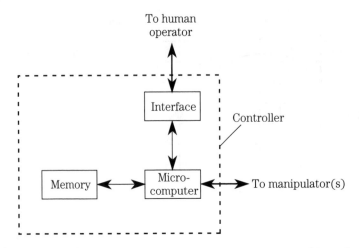

To human
operator

Interface

Controller

Memory

Micro-
computer

To manipulator(s)

CONTROLLER Consists of microcomputer, memory, and interface.

A functional block diagram of a controller is shown in the drawing. A micro-computer is usually central to the system. Movement instructions are stored in memory.

The interface does several things. Mainly, it allows the microcomputer to "communicate" with a human operator or supervisor. Through the interface, it might be possible to reprogram the memory to change the movement instructions. The action or functioning of the robot might be displayed on a monitor screen. There might also be various malfunction indicators. Some interfaces have a teach box, which lets the human operator reprogram the motions and path of the robot. See also COMPUTER-AIDED MANUFACTURING, INDUSTRIAL ROBOTS, MANIPULATOR, MEMORY, POWER SUPPLY, TEACH BOX.

correspondence

In binocular robot vision, also called "stereo," two different images are seen, one by each robot "eye" (see BINOCULAR ROBOT VISION). How does the robot know when its eyes are both looking at the same thing or the same place? This is called *correspondence*. It usually requires a microcomputer, and sometimes even artificial intelligence, to figure the problem out.

Perhaps you've had your sense of correspondence confused when looking at a grid of dots, or at a piece of quadrille graph paper. The illustration shows how your eyes, or a robot's vision system, might be fooled by such a pattern.

At A, both eyes are looking at the same point; depth is perceived the way it should be, even if the views of the object appear slightly different because of the difference in viewing angle through either eye.

At B, the left eye is looking at one object in the set, while the right eye looks at another. Because all the objects are evenly spaced, they still seem to line up as perceived by the "brain." But this is not the true nature of the scene. If a ro-

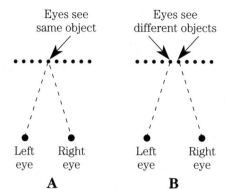

Eyes see same object

Eyes see different objects

Left eye   Right eye

Left eye   Right eye

A   **B**

CORRESPONDENCE
At A, perception is correct. At B, it is fooled.

bot acts on this information, it will make positioning errors. If you've had your depth perception tricked this way, you couldn't tell exactly how far away the pattern was. See also VISION SYSTEMS.

CPU

See CENTRAL PROCESSING UNIT.

crash

Sometimes a computer will "freeze up" without apparent reason. If you own a personal computer (PC), you've probably experienced this. You give a command, or you are typing along, and suddenly the thing stops in its tracks. You press keys, first in a state of confusion, and then in a frenzied attempt to get the computer to do something—anything! But it's no use. This is called a crash.

A crash can be caused by various problems. Usually it is nothing more than an abnormal voltage "spike" or "glitch," causing one of the integrated circuits (ICs) to see a signal that isn't there. In most cases, resetting (rebooting) the computer will normalize things again, although you will probably lose whatever happens to be stored in volatile memory. See also BOOT.

cryptanalysis

Cryptanalysis is the art of breaking ciphers, or codes used to keep unauthorized people from intercepting signals. Ciphers are extensively used in wartime communications. Much effort is spent trying to break the ciphers of the enemy.

An example of a cipher is shown in the table. The alphanumeric sequence *ABC ... XYZ0123456789.,?_* is assigned numbers in order from 1 to 40. Then each character is shifted 13 units downwards in the table. The cipher is transmitted as a string of numerals from the set {1,2,3, . . . 40}. This is an extremely simple cipher, and it would not take a computer very long to break it.

CRYPTANALYSIS: A simple cipher.

Character intended	Code number	Number sent
A	1	28
B	2	29
C	3	30
D	4	31
E	5	32
.	.	.
.	.	.
.	.	.
L	12	39
M	13	40
N	14	1
O	15	2
P	16	3
.	.	.
.	.	.
.	.	.
Z	26	13
0	27	14
1	28	15
2	29	16
3	30	17
.	.	.
.	.	.
.	.	.
9	36	23
period	37	24
comma	38	25
question mark	39	26
space	40	27

With the help of computers, cipher-breaking has become much more sophisticated than it once was. A computer can test different "solutions" to a code much more rapidly than teams of humans ever could. Beyond that, artificial intelligence (AI) can be employed in an attempt to literally figure out what the enemy is thinking. This streamlines the process of cipher-breaking. It lets the cryptanalyst, or code-breaker, get a feel for the general scheme behind a cipher, and in this way, it helps the cryptanalyst understand the subtleties of the code more quickly.

One of the earliest cryptanalysts to use a computer was Alan Turing. Turing is known as a pioneer in AI. During World War II in the early 1940s, the Germans developed a sophisticated machine that encoded military signals. The machine was called *Enigma*. The word *enigma* means something mysterious. And indeed the machine was a mystery to allied cryptanalysts, until Alan Turing designed one of the first true computers to successfully decode the signals.

It has been said that as you improve mouse traps, evolution produces smarter mice. Computers become more powerful, and they can decode more and more complex ciphers every year. But they can also invent ciphers that are harder and harder to decode. It is a vicious circle.

Ultimately, the advantage in cipher-breaking goes to the side with the more advanced AI technology. This is true in almost all aspects of modern warfare. The military is therefore very interested in the field of AI. See also ARTIFICIAL INTELLIGENCE and TURING, ALAN.

cybernetics

The term *cybernetics* refers to the science of goal-seeking, or self-regulating, things. The word itself comes from the Greek "governor." The fields of robotics and artificial intelligence (AI), to which this book is devoted, are subspecialties within the science of cybernetics.

Computer-controlled robots are good examples of cybernetic machines. Their behavior is governed by the programming of a computer, but true cybernetic machines also interact with their environment.

An example of a cybernetic process is pouring a cup of coffee. Suppose you say to your personal robot, "Please bring me a cup of coffee. And be sure it's hot!" In the robot's computer memory, there is data concerning what a coffee cup looks like. Also encoded is the route to the kitchen, the shape of the coffeepot, and, of course, a relative-temperature-interpretation routine, so the robot will know what you mean by "hot." (In this case it's about 110 to 120 degrees Fahrenheit.)

A robot would go through an unbelievably complicated process to get you a cup of coffee. You'll find this out if you try to write down each step in rigorous form (see COMMON-SENSE SUMMER PROJECT). Yet the ultimate cybernetic device (a person) does this without any "brain work" at all.

cyborg

The term *cyborg* is a contraction of the words "*cyb*ernetic" and "*org*anism." It is used in reference to humans whose bodies are largely, or even mostly, made up of robotic parts.

In one sense, every person is a cyborg. We all have goal-seeking and self-regulating minds. If that were not so, this book would never have been written. Neither its author nor its publisher would have been able to coordinate the effort to produce what you're reading right now. Nor would you have any reason to read it.

In robotics, the term *cyborg* refers to a person who is largely, or even mostly, a robot, but is still biologically alive. If a person is given a single robotic hand or arm, it is called a *bionic* body part. Science-fiction carries this notion to the point that a person seriously injured might be "rebuilt" almost entirely of bionic parts. Television producers took this idea and created the show *The Six Million Dollar Man*, an adventure series. The "bionic man" could run as fast as a car on a freeway, throw a football a mile, and win any fight he got into. He had practically unlimited endurance. He was a cyborg.

While technology is a long way from creating cyborgs, the possibility exists that they might someday be commonplace. The cost would be more on the order of six billion dollars, rather than six million, per "unit." See also ANDROID, ARTIFICIAL LIFE, CYBOT SOCIETY.

cybot society

Some people envision a future society comprised of human beings, cyborgs, super-smart robots, and computers. This has been called a *cybot society*. The idea has gained the most popularity in Japan.

A leading Japanese robotics engineer and researcher, Ichiro Kato, has worked with robot substitutes for human body parts. These machines can sometimes function almost as well as the real thing. Kato has made excellent mechanical hands and legs. See KATO, ICHIRO.

How much of a person can be bionic? Kato believes that a human being might be almost entirely bionic. That's almost, but not quite, an android.

While enthusiasm for the idea of a cybot society runs high in Japan, there is somewhat less interest in the United States and Europe. While Americans and Europeans think of robots as serving mainly industrial purposes, the Japanese think of them as being in some sense alive. This might be why the Japanese are so much more active in developing human-like robots. See also ANDROID, ANIMISM, ARTIFICIAL LIFE, CYBORG, PROSTHESIS.

cylindrical coordinate geometry

Cylindrical coordinate geometry is a scheme for guiding a robot arm in three dimensions. A cylindrical coordinate system is a polar system with an extra coordinate added for "height." The drawing at A shows a cylindrical system of coordinates. Using this system, the position of a point can be uniquely determined in three-dimensional space.

In the cylindrical system, a reference plane is used. An origin point is chosen, and also a reference axis, running out away from the origin in the reference plane. In the reference plane, the position of any point can be specified in terms of the radius, or distance from the origin, and the angle measured counterclockwise from the reference axis. You might let the radius be given by x (in centimeters) and the angle by y (in degrees of arc).

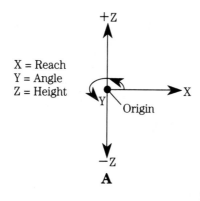

X = Reach
Y = Angle
Z = Height

A

**CYLINDRICAL
COORDINATE GEOMETRY**
Basic scheme (A), and a robot
arm using this geometry (B).

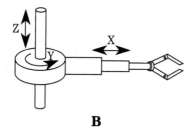

B

The third dimension can be called *height*. It is either positive (above the reference plane) or negative (below it); the height might also be zero (actually in the reference plane). The units for the height, call it z, are the same as the units for radius (x).

The drawing at B shows a robot arm equipped for cylindrical coordinate geometry. The movements x, y, and z are called *reach, base rotation,* and *elevation* respectively. See also CARTESIAN COORDINATE GEOMETRY, POLAR COORDINATE GEOMETRY, ROBOT ARMS, SPHERICAL COORDINATE GEOMETRY.

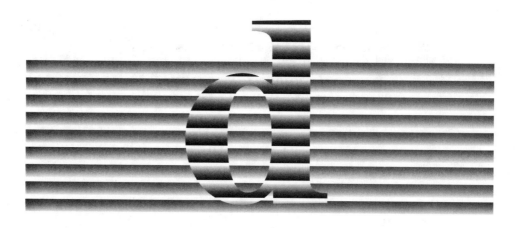

data

The term *data* generally means "digital information" in electronic and computer applications. The information handled and stored by a digital computer is often called data. See the several articles following this one.

Any set of alphanumeric characters (along with punctuation marks and other symbols) representing a specific quantity, is regarded as data. This might be a sentence, chart, or graph, computer program, credit report—anything that can be rendered as digital information.

Some researchers think that all human thought, including emotion, consists of data. Others think there is something more than mere digital information involved in the processes that make us what we are. Are our brain processes really nothing more than lots and lots of data? This question is not just for dreamers; it is important if scientists are to know whether or not artificial intelligence can ever reach human levels.

Data is the name of an android in the popular science-fiction television series *Star Trek: The Next Generation*. In that setting, the answer to the above question is apparently "Yes!" Data looks human, talks like a human, and can even show some emotion. See also ARTIFICIAL INTELLIGENCE, ARTIFICIAL LIFE, ANDROID, COMPUTER, COMPUTER CONSCIOUSNESS, COMPUTER REASONING, DIGITAL.

data-acquisition system

A *data-acquisition system* is a set of electronic circuits intended for gathering data for storage, processing, or direct use.

A personal computer (PC) can be employed as a data-acquisition system. You do this when you connect your PC to a modem, load terminal emulation software into it, and interface it with the telephone lines (see the block diagram). This allows you access to data of many different kinds.

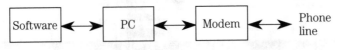

DATA-ACQUISITION SYSTEM
Block diagram of PC system used as a terminal.

Data can be stored in a variety of ways. If the data is transmitted to a diskette or hard disk, it can be kept for an indefinite length of time. Data might be sent to a printer for keeping in the form of hardcopy.

Some data is acquired in the form of software, or programs, which can be run on your PC. But when using software directly from public media, your computer might acquire a virus.

In the future, data-acquisition systems will become more versatile. You might someday be able to access a satellite system with your computer. Amateur radio operators are already exchanging computer data via a mode called *packet radio*.

When many data-acquisition systems are used together, both transmitting and receiving information, the resulting network has a degree of artificial intelligence greater than that of any of the individual computers. See also ARTIFICIAL INTELLIGENCE, COMPUTER, DATA COMMUNICATION, DATA CONVERSION, MODEM, NETWORK, PACKET COMMUNICATIONS, PACKET RADIO, VIRUS.

data communication

Data communication is the transfer of data in both directions between two points, or in all possible ways among three or more points. Each station must have a data-acquisition system. When computers are interconnected so they all work more or less together, the system is called a *network*.

Ideally, data arrives at a terminal exactly as it was sent from the distant terminal. But interference can alter characters. This results in errors. If such an error happens at a critical point in a data transmission, the result can be seriously flawed or misinterpreted data.

Sophisticated data-communication systems have almost completely eliminated errors. This is done by handshaking, in which the receiving terminal analyzes incoming data to see if it is in the right format. If something appears to be out of place, the receiving terminal sends an inquiry signal back to the transmitter, and the transmitter repeats the character(s) in question.

Data communication is usually carried out in digital form. This provides a better signal quality than analog methods. Sometimes data is encoded to prevent unauthorized people from intercepting it. See also ANALOG, DATA-ACQUISITION SYSTEM, DIGITAL, NETWORK, PACKET COMMUNICATIONS.

data conversion

Information can exist in many different forms. It might be analog or digital. In computer practice, it is almost always digital, although some robots and computers must process analog data, such as a voice or a video image.

Digital data can be sent and received in either parallel form or serial form. See ANALOG, DIGITAL, PARALLEL, SERIAL.

Four types of data converters are described in this article.

Analog-to-digital

A voice signal, or any continuously variable signal, can be digitized, or converted into a string of pulses, whose amplitudes can achieve only a finite number of states. This is *analog-to-digital (A/D) conversion*. You can think of A/D conversion as "chopping" or "slicing" a signal into chunks that have only certain sizes.

Resolution

The number of states is always a power of 2, so that it can be represented as a binary-number code. Fidelity gets better as the exponent increases. The number of states is called the *sampling resolution*, or simply the resolution.

You might think the resolution must be large for good reproduction to be possible. But a resolution of $2^3 = 8$ (as shown in Fig. 1) is the standard resolution for commercial digital voice circuits. A resolution of $2^4 = 16$ is adequate for compact disks used in advanced hi-fi systems.

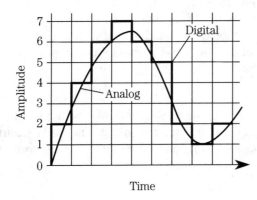

DATA CONVERSION
Fig. 1. Eight-level digital signal (heavy line), along with corresponding analog signal (lighter curve).

Sampling rate

The efficiency with which a signal can be digitized depends on the frequency at which sampling is done. In general, the sampling rate must have a frequency that is at least twice the highest data frequency.

For a voice signal, the commercial standard is 8 kHz, or one sample every 125 us. For music and hi-fi digital transmission, the standard sampling rate is 44.1 kHz, or one sample every 22.7 us.

Robotics/AI applications

In a "smart computer," an analog-to-digital signal might be used to convert your voice, or an image, into digital form that the computer can understand. See SPEECH RECOGNITION, VISION SYSTEMS.

Digital-to-analog

Digital-to-analog (D/A) conversion reverses the process of A/D conversion, so the original analog data is recovered. You can use Fig. 1 to imagine D/A conversion, just by thinking "the other way around" from the way you imagine A/D conversion. The D/A converter "smooths out" the digital signal.

Digital is simple

You might ask, "Why convert a signal to digital form in the first place, if it's going to be changed back to analog anyway?" The reason is that a digital signal is simpler than an analog signal.

It's good to make a signal as simple as possible. This is because the more different a signal is from unwanted noise, the easier it is to separate the data from the noise, and the better is the realizable S/N ratio. Also, computers have an easy time working with digital data, and a very hard time with analog signals.

You might think of signal/noise separation in terms of apples, oranges, and a watermelon. It takes a while to find an orange in a tub of apples. (You'll probably have to dump the tub.) Think of the orange as an analog signal, and the apples as noise. But suppose there's a watermelon in a bushel basket with apples. You'll have no trouble finding the watermelon. Think of the melon as a digital signal, and the apples as noise.

Digital is reproducible

In recent years, digitization has become commonplace not only in data communications, but in music recording and even in video recording. The main advantage of digital recording is that a selection can be recorded, re-recorded, re-re-recorded, etc., and the quality does not diminish.

Robotics/AI applications

In artificial intelligence, digital-to-analog converters can be used in image processing and in speech synthesis. See ARTIFICIAL INTELLIGENCE, SPEECH SYNTHESIS.

Serial-to-parallel

Digital information can be sent in either of two ways: serial or parallel. There are advantages and disadvantages to either method.

When data is sent in series, the bits are all sent along one line, one by one. The advantage of this is that one line suffices to convey the information; the disadvantage is that it takes time to transmit data bit-by-bit.

When data is sent in parallel, groups of bits are sent along several lines at once. This increases the speed of transmission, compared with serial transfer, by a factor equal to the number of lines. The disadvantage is that several (or many) lines must be used.

A serial-to-parallel (S/P) converter gathers bits up in groups from a serial line, and sends them in parallel along several lines. This is shown schematically in Fig. 2. Imagine the data bits as flowing from left to right. The output of a S/P converter can't go any faster than the input, obviously, but the circuit is useful when it is necessary to interface between a serial-data device and a parallel-data device. See PARALLEL, SERIAL.

Parallel-to-serial

A parallel-to-serial (P/S) converter does the opposite of the S/P circuit. It gathers the bits from multiple lines, and transmits them out one at a time, at a regular rate and in a defined sequence.

The output of the P/S converter must, over a period of time, keep up with the input. If the output is slower than the input, bits will accumulate in the converter. A small buffer memory stores the bits from the parallel lines while they are awaiting transmission along the serial line, but this memory cannot have unlimited storage capacity.

You can think of the operation of a P/S converter by looking at Fig. 2 and imagining the bits moving from right to left.

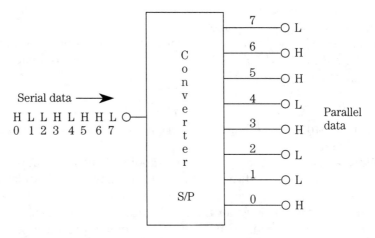

DATA CONVERSION Fig. 2. A serial-to-parallel converter. High = *H*, Low = *L*; numbers are for sequence reference.

data storage media

In a computer system, data is retained in various kinds of electronic memory circuits, and also on magnetic disks and compact disks. A summary of each of these media follows.

Integrated circuits

Binary digital data, in the form of high and low levels (logic ones and zeros), can be stored in ICs. In ICs, memory can take various forms.

RAM A *random-access memory (RAM)* stores binary data in arrays. The data can be addressed (selected) from anywhere in the matrix. Data is easily changed and stored back in RAM, in whole or in any part. A RAM is sometimes called a read/write memory.

An example of RAM is a word-processing computer file that you are actively working on. This book was written in semiconductor RAM before being stored on disk (another kind of RAM) and ultimately printed on the paper now before you.

ROM In contrast to RAM, read-only memory (ROM) can be accessed, in whole or in any part, but not written over. A standard ROM is programmed at the factory. This permanent programming is known as *firmware*. But there are also ROMs that you can program and reprogram yourself.

EPROM An erasable programmable ROM (EPROM) is an IC whose memory is of the read-only type, but that can be reprogrammed by a certain procedure.

Bubble memory Bubble memory uses magnetic fields within ICs. The scheme is especially popular in computers, because a large amount of data can be stored in a small physical volume.

IC capacity This book was written with word-processing software on a PC with 640 kilobytes (640 KB) RAM. This is about 640 typewritten pages—an average novel.

Magnetic media

For permanent storage of information, magnetic media are commonly used. These include tape and disks.

Audiotape The earliest computers used audiotape to store data. This is still possible, but disks are more often used in modern computers. Nowadays, tape is used mainly for storing voice, music, and/or image data.

Videotape Videotape is wider than audiotape, and must deal with a much larger amount of data per unit time. The most common videocassette tape is called *VHS*. The tape itself comes in cassettes that measure 7⅜ × 4 inches, and are about an inch thick. The tape is ½ inch wide. Videotape, like audiotape, must be protected from magnetic fields and from excessive heat.

Magnetic disks Personal and commercial computers almost always use magnetic disks. They come in two forms: the hard disk and the floppy disk or diskette.

The hard disk in a typical PC can hold more than 20 Mb of data. This figure doubles every few years, and the trend can be expected to continue until the point of diminishing returns is reached. Hard disks are permanently installed in the computer.

Diskettes can be interchanged in seconds, so there is no limit to how much data you can sock away on them. A full-wall bookcase of floppies could have more manuscript than you'd write in 1,000,000 years.

Compact disks

The main asset of CDs is their superior reproduction quality. There's no limit to the number of times a CD can be replayed, because laser beams are used to recover the sound. The lasers bounce off tiny pits on the disk. Light beams, of course, do not scratch the CD; nothing mechanically rubs against the disk. A CD won't jam because it does not move while it is being replayed. The data can't be distorted by stretching, as can happen with magnetic tape.

For further information

For more detailed information on various forms of data storage media, consult the articles: COMPACT DISK, DISKETTE, HARD DISK, INTEGRATED CIRCUIT, MEMORY.

debugging

Debugging is a process by which engineers get rid of the flaws in a circuit, machine, or computer program. This includes getting the device to work as efficiently as possible.

Usually, when a program is first run after being written, it doesn't work, or else it works inefficiently. The same is often true for electronic circuits and mechanical devices. Debugging can sometimes be quick and easy; often it is a time consuming and difficult process.

Sometimes debugging means that a program or design must be scrapped, and the engineer must start all over again from a new angle.

decimal number system

The *decimal number system* is the scheme that people most commonly use to denote numerical values. This system is sometimes called *modulo-10*, *base-10*, or *radix-10*, although the number 10 has a value that varies depending on the actual counting base. Maybe it would be better to say that the decimal number system is *modulo xxxxxxxxxx* (that is, to actually write out the quantity as a character, in this case x). See MODULO.

There are ten different digits in the decimal system, representable by the set {0,1,2,3,4,5,6,7,8,9}. Digits are written in a certain sequence to get a number. Depending on the position of a digit in the sequence, its value is multiplied by some power of 10.

In computers, the decimal number system is awkward. It is easier for a computer to work with powers of 2, than with powers of 10. Therefore, computers most often employ binary numbers. This is *modulo 2*. Sometimes computers use *modulo 8* or *modulo 16*, known as *octal* and *hexadecimal* number systems respectively.

Even though a computer doesn't function in decimal notation, it converts the decimal numbers you give it into its own system (usually binary), and also converts its own numbers into decimal form before they appear on the monitor or printer. Therefore, as far as you're concerned, the computer understands modulo 10. See also BINARY NUMBER SYSTEM, HEXADECIMAL NUMBER SYSTEM, OCTAL NUMBER SYSTEM.

decoding

Decoding is the process of converting a message, received in some form of code, into plain language. In a computer, this is done by the software. You see words on the monitor screen, or at the printer output; these are made up of characters that are represented by groups of binary digits (bits). The software converts the binary data into an image, or a set of commands that instructs the printer how to arrange the ink on the paper.

In a machine with speech synthesis, the software converts the data bits into groups of audio waves at just the right frequencies, and with just the right timing, to pronounce recognizable words. See SPEECH SYNTHESIS.

With encrypted data, or information that has been deliberately altered to make it recognizable only to certain people, the decoding process is known as *deciphering*. See also ENCODING, CRYPTANALYSIS.

degrees of freedom

The term *degrees of freedom* refers to the number of different ways in which a robot arm can move. Most robot arms move in three dimensions, but often they have more than three degrees of freedom.

You can use your own arm to get an idea of the degrees of freedom that a robot arm might have.

Extend your right arm straight out toward the horizon. Extend your index finger so it is pointing. Keeping your arm straight, move it from the shoulder. You can move in three ways. Up-and-down movement is called *pitch*. Movement to the right and left is called *yaw*. You can also rotate your whole arm as if you were using it as a screwdriver. This is called *roll*. Your shoulder has three degrees of freedom: pitch, yaw, and roll.

Now move your arm from the elbow only. This is rather hard to do without also moving your shoulder. Holding your shoulder in the same position constantly, you will see that your elbow joint has the equivalent of pitch in your shoulder joint. But that is all. Your elbow, therefore, has one degree of freedom.

Extend your arm toward the horizon again. Now move only your wrist. Try to keep the arm above the wrist straight and motionless. Your wrist can bend up and down, side to side, and it can also twist a little. Your lower arm has the same three degrees of freedom that your shoulder has. (Most of the roll takes place all along your arm below the elbow.)

In total, your arm has seven degrees of freedom: three in the shoulder, one in the elbow, and three in the arm below the elbow.

Three degrees of freedom are sufficient to bring the end of a robot arm to any point within its work envelope, or work space, in three dimensions. Thus, in theory, you might think that a robot should never need more than three degrees of freedom. But the extra possible motions, provided by multiple joints, give a robot arm versatility that it could not have with just three degrees of freedom. See also CARTESIAN COORDINATE GEOMETRY, CYLINDRICAL COORDINATE GEOMETRY, ROBOT ARMS, SPHERICAL COORDINATE GEOMETRY, WORK ENVELOPE.

degrees of rotation

Degrees of rotation are a measure of the extent to which a robot joint, or a set of robot joints, is turned. Some reference point is always used, and the angles are given in degrees with respect to that joint.

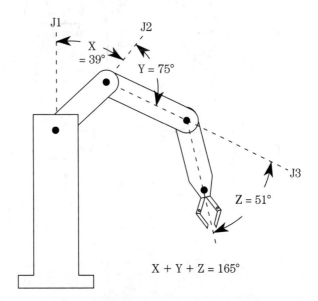

DEGREES OF ROTATION Angles X,
Y, and Z are measured relative to axes J1, J2, and J3.

Rotation in one direction (usually clockwise) is represented by positive angles; rotation in the opposite direction is specified by negative angles. Thus, if angle X = 58 degrees, it refers to a rotation of 58 degrees clockwise with respect to the reference axis. If angle Y = –74 degrees, it refers to a rotation of 74 degrees counterclockwise.

The drawing shows a robot arm with three joints. The reference axes are J1, J2, and J3, for rotation angles X, Y, and Z. The individual angles add together.

To move this robot arm to a certain position within its work envelope, or the region in space that the arm can reach, the operator enters data into a computer. This data includes the measures of angles X, Y, and Z. The operator has specified X = 39, Y = 75, and Z = 51. In the drawing, no other parameters are shown (this is to keep the illustration simple). But there would probably be variables such as the length of the arm sections, the base rotation angle, and the position of the gripper. See also DEGREES OF FREEDOM, ROBOT ARMS, WORK ENVELOPE.

derivative

The term *derivative* refers to the rate of change of a mathematical function. For example, speed (or velocity) is the derivative of displacement (position). Acceleration is the derivative of velocity.

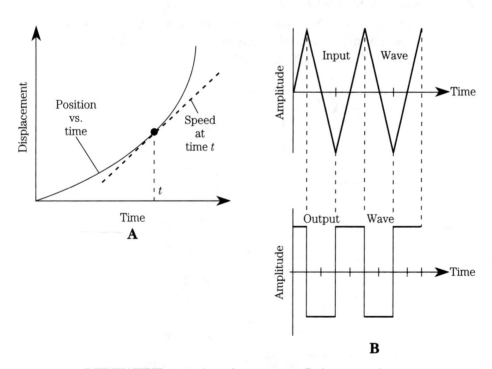

DERIVATIVE At A, for a drag racer; at B, for a waveform.

Drawing A shows a graph of displacement as a function of time. This function appears as a curve. You might think of it as a graph of the distance traveled by a dragster in a race, in which case the displacement might be given in feet and the time in seconds. At a certain time, call it t, the speed is equal to the slope of the line tangent to the curve at that time. For the dragster, the speed would be given in feet per second.

In digital electronics, a circuit that continuously takes the derivative of an input wave is called a *differentiator*. An example of the operation of a differentiator is shown in drawing B. The input is a triangular wave. The output follows the slope, or derivative, of this wave, alternately positive (upwards to the right) and negative (downwards to the right). This derivative is a square wave. See also INTEGRAL.

desktop computer

See COMPUTER.

Devol, George C., Jr.

George Devol is sometimes called the "grandfather of industrial robotics." He engineered many different devices and processes for computer-controlled machines as early as the 1940s. He did much of his work with another pioneer in the field, Joseph Engelberger. Engelberger helped to put Devol's ideas into practice. See ENGELBERGER, JOSEPH F.

Devol was concerned that factory work made people serve machines, when it was supposed to be the other way around. (Some social scientists think this is still a problem.) Much factory work in the early part of the 20th century was extremely tedious. As many as half of the people working in factories might, Devol thought, be replaced by machines. A machine can't get bored, and it breaks down less often than a person gets sick. If the mundane jobs were done by machines, then people could do work that was more rewarding and more interesting.

One of Devol's most significant achievements was the development of a primitive "teach box," or a method of programming a robot to move in a certain way. At first, companies were not very interested in this idea. The corporations had engineers and plans of their own, and Devol's ideas were too radical. But eventually, Devol's methods gained acceptance, because they worked better than other designs. See AUTOMATED INTEGRATED MANUFACTURING SYSTEM, CONTROLLER, INDUSTRIAL ROBOTS, TEACH BOX.

dichotomizing search

See BINARY SEARCH.

differential amplifier

A circuit that responds to the difference in amplitude between two signals, often producing *gain* also, is called a *differential amplifier*. The output is proportional to the difference between the input signal levels. If the inputs are identical, then the output is zero.

The nomograph shows how the instantaneous output of a differential amplifier varies as the instantaneous input values change. To find the output, place a ruler so its edge passes through the two input points; the output will be the point on the center scale through which the ruler passes. In this example there is no gain in the amplifier. You might multiply all the center values by a factor of 10 to illustrate the behavior of a differential amplifier with gain.

Differential amplifiers are sometimes used in robotic sensing systems. The output can be used as an "error signal," so that a robot will follow a path along which two reference sounds or radio waves are exactly in phase, for example. See also DIFFERENTIAL TRANSDUCER.

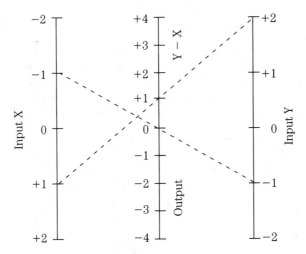

DIFFERENTIAL AMPLIFIER The output is proportional to the difference between the two inputs.

differential transducer

A *differential transducer* is a sensing device with two inputs and one output. The output is proportional to the difference between the input signal levels. An example is a differential pressure transducer, which responds to the difference in mechanical pressure at two points.

Any pair of transducers can be connected in a differential arrangement. Usually, this involves connecting the transducers to the inputs of a differential amplifier.

When the two variables have the same magnitude, the output of the differential transducer is zero. The greater the difference in the magnitudes of the sensed effects, the greater the output. The most output occurs when one of the sensed effects is intense, and the other is zero or near zero. See also DIFFERENTIAL AMPLIFIER.

differentiation

See DERIVATIVE.

digital

A circuit or device is *digital* when it can attain only a finite number of levels or states. Usually, this number is a power of 2. This is in contrast to analog quantities, which can vary continuously over a range of values.

In computers, circuits usually operate in the binary system. There are only two possible levels, called *high* and *low*. Sometimes these levels are assigned numerical values 1 and 0, or truth values T and F, or statements "yes" and "no." All the data in a personal computer is stored in digital form, as a vast array of logic highs and lows.

In robotics, digital motion refers to a device that can stop only at certain positions within its work envelope. These positions can be programmed into the

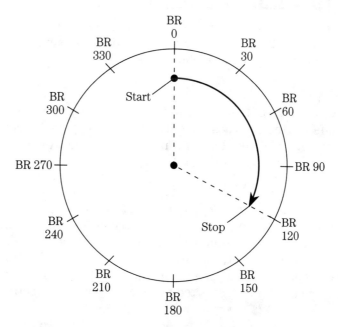

DIGITAL Digitally programmable
base rotation for "BR 120" (see text for discussion).

computer that controls the robot. For example, the base of a robot arm might rotate to any multiple of 30 degrees in the complete circle of 0 to 360 degrees (see drawing). The angle, measured as a compass bearing clockwise from true north, would be programmed into the computer. A command such as "BR 120" (base rotation: 120 degrees) would cause the machine to move to the proper position and stop there. See also ANALOG, BINARY.

digital electronic automation

Digital Electronic Automation (DEA) is a research and manufacturing company located in Turin, Italy. The company is known for its industrial robots. The robots produced by DEA are noted for their high speed, and also for their use of tactile sensors.

Using tactile sensors, a robot arm can tell whether or not a part being installed is the correct one, whether it is of the required quality, and whether it has been placed in the right position. This is invaluable in assembly and quality control. Robots made by DEA are in wide use in consumer-appliance factories throughout Europe. In the United States, General Electric has purchased some design rights from DEA. See also ASSEMBLY ROBOTS, INDUSTRIAL ROBOTS, TACTILE SENSORS.

digital integrated circuit

A *digital IC*, also sometimes called a *digital-logic IC*, operates using two states, called *high* and *low*. These are sometimes called *logic 1* and *logic 0*, respectively. Several different digital IC technologies exist.

Bipolar types

Digital ICs consist of gates that perform logical operations at high speeds. There are several different technologies, each with its own unique characteristics. Digital-logic technology might use bipolar transistors or metal-oxide-semiconductor (MOS) devices.

TTL

In *transistor-transistor logic (TTL)*, arrays of bipolar transistors, some with multiple emitters, operate on dc pulses. This technology has several variants, some of which date back to around 1970. The hallmark of TTL is immunity to noise pulses. A simple TTL gate is illustrated in the drawing. The transistors are either all the way on (saturated), or else all the way off (cut off). This is the reason why TTL is not very much affected by external noise "distractions."

ECL

Another bipolar-transistor logic form is known as *emitter-coupled logic (ECL)*. In ECL, the transistors are not operated at saturation, as they are with TTL. This

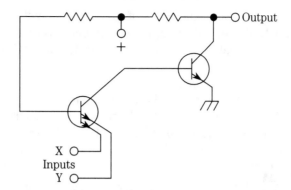

DIGITAL INTEGRATED CIRCUIT A TTL gate.

increases the speed of operation of ECL compared with TTL. But noise pulses have a greater effect in ECL, because unsaturated transistors amplify as well as switch signals. An ECL gate is somewhat more complex than a TTL gate, using four bipolar transistors.

MOS types
Several variants of MOS technology have been developed for use in digital devices. They all offer superior miniaturization and reduced power requirements as compared with bipolar digital ICs.

CMOS
Complementary-metal-oxide-semiconductor (CMOS), pronounced "seamoss" (and sometimes written that way by lay people who have heard the term but never seen it in documentation), employs both N type and P type silicon on a single chip. The main advantages of CMOS technology are extremely low current drain (and therefore a low-power requirement), high speed, and immunity to noise. A diagram of a CMOS gate is shown in the article COMPLEMENTARY-METAL-OXIDE-SEMICONDUCTOR TECHNOLOGY.

NMOS/PMOS
N-channel MOS (NMOS) offers simplicity of design, along with high operating speed. P-channel MOS is similar to NMOS, but the speed is slower.

Trends
Research is constantly being done to find ways to increase the speed, increase the memory capacity, and reduce the power requirements of digital ICs. The motivation arises mainly from constant consumer pressure for more sophisticated and portable computers.

As technology advances, you can expect to see personal computers (PCs), both of the desktop and the laptop variety, with gigabyte-level memory and working speed comparable to today's industrial machines. With improved CMOS devices and also with better batteries, power supplies will last longer between chargings. It is possible that some PCs will be able to work off solar cells, even with artificial light.

The ultimate goal is a PC, perhaps even a laptop, with a high degree of artificial intelligence (AI). Some researchers joke that if we can make a smart enough portable PC, we won't have to think anymore, anywhere we go—the PC will do that for us. See also ARTIFICIAL INTELLIGENCE, COMPUTER.

digital logic

See LOGIC.

digital signal processing

A new, and rapidly advancing, communications technique, *digital signal processing (DSP)* promises to revolutionize voice, digital, and image communications. This is important in "smart computers" because it will improve the quality of speech recognition, speech synthesis, and vision systems. See SPEECH RECOGNITION, SPEECH SYNTHESIS, VISION SYSTEMS.

In analog modes, DSP works by converting the received voice or video signal input into digital data by means of an analog-to-digital (A/D) converter. The digital signal is processed, and is reconverted back to the original voice or video via a D/A converter. See ANALOG, DATA CONVERSION.

In circuits that use only digital modes, A/D and D/A conversions are not necessary, but DSP can still be used to "clean up" the signal. This reduces the number of errors. See DIGITAL.

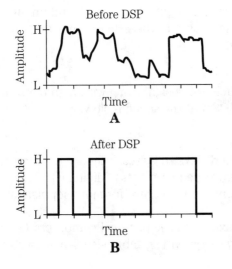

DIGITAL SIGNAL PROCESSING
Before processing (A) and after (B).

It is in the digital part of the DSP circuit that the signal enhancement takes place. Digital signals have a finite number of discrete, well-defined states. It is easier to process a signal of this kind than to process an analog signal, which has a theoretically infinite number of possible states. The DSP circuit gets rid of confusion between digital states. It "cleans up" a digital signal, as shown in the drawing. The signal before processing is on top (A); the signal after processing is below (B). The result is data that is essentially interference-free.

directional transducer

A *directional transducer* is a device that senses some effect to an extent that depends on the direction from which the effect comes. Transducers are extensively used in robotic sensing and guidance systems.

Probably the simplest example of a directional transducer is a common tape-recorder microphone. These are almost always unidirectional, that is, responding best in one direction (see the drawing, at A).

An example of a bidirectional transducer is a horizontal radio antenna known as a *dipole*. The directional pattern in the horizontal plane looks like a figure 8, as shown at B.

Some transducers are omnidirectional in a specified plane. A vertical radio antenna is omnidirectional in the horizontal plane (see the figure at C). It works equally well in all horizontal directions. But its sensitivity varies in vertical planes.

Some types of microphones are equally sensitive in all possible directions; the directional pattern for such a transducer is a sphere in three dimensions. This is a truly nondirectional transducer.

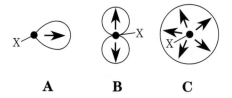

DIRECTIONAL TRANSDUCER
Unidirectional (A), bidirectional
(B), and omnidirectional (C)
patterns. Transducer is at point X.

A   B   C

direct-drive robots

A *direct-drive robot* is a robot that uses special motors, so that gears are almost completely eliminated. This improves the efficiency compared with designs using more gears. The first direct-drive robotic device was developed in the early 1980s at Carnegie-Mellon University. The first practical model was manufactured in 1984 by Adept Technology. Today, direct-drive robots are in use in Japan as well as in the United States.

Direct-drive robots were developed along the same lines as the selective compliance assembly robot arm, or SCARA. See also SCARA.

direction finding

There are two different kinds of robotic direction finding. The first involves a robot figuring out where it is, and what direction it is going. The second involves a robot determining the direction in which a destination point lies.

A "smart robot" can find its position by comparing the signals from two stations whose positions are known, as shown at A. By adding 180 degrees to the bearings of the sources X and Y, the robot (triangle) obtains its bearings as "seen" from the sources (dots).

Drawing B is a block diagram of a radio direction finder. The receiver has a signal-strength indicator and a servo that turns the antenna. The antenna has a sharp null in its response. When the antenna is turned so that the signal source is in this null, the signal strength drops dramatically. Then the robot's artificial intelligence (AI) system, or "brain," gets the bearing by some means, such as comparing the orientation of the antenna with the indication of a magnetic compass. The robot can then, if it is so instructed, seek out one source or the other. Or, by means of a computer map, it can move to any place whose position has been programmed into its AI system. See also ARTIFICIAL STIMULUS, DISTANCE MEASUREMENT.

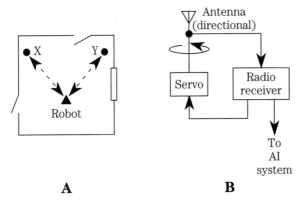

A   **B**

DIRECTION FINDING At A, robot finds its position in a room. At B, block diagram of radio direction finder.

direction resolution

The ability of a robot to separate two objects, located in almost the same direction, is called *direction resolution*. Sometimes it is also called *azimuth resolution*. It is given in degrees of arc.

Two objects might be so nearly in the same direction that the robot "sees" them as being one and the same object. But if they are at different radial distances, the robot can tell them apart by distance measurement. See also DISTANCE MEASUREMENT, DISTANCE RESOLUTION, VISION SYSTEMS.

diskette

A *diskette* is a small magnetic disk used for storing computer data and software. Diskettes are the most common storage medium for personal computers (PCs).

Diskettes are almost always found in either of two standard sizes. One is 5.25 inches in diameter, and is housed in a flexible, thin, square case. Because the case is flexible, this type of diskette is sometimes called a *floppy disk*. The other common size is 3.5 inches in diameter. It has a rigid, square case.

A typical high-density, 5.25-inch floppy disk can store up to 1.2 megabytes (Mb) of data. The 3.5-inch diskette, while physically smaller than its cousin, can store slightly more data, about 1.44 Mb.

The data is stored onto, and read from, a diskette by a magnetic read/write head. No two bits of data can be separated by more than the diameter of the diskette; this is the main advantage of diskettes over tapes, which can be hundreds of feet long. The arrangement of the head and disk in a diskette drive resembles an old-fashioned turntable. The diskette spins, and the head moves back and forth, so the head can reach any point on the surface. The head also moves up and down, in the same way the arm of a turntable lifts and drops.

The illustration shows a magnified, cut-away view of a diskette as it moves underneath a read/write head. In this drawing, the head can move toward or away from you (radially on the disk), and also up (to prevent unwanted reading and writing) and down (for reading and writing on the disk).

A diskette stores as much data as a medium-sized book, in less than one-tenth the volume. You can keep a library of diskettes for your PC. Imagine how much data could be stored on a full-wall set of shelves, in the form of diskettes!

A "smart robot" might have a diskette drive built-in, allowing different programs to be used. For example, you might have a diskette with a computer map of your house, and another diskette with a map of your friend's house, and still another with a map of your workplace or school. Yet another diskette could cause the robot to give a lecture about astrophysics, or to play music on a piano. See also COMPUTER MAP, HARD DISK.

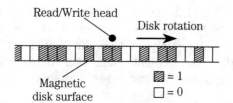

DISKETTE
Cross-sectional view of a diskette as it moves beneath a magnetic head.

disk operating system

All computers that are IBM (International Business Machines) compatible make use of a system known as the *disk operating system (DOS)*. There are two variations, called *PC-DOS*, marketed by IBM, and *MS-DOS*, marketed by Microsoft.

In DOS, you work with commands. That is, you type certain things on a keyboard to give the computer its instructions. To call up the file in which this material was written, the first command given was "EDITOR" (to load the software for word processing). The next command was "LOAD PANSONIC.PRN" (load software for Panasonic printer). After that, the next command was "CALL RAI.D" (to gain access to the file called RAI.D, containing all the text for the letter D of this book).

Microsoft DOS has been improved over the years. You will therefore hear about versions 2.0, 3.0, 3.1, 4.0, 5.0, and so on. Generally, the higher the number, the more "powerful" the software, in the sense that it can run programs faster and with more memory.

As computer memory capacity expanded, Microsoft developed Windows. This operating system might be thought of as the next step past MS-DOS 5.0. In computers that begin to show true artificial intelligence (AI), the more powerful operating systems are obviously preferable to earlier versions. In the future, systems will be developed with consumer AI devices in mind. See also ARTIFICIAL INTELLIGENCE, COMPUTER, SOFTWARE, WINDOWS.

displacement transducer

A *displacement transducer* is a device that measures a distance or angle traversed, or the distance or angle separating two points. Or, it might convert a signal into movement over a certain distance or angle. A transducer that measures distance in a straight line is a linear displacement transducer. If it measures an angle, it is an angular displacement transducer.

Suppose you want a robot arm to rotate 28 degrees in the horizontal plane—no more and no less. You give a command to the robot's computer, such as "BR 28" (base rotation 28 degrees). The computer then sends a signal to the robot arm, so that it rotates clockwise. An angular distance transducer keeps track of the angle of rotation, sending a signal back to the computer. This signal increases in linear proportion to the angle that the arm has turned (see the graph).

By issuing the command "BR 28" you told the computer two things: first, to start the base of the arm rotating, and second, to stop the rotation when it has moved through 28 degrees of arc. This second command set a threshold level for the return signal. As the signal from the displacement transducer increases, it reaches this threshold at 28 degrees of rotation. The computer is programmed to stop the arm at this time.

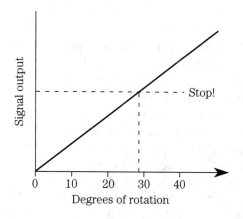

DISPLACEMENT TRANSDUCER
Rotation stops when output reaches threshold level.

There are other ways to get a robot arm to move, besides using distance transducers. The figure above is just one example of how a distance transducer might be used in a robotic system. See also ROBOT ARMS, TEACH BOX.

distance measurement

Distance measurement, also called *ranging*, is a method for a robot to navigate in its environment. It also allows a central computer to keep track of the where-abouts of insect robots. There are several ways for a "smart robot" to measure the distance between itself and some object.

Sonar literally uses sound or ultrasound, bouncing the waves off of things around the robot and measuring the time for the waves to return. If the robot senses that an echo delay is extremely short, it knows that it is getting too close to something (see illustration). Sound travels about 1100 feet per second in air.

Radar works just like sonar, but uses radio waves rather than sound waves. Light beams can also be used. But radio and light beams travel at such high speeds that it is hard to measure delay times for nearby objects. Also, the objects might not reflect the waves very well.

Beacons of various kinds can be used for distance measurement. These devices can use sound, radio waves or light waves. See also ARTIFICIAL STIMULUS, AUTONOMOUS ROBOTS, BEACON, DIRECTION FINDING, INSECT ROBOTS, PROXIMITY SENSING.

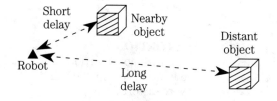

DISTANCE MEASUREMENT
Object distance is proportional to sound-return delay.

distance resolution

Distance resolution is the precision of a robotic distance measurement system. It is the ability of the system to differentiate between two objects that are almost, but not quite, the same distance away from the robot. It can be measured in feet, inches, meters, millimeters, or even smaller units.

When two objects are very close to each other, a distance-measuring system will see them as a single object. But as the objects get farther apart, they become distinguishable. The minimum radial separation of objects, for a "smart robot" to tell them apart, is the distance resolution for the ranging system.

Distance resolution sometimes depends on how far away the objects actually are. Often, nearby things can be resolved better than ones far away (see the drawing). Suppose two objects are separated radially by one foot. If their mean (average) distance is 10 feet, their separation is $\frac{1}{10}$ (10 percent) of the mean distance. But if their mean distance is 1000 feet, their separation is $\frac{1}{1000}$ (0.1 percent) of the mean distance. If the distance resolution is 1 percent of the mean distance, then the robot can tell the nearer pair of objects apart, but not the more distant pair.

If the radial distances to two or more objects are all the same or nearly the same, then their distances can't be resolved, no matter how far apart they actually are (see DIRECTION RESOLUTION).

Distance resolution depends on the type of ranging system used. The most sensitive methods compare the phases of laser beams arriving from, or reflected by, beacons. See also BEACON, DISTANCE MEASUREMENT, PROXIMITY SENSING, VISION SYSTEMS.

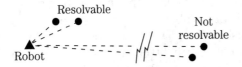

DISTANCE RESOLUTION
Nearby objects are often easier to resolve than distant objects.

distortion

When a digital pulse travels over a long distance, and/or goes through many circuits getting from its origin to its destination, there is always some distortion. This renders the pulse less and less like the original, until eventually it is hard to distinguish it from background noise (see the drawings).

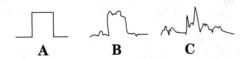

DISTORTION
Pulse with no distortion (A), moderate distortion (B), and severe distortion (C).

Distortion is much less of a problem with modern digital devices, as compared with old-fashioned analog systems. The main reasons for this are superior wiring methods, digital signal processing and fiberoptic data transmission methods. See also DIGITAL SIGNAL PROCESSING, FIBEROPTIC DATA TRANSMISSION.

diving machine

See SUBMARINE ROBOTS.

domain of function

The *domain* of a mathematical function is the set of values for which the function is defined. Every x in the domain of a function f is mapped by f onto a definite, single value y. Also, any x not in the domain is not mapped onto anything by the function f.

Suppose you are given the function $f(x) = +x^{1/2}$. (The ½ power is the square root). The graph of this function is shown in the drawing. Note that the function is not defined for negative values of x, and is also not defined, as shown in the drawing, for $x = 0$. The function $f(x)$ has values y if, and only if, $x > 0$. Therefore the domain of f is the set of positive real numbers.

Computers work extensively with functions, both analog and digital. Functions are important in robotic navigation, location and measurement systems. See also FUNCTION, RANGE OF FUNCTION.

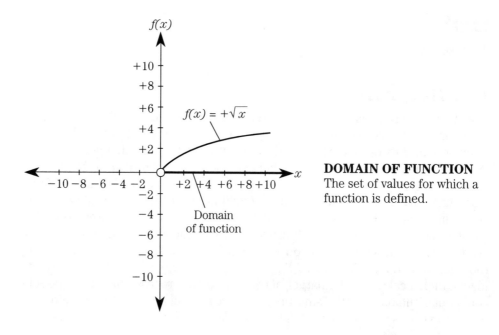

DOMAIN OF FUNCTION
The set of values for which a function is defined.

don't-care state

In a logic function, some states do not affect the outcome. This happens when a logic function is undefined for some input combinations. When this occurs, the states that have no effect on the output are called *don't-care states*.

Suppose, for example, you want to represent the decimal numbers 0 through 9 in binary form. Then you will need four binary places, obtaining:

$$
\begin{array}{ll}
0 = 0000 & 5 = 0101 \\
1 = 0001 & 6 = 0110 \\
2 = 0010 & 7 = 0111 \\
3 = 0011 & 8 = 1000 \\
4 = 0100 & 9 = 1001
\end{array}
$$

This does not use all possible four-digit binary numbers. There are six left over: 1010, 1011, 1100, 1101, 1110, and 1111. (These correspond to decimal numbers 10 through 15.) These numbers are undefined, unless they are defined by default.

Suppose that binary 1010, 1100, and 1110 are assigned to 8, and binary 1011, 1101, and 1111 are assigned to 9. This prevents confusion that might occur when some binary numbers are left undefined.

The values of the two middle digits are don't-care states in these default definitions. When the leftmost digit is 1, the last digit alone is sufficient for the computer to know whether the decimal number is 8 or 9. See also BINARY, BINARY NUMBER SYSTEM.

DOS

See DISK OPERATING SYSTEM.

Drexler, Eric

Imagine robots so small that they could actually fit inside a single living cell, or travel around in the bloodstream, navigated by on-board computers. Is such a thing possible? Eric Drexler thinks so, and is working toward that goal.

What would such microscopic robots be good for? According to Drexler and his colleagues, they might repair damaged DNA (deoxyribonucleic acid). Some researchers have suggested that robots could be designed specifically to act as antibodies against diseases that defy biologists' attempts to find cures. As of this writing, the most important such disease is AIDS (acquired immunodeficiency syndrome). Plagues that people once thought were eradicated for good, such as tuberculosis and malaria, are evolving new strains that resist conventional treatments. Biological research is largely a trial-and-error process. But what if people could build millions of little "smart robots," programmed to go after certain bacteria and viruses and kill them? "Entirely possible," researchers like Drexler say.

Drexler imagines building a molecular computer from individual atoms of carbon, a fundamental ingredient of all living matter. These computers would store data in much the same way that DNA does, but they would be programmed by people rather than nature. These computers could be as small as 0.1 micron (10^{-7} meter or 0.0001 millimeter) in diameter. Even an object this small has enough carbon atoms to make a central processing unit (CPU) equivalent to that of a typical personal computer (PC).

The building of computers and robots molecule-by-molecule might seem like an impossible task. But not to scientists in the field of artificial life. Compare the technology of today with that of, say, 1948, when a machine equivalent to a modern PC would have consumed the power of a city the size of Rochester, Minnesota, and would have needed all the downtown buildings for its relays and vacuum tubes. As if that isn't ridiculous enough, a computation that takes your notebook computer five seconds to solve would have taken that huge, bumbling thing a couple of weeks. Now extrapolate: 1948 is to today, as today is to—what? See also ARTIFICIAL LIFE, BIOCHIP, BIOLOGICAL ROBOTS, NANOTECHNOLOGY.

drop-in and drop-out

When digital data is stored on, and retrieved from, magnetic media, errors occasionally occur. A logic 1 might get changed to 0, or vice-versa. Or a bit might get accidentally inserted or dropped. The generation of extra bits is called *drop-in*. The loss of a bit is called *drop-out*.

Drop-in and drop-out can occur because of dust on a disk, because of stray electrical impulses, or as a result of a defect in a disk. Errors can also take place for no apparent reason, other than the imperfect nature of physical things. The probability of errors increases as data files are copied over and over.

Modern digital systems have almost entirely eliminated the problems of noise and distortion that plagued older, analog systems. Nevertheless, an error in just the wrong place, at just the wrong time, can cause a major malfunction in an artificial intelligence (AI) system or in a "smart robot."

Drop-in and drop-out can be minimized by keeping the temperature and humidity within reasonable limits, and by keeping the air as free as possible from dust, salt spray, and corrosive pollutants. See also ANALOG, DIGITAL, DIGITAL SIGNAL PROCESSING, DISTORTION.

dual inline package

The *dual inline package (DIP)* is a common housing for integrated circuits (ICs). A flat, rectangular box containing the chip is fitted with pins along either side, as shown in the drawing. The number of pins varies from three or four to a dozen or more on each side.

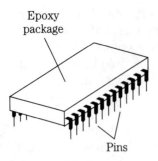

Epoxy
package

Pins

DUAL INLINE PACKAGE
A common housing for integrated
circuits.

The DIP package makes it easy to install and replace ICs. Sockets can be used for this purpose. Sometimes DIPs are soldered directly onto a printed circuit (PC) board. It is somewhat more difficult to remove and replace ICs that are soldered to the board, but there is less chance of intermittent failures with this method. See also FLATPACK, INTEGRATED CIRCUIT, SINGLE INLINE PACKAGE.

duplex

The term *duplex* can refer to either of two different things in computer communications.

When you connect your personal computer (PC) to a modem and use it with the telephone line, for example, to access an on-line network, you can "talk" with the central computer, or with other computers, almost as if you were in a normal conversation. Data goes in both directions, more or less simultaneously. This is duplex communication. It can be done with radio and fiberoptic links, as well as on the telephone lines.

When two different signals are sent over a single circuit, for example a picture and a voice signal, the mode is sometimes called duplex. See also MODEM, NETWORK.

duration time

Duration time is the length of a data pulse, usually representing one bit, from the instant of turn-on till the instant of turn-off. If a pulse is perfectly rectangular (as it should ideally be), then duration time is easy to determine, as shown in the drawing.

Pulses are often not perfectly rectangular. There might be distortion in the circuit. Then, the duration of a pulse must be determined by an approximation method. An equivalent rectangular pulse, having the same area beneath its graph, and having the appropriate maximum amplitude, is constructed mathematically, and its duration is then calculated.

Duration time is inversely proportional to data speed. That is, the faster the data rate, the shorter the pulse duration. See also BAUD RATE, BIT, DISTORTION.

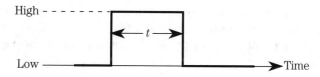

DURATION TIME The time t,
during which a digital pulse is in the high condition.

duty cycle

Duty cycle is the proportion of time during which a circuit, machine, or component is operated.

Suppose that a motor is run for one minute, then is shut off for two minutes, then is run for a minute again, and so on. Then the motor runs for one out of three minutes, or ⅓ of the time, and its duty cycle is therefore ⅓, or 33 percent.

If a device is observed for a length of time t_o, and during this time it runs for a total time t (in the same units as t_o), then:

$$\text{duty cycle (\%)} = 100\frac{t}{t_o}$$

as shown in the illustration.

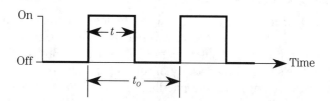

DUTY CYCLE Proportion
of time t/t_o, during which device is working.

When determining the duty cycle, it is important that the observation time t_o be long enough. In the case of the motor described above, any value of t_o less than three minutes would be too short to get a complete sample of the data.

The more a circuit, machine, or component is used, the sooner it will wear out. That is, the higher the duty cycle, the shorter the useful life. This effect is most pronounced when a device is worked hard. Also, the rating of a device often depends on the duty cycle at which it is expected to be used.

Suppose the motor described above is rated at a torque of 10 newton-meters (10 nm) for a duty cycle of 100 percent. If it must constantly provide a torque of 9.9 nm, then it will be taxed to its utmost. If it must constantly turn a load of 12 nm, it might break down long before it should. For a duty cycle of 33 percent, however, it might be rated at 15 nm, as long as any single working period does not exceed two minutes. If it only needs to turn 0.5 nm, however, the motor cannot only run continuously, but it will probably last longer than its expected life. But such over-engineering would represent a needless expense.

Devices such as robot motors can be protected from overwork (either momentary or long-term) by means of back pressure sensors. See BACK PRESSURE SENSOR.

dynamic stability

Dynamic stability is a measure of the ability of a robot to maintain its balance while in motion.

A robot with two or three legs, or that rolls on two wheels, can have excellent stability while it is moving. But when it comes to rest, it is unstable. A two-legged robot can be pushed over easily when it is standing still. This is one of the major drawbacks of biped robots. It is very difficult to program a good sense of balance, of the sort you take for granted, into a machine.

Robots with four or six legs have good dynamic stability. But they are usually slower in their movements, compared with machines having fewer legs. See also BIPED ROBOTS, INSECT ROBOTS, STATIC STABILITY.

dynamic transducer

A *dynamic transducer* is a coil-and-magnet device that converts mechanical vibration into electrical currents, or vice-versa. The most common examples are the dynamic microphone and the dynamic loudspeaker. But dynamic transducers can be used as sensors in a variety of robotic applications.

The illustration shows a functional diagram of a dynamic transducer suitable for converting sound waves into electric currents, or vice-versa. A diaphragm is attached to a permanent magnet. The magnet is surrounded by a coil of wire.

Sound vibrations cause the diaphragm to move back and forth; this moves the magnet, which causes fluctuations in the magnetic field within the coil. The result is alternating-current (ac) output from the coil, having the same wave-shape as the sound waves that hit the diaphragm.

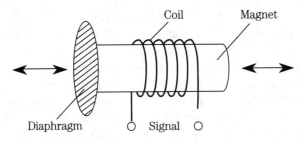

DYNAMIC TRANSDUCER The magnet moves relative to the coil.

If an audio signal is applied to the coil of wire, it creates a magnetic field that produces forces on the permanent magnet. This causes the magnet to move, pushing the diaphragm back and forth, and therefore creating sound waves in the air that exactly follow the waveform of the signal.

Dynamic transducers are commonly used in speech-recognition and speech-synthesis systems. See also SPEECH RECOGNITION, SPEECH SYNTHESIS.

economic effect of robotics

Robots allow production of more goods at a lower cost than is possible without them. A robot won't break down as often as a human worker gets sick. Robots can be used in dangerous jobs, saving human lives (and lowering medical bills). Robots can't get bored, so they can do jobs that would numb people's minds with monotony. Many scientists and writers think that the future success of industrialized economies will depend on robotization. Nations that employ robots might prosper; nations that do not use robots will never become major economic powers.

All these great things are meaningless to the person who is out of work, having been displaced by a robot. Sometimes such workers feel insulted as well as injured: "I was replaced by a *machine*!" This problem can be solved, however, because robots help the economy more than they hurt it. One solution would be to set up schools, paid for with some of the profits resulting from robotization. These schools would retrain people who have been put out of work by robots, so that they could find jobs that would make better use of their human talents. This would in turn help the economy still more—and the people would be happier too.

As economies become less industrial and more information-based, artificial intelligence (AI), as well as robotics, promises an expanding market of well-paid, interesting work. Ironically, robots and computers might be the key to making training affordable to more people.

See also ARTIFICIAL INTELLIGENCE, ASSEMBLY ROBOTS, AUTOMATED INTEGRATED MANUFACTURING SYSTEM, COMPUTER-ASSISTED INSTRUCTION, EDUCATIONAL ROBOTS.

edge connector

An *edge connector* is a plug-and-socket device used to connect a printed-circuit (PC) board in an electronic system. The connector gets its name from the fact that the "male" portion is actually contained right along the edge of the circuit board, as shown in the drawing. The "female" connector is a receptacle, into which the edge of the board snugly fits. There are often dozens of contacts.

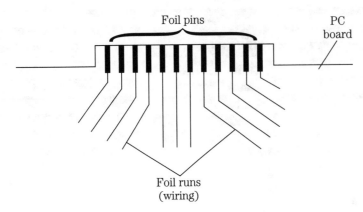

Foil pins

PC
board

Foil runs
(wiring)

EDGE CONNECTOR Pins are contained right in the circuit board.

Edge connectors are ideally suited to modular construction of electronic equipment. In troubleshooting, a computer identifies which board most likely contains the problem. That board can be pulled out easily, and plugged into a service machine. The machine locates the faulty component. A technician replaces the bad part, tests the board again, and then puts it back into the original machine, or into storage for future use in the field. See also MODULAR CONSTRUCTION, PRINTED CIRCUIT.

edge detection

The term *edge detection* refers to the ability of a robotic vision system to locate boundaries. It also refers to the robot's knowledge of what to do with respect to those boundaries.

A robot car, for example, uses edge detection to see the edges of a road, and uses the data to keep itself on the road. But it also needs to stay a certain distance from the right-hand edge of the pavement, so that it doesn't cross over into the lane of oncoming traffic (see the drawing). It must stay off the road shoulder. Thus, it must tell the difference between pavement and other surfaces, such as gravel, grass, sand, and snow. The robot car could use beacons for this purpose, but this would require the installation of the guidance system beforehand. That would limit the robot car to roads that are equipped with such navigation aids.

A personal robot, of the type people often imagine doing work similar to that of a maid or butler, would need to see the edges of a door before going through the door. Otherwise it might run into walls, or into closed doors, or into windows. It had better know when it is coming near a flight of stairs, too. See also ARTIFICIAL INTELLIGENCE, VISION SYSTEMS.

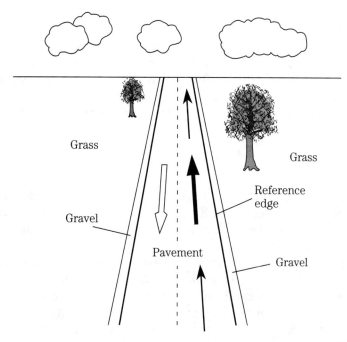

EDGE DETECTION In this example,
the robot must keep track of the right-hand edge of a pavement.

educational robots

The term *educational robot* applies to any robot that causes its user(s) to learn something. Especially, this term applies to robots available for consumer use. Robots of this kind have become popular among children, particularly in Japan, but increasingly in the U.S. While these machines are toys in the sense that children have fun using them, they often are excellent teachers as well. Children learn best when they're having fun at the same time. Kids and robots do well together.

An instructional robot is an educational robot that is intended only as a teacher. Robots of this kind can be purchased for use in the home, but more often they are found in schools, especially at the junior-high and senior-high level (grades 7 through 12).

Robots are intimidating to some students. But once a child or young adult gets used to working or playing with machines, robots can become companions, even friends—especially if there is some measure of artificial intelligence (AI). See also ARTIFICIAL INTELLIGENCE, COMPUTER-ASSISTED INSTRUCTION, PERSONAL ROBOTS.

efficiency

Efficiency is the ratio, usually expressed as a percentage, between the output power produced in the desired form by a device, and the total input power to that device. Efficiency can also be given as an energy ratio over a period of time.

If P_{out} is the output power or energy in the desired form and P_{in} is the total input power or energy, then the efficiency, Eff, in percent is given by the formula:

$$Eff = \frac{100P_{out}}{P_{in}}$$

All machines have an efficiency greater than zero (absolutely useless) but less than 100 percent (perfect). Of course, when designing and building machines of all kinds, engineers strive for the greatest possible efficiency, while keeping the cost within reason.

The term *efficiency* is sometimes used in a general, nonmathematical sense, as an expression for how well a machine, program or process does a certain job. You might say, "That is an inefficient computer program," meaning that it contains far more steps than it needs. Or you might say, "Six-legged robots are efficient in rough terrain," meaning that they work better than most other robot schemes in that environment.

elastomer

An *elastomer* is a flexible substance resembling rubber or plastic. In robotic tactile sensors, elastomers can be used to detect the presence or absence of mechanical pressure.

The drawing shows how an elastomer might be used to detect, and locate, a pressure point. The elastomer conducts electricity fairly well, but not perfectly. It has a foam-like consistency, so that it can be compressed. An array of elec-

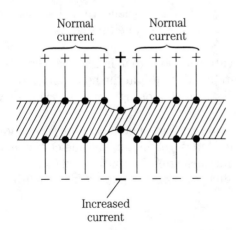

ELASTOMER
Where the material is compressed, the current increases between electrodes.

trodes is connected to the top of the elastomer pad; an identical array is connected to the bottom of the pad. These electrodes run to the robot's artificial intelligence (AI) system, or "brain."

When pressure appears at some point in the elastomer pad, the material compresses, and this lowers its electrical resistance in a small region. This is detected as an increase in the current between electrodes in the top pad and the bottom pad, but only within the region where the elastomer is being compressed. The data is sent to the AI system, which determines where the pressure is taking place, and how intense it is. See also TACTILE SENSORS.

electric eye

An *electric eye* optically senses an object and then actuates a device. For example, it might be set up to detect anything passing through a doorway. This can count the number of people entering or leaving a building. Another example is the counting of items on a fast-moving assembly line; each item breaks the light beam once, and a circuit counts the number of interruptions.

Usually, an electric eye has a light source and a photocell; these are connected to an actuating circuit as shown in the block diagram. When something interrupts the light beam, the voltage or current from the photocell changes dramatically. It is easy for electronic circuits to detect this voltage or current change. Using amplifiers, even the smallest change can be used to control large and powerful machines.

Electric eyes don't always use visible light. Infrared, or energy with a wavelength somewhat longer than visible red, is commonly used in optical sensing devices. This is ideal for use in burglar alarms, because an intruder cannot see the beam, and therefore cannot avoid it.

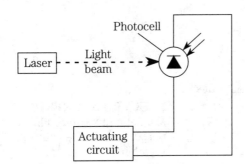

ELECTRIC EYE
Consists of light source,
photocell, and actuating circuit.

electrodynamics

Electrodynamics is a branch of science concerned with the interaction of electrical and mechanical effects. The attraction of opposite charges, the repulsion of like charges, the attraction of opposite magnetic poles, and the repulsion of like magnetic poles all fall in the category of electrodynamic effects. The term arises from the concept of "moving electricity."

Motors and generators are examples of electrodynamic devices. These operate according to the principles of magnetism. Moving-needle meters (analog) also work according to the principles of electrodynamics. The servo systems used by robots are examples of more refined electrodynamic machines. See also GENERATOR, MOTOR, SERVOMECHANISM, SERVO ROBOTS, SERVO SYSTEM.

electromagnetic field

An *electromagnetic (EM) field* is a force field that has both electrical and magnetic components. Such fields are always produced when charged objects, such as electrons in a wire, are accelerated.

The controlled, constant acceleration of electrons is responsible for radio waves. These are EM fields having frequencies ranging from about 3000 Hertz to 3000 gigahertz (3,000,000,000,000 Hz). The 60-Hz alternating-current (ac) utility produces an EM field whose frequency is lower than that of radio waves. Infrared, visible light, ultraviolet, X rays, and gamma rays are forms of EM energy with frequencies higher than those of the radio waves.

All EM fields consist of an electric component whose flux is at right angles to a magnetic component. The fields are constantly alternating back and forth. As a result, they tend to travel through space, the ac electric component producing an ac magnetic component of identical frequency, which in turn gives rise to another ac electric component, and so on indefinitely. The direction in which the EM field travels is at right angles to both the electric and magnetic lines of flux (see the drawing).

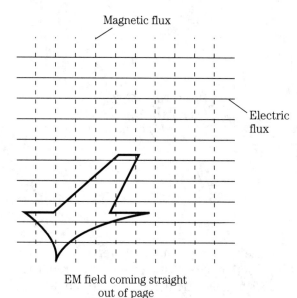

Magnetic flux

Electric flux

EM field coming straight out of page

ELECTROMAGNETIC FIELD
Magnetic flux (dotted lines) is perpendicular to electric flux (solid lines).

Electromagnetic fields are used extensively for data communications. Microwave links, fiberoptic systems, and satellite links are the most common examples. But there are situations in which these fields can upset the functioning of electronic equipment, particularly computers. For this reason, sensitive digital equipment, such as artificial intelligence (AI) systems and robot controllers, must be shielded from stray EM fields. See also ELECTROMAGNETIC SHIELDING, FIBEROPTIC DATA TRANSMISSION, MICROWAVE DATA TRANSMISSION, SATELLITE DATA TRANSMISSION.

electromagnetic shielding

Electromagnetic shielding is a means of preventing computers and other sensitive equipment from being affected by stray electromagnetic (EM) fields. Computers also generate EM energy of their own, and this can cause interference to other devices, especially radio receivers, unless shielding is used.

The simplest way to provide EM shielding for a circuit is to surround it with metal, usually copper or aluminum, and to connect this metal to electrical ground. Because metals are good conductors, an EM field sets up electric currents in them. These currents oppose the EM field, and if the metal enclosure is grounded, the EM field is in effect shorted out.

It isn't always easy to completely enclose a circuit with a metal shield. In a computer, the keyboard is generally made of plastic. Fortunately, small gaps in the enclosure can be tolerated. The keyboard housing can be metal, with holes for the keyswitches.

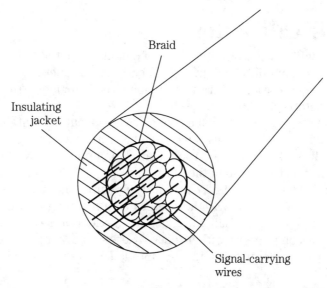

ELECTROMAGNETIC SHIELDING A shielded, multiconductor cable.

In addition to metal enclosures, interconnecting cables must be shielded. This is done by surrounding all the cable conductors with a copper braid (see drawing). Cable of this kind is sometimes called *shielded cable*; the braid is electrically grounded via the connectors at the ends of the cable.

One of the biggest advantages of fiberoptic data transmission is the fact that it does not need EM shielding. Fiberoptic systems are immune to the EM fields produced by radio transmitters and alternating-current (ac) utility wiring. Fiberoptic systems also work without producing external EM fields. See also ELECTROMAGNETIC FIELD, FIBEROPTIC DATA TRANSMISSION.

electromechanical transducer

Any device that converts electrical energy into mechanical energy, or vice-versa, is an *electromechanical transducer*. Motors and generators are what you probably think of first. A motor works via magnetic forces produced by electric currents; the generator produces electric currents as a result of motion of an electric conductor in a magnetic field.

Devices that convert sound into electricity, or vice-versa, are electromechanical transducers. Speakers and microphones are the most common examples. They usually work via dynamic principles, but some work electrostatically. See DYNAMIC TRANSDUCER, ELECTROSTATIC TRANSDUCER.

Analog meters are electromechanical transducers. They convert electric current into displacement. In recent years, digital meters have largely replaced analog meters. This is because digital devices do not have moving parts to wear out.

Robots use electromechanical transducers in many ways, particularly in servo devices. See also SERVOMECHANISM, SERVO ROBOTS, SERVO SYSTEM.

electronic calculator

An *electronic calculator* is a primitive form of digital computer. Calculators can always add, subtract, multiply, and divide numbers. Other functions are often included, such as square root, logarithms, exponential functions, trigonometric functions, and factorials.

The drawing shows the arrangement of a typical calculator suitable for balancing a checkbook. A unit like this can be bought in a department store for less than 20 dollars. It works in ordinary room light, being powered by photovoltaic (solar) cells. It has no battery to bother with.

The most advanced electronic calculators are programmable, and might better be called *nanocomputers* (the prefix *nano* meaning, in this context, smaller than micro but bigger than nothing). The distinction between large calculators and microcomputers is rather fuzzy. There's no exact point where you can say, "This is a calculator, but add one more logic gate and it will become a microcomputer."

The earliest digital computers were actually electronic calculators. But what a difference there is between those lumbering monstrosities and the pocket calculators of today! The calculator shown in the drawing would, in 1946, have taken up the better part of a room and needed a special utility line all for itself. See also COMPUTER, ELECTRONIC NUMERICAL INTEGRATOR AND CALCULATOR.

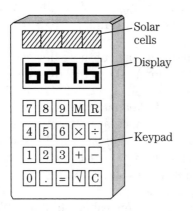

ELECTRONIC CALCULATOR
A typical calculator for doing basic arithmetic.

electronic mail

Electronic mail, also called *E-mail*, is a communications system that lets people leave written messages for each other. This requires a personal computer (PC), special software called terminal emulation software, and a modem for interconnection with the telephone line. Amateur radio operators use the airwaves instead of "landline."

To use E-mail, you pay a flat fee for up to a certain length of time on line, or for a certain number of messages, per month. Ham radio operators, of course, don't need to pay anything to send E-mail via the radio. See also BULLETIN BOARD, MODEM, PACKET COMMUNICATIONS, PACKET RADIO, TERMINAL EMULATION SOFTWARE.

Computers can use E-mail to exchange data. By interconnecting computers in a network, a "smarter" system can be obtained than is possible with one computer. In the future, networking will become more and more important, because of the increasing demand for artificial intelligence (AI) that is available to a large number of subscribers all at once. See ARTIFICIAL INTELLIGENCE, NETWORK.

electronic memory

See MEMORY.

electronic music

See COMPUTER MUSIC.

electronic numerical integrator and calculator

The first digital computers made use of vacuum tubes or relays for switching, because transistors hadn't been invented yet, and integrated circuits (ICs) were nothing but the ideas of dreamers. One such computer, called the *Electronic Numerical Integrator and Calculator (ENIAC)*, was first put to use in the mid-1940s at the University of Pennsylvania.

The ENIAC had about 19,000 vacuum tubes. Each of these devices needed a source of power to drive its filament, in addition to the actual digital signal voltages. The ENIAC weighed 60,000 pounds and took up a good part of a building. It needed a large air-conditioning system just to keep it from overheating.

Today's modern personal computers (PCs), such as you can buy in department stores for a few hundred dollars, contain ICs with silicon chips more powerful than ENIAC. A typical silicon chip is so small that you'd need a good magnifying glass to tell it from a grain of sand. It can run off a small battery. It produces practically no heat. It needs no maintenance.

The transition from ENIAC to modern PC took only about 40 years. What will happen in the next 40 years? See ARTIFICIAL INTELLIGENCE, ARTIFICIAL LIFE, COMPUTER, INTEGRATED CIRCUIT.

electronic warfare

Electronic warfare is the use of electronic equipment and techniques in military defense or aggression. The extent of electronic automation can vary from simple communications systems to "star wars" fought among machines.

Military minds have long dreamed of ways to alleviate the human suffering involved in war. This was the major motivation behind the so-called Strategic Defense Initiative (SDI) that received so much attention during the presidency of Ronald Reagan. The *Star Wars* program, as SDI was often called, would employ computer-controlled "robot spacecraft" to intercept and destroy enemy missiles while they were in space, dozens or hundreds of miles above the surface of the earth.

Some military experts think it is possible to build robot soldiers. These non-human warriors could not only fight on the ground, but they could pilot aircraft, helicopters, and boats. Presumably, they would not feel the pain, boredom, loneliness, sorrow, and other agonies that go along with fighting in a war. The use of robots in warfare is probably inevitable (see CAPEK, KAREL). Artificial intelligence (AI) could help generals plan their strategies, and could control whole armies of insect robots (see INSECT ROBOTS).

What will all this come to? Maybe someday, the leaders of the world will agree to do away with military hardware altogether, because it costs too much money. Wars could be made completely electronic. Conflicts between nations could be resolved by having the generals play video games. See also BIOLOGICAL ROBOTS, COLOSSUS, COMPUTERIZED DEFENSE SYSTEM, MILITARY ROBOTS.

electrostatic transducer

An *electrostatic transducer* is a device that changes mechanical energy into electrical energy or vice-versa, by taking advantage of electrostatic forces. The most common types involve conversion between sound waves and audio-frequency electric currents.

The drawing is a functional diagram of an electrostatic microphone. Incoming sound waves cause vibration of the flexible plate. This produces rapid (although small) changes in the spacing, and therefore the capacitance, between the two plates. A direct-current (dc) voltage is applied to the plates, as shown. As the capacitance changes between the plates, the electric field between them fluctuates. This produces variations in the current through the primary winding of the transformer. Audio signals appear across the secondary winding.

An electrostatic speaker works in just the opposite way. Currents in the transformer produce changes in the voltage between the plates. This change results in electrostatic force fluctuations, pulling and pushing the flexible plate in and out. The motion of the flexible plate produces sound waves.

Electrostatic transducers can be used in most applications where dynamic transducers are employed. This includes speech recognition and speech synthesis systems. Advantages of electrostatic transducers include light weight and good sensitivity. They also work with extremely low electric current. See also DYNAMIC TRANSDUCER, SPEECH RECOGNITION, SPEECH SYNTHESIS.

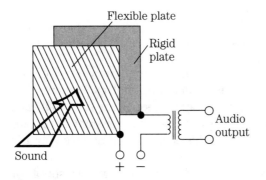

Flexible plate

Rigid plate

Audio output

Sound

+ −

ELECTROSTATIC TRANSDUCER
Sound waves cause changes in capacitance, producing audio-frequency current.

ELIZA

One of the more controversial developments in artificial intelligence (AI) involved a program called *ELIZA*. This program was put together in the 1960s by Joseph Weizenbaum of the Massachusetts Institute of Technology (MIT).

The purpose of ELIZA was to simulate a psychoanalyst (a doctor who helps people work out their problems by talking with them). The "patient" would sit at a computer terminal and "converse" with the "doctor" by typing sentences on a keyboard. The ELIZA program was, in fact, sometimes called "DOCTOR."

Suppose you were the "patient," and you sat down to the computer to talk with ELIZA. You would see, on the screen:

Speak up!

You might then type:

I'm upset.

The computer might then respond with:

Why are you upset?

to which you might reply:

I don't know. That's why I'm here.

The conversation would then proceed, with ELIZA asking questions, and the "patient" giving answers or asking other questions. The program would never really commit itself by saying that's wrong or don't ever do that again. The "doctor" would just make phrases, some from its own memory and some stored from things the "patient" said earlier. Nevertheless, ELIZA often behaved so much like a real psychiatrist that some people actually suggested that it was just as good as a human doctor.

Weizenbaum was disturbed by the reactions and the controversy ELIZA caused. The program was not really very "smart," especially by standards of the 1990s. The ELIZA program could not then, as computers still cannot, have any feeling or concern for human beings. See also ARTIFICIAL INTELLIGENCE.

embedded path

An *embedded path* is a means of guiding a robot along a specific route. Automated guided vehicles (AGVs) employ this guidance scheme.

One common type of embedded path is a buried, current-carrying wire. The current in the wire produces a magnetic field that the robot can follow. This method of guidance has been suggested as a way to keep a car on a highway, even if the driver isn't paying attention. The wire needs a constant supply of electricity for this guidance method to work. If this current is interrupted for any reason, the robot will lose its way.

Alternatives to wires, such as colored paints or tapes, do not need a supply of power, and this gives them an advantage. Tape is easy to remove and put somewhere else; this is difficult to do with paint, and practically impossible with wires embedded in concrete. See also AUTOMATED GUIDED VEHICLE, GUIDANCE SYSTEMS.

empirical design

Empirical design is an engineering technique, in which experience and intuition are used in addition to theory. The process is largely trial-and-error. The engineer starts at a logical point, based on theoretical principles. But experimentation is necessary in order to get the device or system to work just right.

Robots are ideally suited to empirical design techniques. You can't just make up plans for a robot, no matter how detailed or painstaking the drawing-board process might be, and expect the real machine to work perfectly on the first trial. A prototype is built and tested, noting the flaws. The engineer goes back to the drawing board and revises the design. Sometimes it's necessary to start all over from scratch; more often, small changes are made. The machine is tested again, and the problems noted; another drawing-board round follows. This process is repeated until the machine works the way the engineer wants it to.

Computer programs are perfected in a similar way. A program rarely works perfectly the first time it is run. Debugging is necessary, somtimes requiring several rounds of rewriting and rerunning. See COMPUTER PROGRAMMING, DEBUGGING.

In artificial intelligence (AI), "perfection" is an ideal that is never achieved totally. This is because, no matter how smart a computer becomes, it can always be made a little smarter—and wiser too! Computer consciousness is always advancing. In some ways, the behavior of AI systems evolves on its own, as if it were independent of the minds of the engineers involved. See ARTIFICIAL INTELLIGENCE, COMPUTER, COMPUTER-AIDED DESIGN, COMPUTER CONSCIOUSNESS, COMPUTER EMOTION, COMPUTER GRAPHICS, COMPUTER-INTEGRATED MANUFACTURING, COMPUTER REASONING.

encoding

Encoding is the process by which a machine translates a spoken or written language, like English or Japanese, into code. When you type something on the keyboard of a computer, for example, the characters (letters, numerals, and punctuation) are converted into a stream of data bits. If your computer has speech recognition, it converts audible sounds into data bits. The binary data stream, in these cases, is the code.

In a robot arm, the term *encoding* refers to the method by which the arm sets or determines its position. This is done in either of two ways.

In absolute position encoding, the angle of the robot arm, or of a certain section of the arm relative to the section ahead of it, is translated directly into binary code. The angle might vary from 0 to 90 degrees, for example, encoded with a resolution of one degree. This would need seven bits, ranging from 0000000 (decimal 0) to 1011010 (decimal 90), as shown in the drawing. Some other variable, such as linear displacement, might be used. Then the output would be in units like millimeters or inches. This method could be employed for a telescoping robot arm.

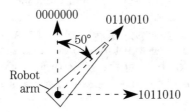

ENCODING
Digital encoding for robot arm from 0 to 90 degrees. The arm is at 50 degrees.

In incremental position encoding, a binary signal is sent whenever a robot arm or section actually moves. The value of the digit depends on the extent of the increment; that is, it indicates the angle or distance that the device has moved. To get the absolute angle or position, it is necessary to add up increments over time. See also BIT, DATA CONVERSION, ROBOT ARMS.

end effector

The device or tool that is connected to the end of a robot arm depends on what the arm is supposed to do. This tool is called an *end effector*.

If a robot is designed to set the table for supper, "hands" would be attached to the ends of the robot's arms. The end effectors would be robot grippers. In an assembly-line robot designed to insert screws into cabinets, a rotating-shaft device and screwdriver head would be attached at the end of the arm. This might be changed to a bit for drilling holes, or an emery disk for sanding wood.

A given type of robot arm can usually accommodate only certain kinds of end effectors. You couldn't take the table-setting robot, directly replace one of its grippers with a screwdriver, and then expect it to tighten the screws on the kitchen-door hinges. You would have to change the software in the robot, so that it would be in "handyperson mode" rather than "waitperson mode." You would also need to change the hardware to twist, rather than grip. See also ROBOT ARMS, ROBOT GRIPPERS.

endless loop

In computers and artificial intelligence (AI) systems, the term *endless loop* refers to a situation in which the machine gets "hung up" in a circle of reasoning that takes it nowhere. In programming this is sometimes called an *infinite loop*. It causes a computer to waste some, or perhaps even all, of its effort simply "chasing its tail."

The simplest type of endless loop involves a bug in a program, where the computer is told to move from one step to another, then back to the one. Line 250 might say, "GO TO 720" and line 720 would then say "GO TO 250." Once the computer reaches line 250 in the execution of this program, the program will stop running for all practical purposes. Of course, the solution to this problem is to debug the program.

A smart robot might be told, "Put the glass on my chair at the table," but for some reason hear this command as "Put the glass on my chair *and* the table." Unless the programming were savvy enough to know that this was an impossible command, the robot would probably take the glass, go to your place at the table, and proceed to oscillate its arm back and forth from table to chair to table to chair, never succeeding in placing the glass on both at the same time.

In more complex systems, endless loops can cause degradation in a system, without a complete breakdown. An excellent example can be found in neurotic human behavior of the type called "obsessive-compulsive." Suppose you're a compulsive buyer of clocks and lamps. Your house might look like a store where these items are sold, because you have so many clocks and lamps; however, when you see an interesting new clock or a trendy lamp, you can't help yourself. You keep repeating this behavior even though it has brought you to the brink of bankruptcy.

Some researchers think we humans are super-smart robots. They say that when AI systems become advanced, they will develop neuroses like we have. Then there might be "electronic psychiatrists," who are expert program debuggers, trained to get rid of behavioral hangups in our smart robots. See ARTIFICIAL INTELLIGENCE.

In a servo system, a condition called *hunting* sometimes occurs when the machine repeatedly overcompensates. This is a form of endless loop. See HUNTING.

endoradiosonde

An *endoradiosonde* is a tiny sensing and transmitting device that can be swallowed, so that it goes through the human stomach and intestines. Transducers monitor conditions within the digestive system. A radio transmitter, modulated by the output of the transducers, sends a low-power signal to an instrument bank that a doctor observes.

The endoradiosonde is a primitive type of device that might best be called an endorobot: a robot that works inside of a human body. Some researchers think that future endorobots might not only monitor body functions, but modify them. An example would be robot antibodies that could go around and kill bacteria and viruses. See also MEDICAL ROBOTS.

Engelberger, Joseph F.

Joseph F. Engelberger has been called the "father of industrial robotics," and was one of the first research-and-development (R&D) scientists to design and build workable robots. He founded Unimation, Inc., a major manufacturer of industrial robots, in 1961.

In the early days of robotics, people didn't like the term *robot*. It smacked of science fiction. There was pressure to call the devices "programmable industrial manipulators" and other creative names. But Engelberger called his machines what he wanted them to be called: robots. That word has only two syllables, and it's easy to remember. And, to Engelberger, it was just plain cool. So robots they were.

Engelberger got his education in the newly developed field of servo systems in the late 1940s, at about the same time the first digital computer was built. He went to work in the aerospace industry. In 1956 he met George Devol (see DEVOL, GEORGE C., JR.), who believed that factory work was dehumanizing, and perhaps better done by automatic machines than by people. Thus Engelberger became Devol's colleague, and together they developed, refined, and marketed the first true industrial robots.

Technology evolved more slowly than ideas, especially at first. The manufacturing industry was slow to respond to new ideas; they didn't want to take unnecessary risks. Bankers were hard to impress. But Devol and Engelberger didn't give up. They believed in their concept, and were confident that robots would eventually prove themselves. Fortunately, some manufacturers were willing to give robots a try. As a result of the persistence of a few open minds, industrial robots have become commonplace. See also AUTOMATED INTEGRATED MANUFACTURING SYSTEM, INDUSTRIAL ROBOTS.

ENIAC

See ELECTRONIC NUMERICAL INTEGRATOR AND CALCULATOR.

entry-level system

An *entry-level system* is the least sophisticated computer you can use for the things you want to do. For example, if you want to do word processing and simple graphics, you can use a personal computer (PC) with an 80386 microprocessor, a 40-megabyte hard disk, one or two diskette drives with 1.2 or 1.44 megabytes of memory each, several hundred kilobytes of random-access memory (RAM), a color monitor, and a dot-matrix printer. (This is true as of 1994.)

In 25 years, the system just described will seem primitive to the average PC user, but it will still be adequate for many people's needs. Nevertheless, as computers get smarter and faster, people tend to take each improvement for granted almost immediately. That is to say, we are easily spoiled. Once you get used to a computer that will talk to you and listen to you, you'll never be satisfied with keyboard-only computer operation again!

A typical system in the year 2020 might have an 809986 microprocessor (or whatever it's called), a 32-gigabyte hard drive, four diskette drives with 1.2 gigabytes of memory each, 16 gigabytes of RAM, a high-resolution, 8192-hue color monitor, and a printer that can spit out 1,500 pages per minute. For today's high-school computer programming or English composition, such a system would be overkill. But in the year 2020, 16-year-olds might be conversing with their PCs about the merits of Einstein's concept of time as a dimension, or about what adjective fits best in a certain sentence.

The most important feature of computer evolution is memory capacity, and the speed with which data can be stored and retrieved. This is because, as software gets "smarter," it needs more memory, and this memory must be usable at higher speeds.

Eventually, home computers will have artificial intelligence (AI). Entry-level systems will seem powerful, indeed, when you look at the specifications alone. But the real difference will be in the software, which will be so "smart" that even the most inexperienced people will have no problems using it. In that sense, the impression will not be one of awesome power, but one of simple user-friendliness. See also ARTIFICIAL INTELLIGENCE, BYTE, COMPUTER, DISKETTE, HARD DISK, HUMAN ENGINEERING, MEMORY, MONITOR, PRINTER, SOFTWARE.

epipolar navigation

Epipolar navigation is a means by which a machine can locate objects in space. It can also navigate, and figure out its own position and path in three dimensions. Epipolar navigation works by evaluating the way an image seems to change as viewed from a moving point of view. Your eyes and brain do this without your having to think, although they are not very exact. Robot vision systems, along with artificial intelligence (AI), can do it with extreme precision.

To illustrate epipolar navigation, imagine you're piloting an airplane over the ocean. The only land you see is a small island. You have an excellent map that shows the location, size, and exact shape of this island (see the drawing). For instrumentation, you have only a computer, a good video camera, and lots of AI software. How can you figure out your coordinates and altitude using just these devices?

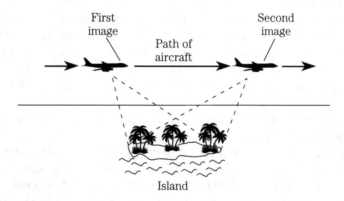

EPIPOLAR NAVIGATION Works by means of perspective.

You let the computer work with the image of the island. As you fly along, the island seems to move underneath you, although you know it's stationary on the ocean surface. You aim the camera at the island and keep it there. The computer sees an image that constantly changes shape. The computer has the map data, so it knows the true size, shape, and location of the island. The computer compares the shape/size of the image it sees, from the vantage point of the aircraft, with the actual shape/size of the island, that it knows from the map data. From this alone, it can tell you:

- your altitude
- your speed relative to the surface
- your exact latitude
- your exact longitude.

Does this seem impossible? The key is that there exists a one-to-one correspondence between all points within sight of the island, and the size/shape of the island's image. The correspondence is far too complex for you to memorize exactly, although, if you've flown in that area many times, you would have an idea of your position and speed based on what the island looked like. But the AI computer has it all in its data banks. For it, matching the image it sees with a particular point in space is simplicity itself.

Epipolar navigation works on any scale, for any speed. It is a method by which robots can find their way without triangulation, direction finding, beacons, sonar, or radar. It is only necessary that the robot have a computer map of its environment. See also ARTIFICIAL INTELLIGENCE, COMPUTER MAP, LOG POLAR NAVIGATION, VISION SYSTEMS.

error accumulation

When measurements are made in succession, the possible error adds up. This is called *error accumulation*.

Analog error accumulation can be illustrated by a measurement example. Suppose you want to measure a long piece of string (about 100 meters, for instance) using a meter stick marked off in millimeters (mm). You must place the stick along the string over and over again, about 100 times. If your error is up to plus-or-minus 1 mm with each measurement (that's doing very well), then after 100 repetitions, the possible error is up to plus-or-minus 100 mm, or 0.1 meter.

In digital systems, errors occur as misinterpreted bits. A machine might see a logic low when it should see high, or vice-versa. Suppose that, on a computer diskette, there are an average of 3 digital errors introduced each time a copy is made. If you copy the diskette repeatedly, making a copy of a copy of a copy of a . . . of a copy (n times), then for the nth-generation copy, there will be an average of 3n digital errors. A 100th-generation copy would have an average of 300 mistaken bits. See also ANALOG, DIGITAL.

error correction

Error correction is a computer operating system in which certain types of mistakes are corrected automatically. An example is a program that maintains a large dictionary of English words. The operator of a word-processing machine might misspell words or make typographical errors. Running the error-correction program will cause the computer to single out all peculiar-looking words, bringing them to the attention of the operator. The operator can then decide whether the word is correct or not. With modern computers, huge vocabularies are easily stored. An error-correction program of this kind is a primitive form of artificial intelligence (AI). See ARTIFICIAL INTELLIGENCE.

When robots must keep track of variables like position and speed, error correction can be used when an instrument is known to be inaccurate, or when values depart from the reasonable range. A computer can keep track of error accumulation, periodically checking to be sure that discrepancies aren't adding up too much. See ERROR ACCUMULATION.

In robotic navigation systems, error correction refers to the processes that keep the device on its intended course. See ERROR-SENSING CIRCUIT.

In a servo system, error correction is done by means of feedback. See ERROR SIGNAL, SERVOMECHANISM, SERVO SYSTEM.

error-sensing circuit

An *error-sensing circuit* produces a signal when two inputs are different, or when a variable deviates from a chosen value. If the two inputs are the same, or if the variable is at the chosen value, the output is zero. This type of circuit is also sometimes called a *comparator*.

Suppose that you want a robot to "home in" on some object. The object has a radio transmitter that sends out a beacon signal. The robot has direction finding (DF) equipment built-in. When the robot is heading in the right direction, the beacon is in the DF null, and the received signal strength is zero. If the robot turns off course, the beacon is no longer in the DF null, and a signal is picked up by the DF receiver (see the drawing). This signal goes to the robot's controller, which steers the robot to the left and right until the beacon signal once again falls into the DF null. See also BEACON, DIRECTION FINDING, ERROR CORRECTION, ERROR SIGNAL, SERVOMECHANISM, SERVO ROBOTS, SERVO SYSTEM.

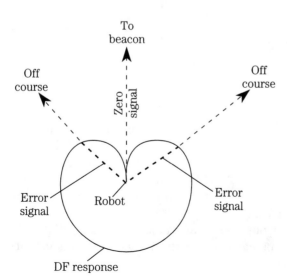

ERROR-SENSING CIRCUIT
Operation of robotic direction-finding (DF) device.

error signal

An *error signal* is a voltage put out by an error-sensing circuit. This signal occurs whenever the output of the device differs from a reference value.

In the direction-finding (DF) device described under ERROR-SENSING CIRCUIT, the output might look like the graph as shown in the drawing. If the robot is pointed on course, the error signal is zero. If it is off course, either to the left or the right, a positive error signal voltage is generated. This voltage depends on how far off course the robot is headed.

The DF circuit is designed to seek out, and maintain, a heading such that the error signal is always zero. To do this, the error signal is itself used by the robot's controller to change the heading. This is the same principle by which a hidden radio transmitter is found. See also BEACON, DIRECTION FINDING, ERROR CORRECTION, SERVOMECHANISM, SERVO ROBOTS, SERVO SYSTEM.

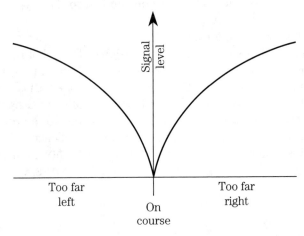

ERROR SIGNAL Received signal strength is zero only when robot is on course.

ethical slave

One of the most fascinating, and also controversial, ideas in robotics and artificial intelligence (AI) research deals with so-called "flesh-and-blood robots." In its most advanced form, such a machine is an android. An example from television is *Data* on the series *Star Trek: The Next Generation*.

Suppose that an android has been developed, and is ready to be mass produced. How would we humans use them?

The military would have an obvious role for androids: as soldiers, pilots, and sailors. This would spare human beings ever having to actively fight in wars again. See MILITARY ROBOTS.

Androids could be used for mining; they could work in radioactive places. They could work in labs with deadly bacteria and viruses. No person would ever again have to get sick, or get injured, or die doing such jobs. Androids might get hurt or killed; but being machines, not people, the loss would not be too great. They would, after all, feel no pain. They would not get sad or depressed or bored.

Would they?

Some scientists think androids are the perfect solution to dangerous, monotonous work now done by humans. But others do not agree. It's easy to imagine androids as slaves. If they aren't really human, aren't really alive, then we don't have to feel guilty about mistreating them. If we yell at them, call them names, order them around, work them without pity, they won't suffer, because they are not really alive. So there's nothing wrong with being cruel to an android.

Is there?

Some psychologists argue that androids could take the abuse we now heap onto human beings, improving the way people treat each other. "We're nice to our pets, our dogs and cats," they say. "So we'll treat androids all right too." But other psychologists warn that androids might bring out the worst kinds of human behavior. "Read some history books!" they say. "We'll be creating ethical slaves. We'll be doing the same thing we've done for centuries: mistreating and degrading living things, and making up excuses that it's okay."

Would an android be alive? Maybe we can let our own choice of words help us decide. The term *artificial life*, used in reference to the most advanced forms of AI, suggests that the answer is "Yes." See also ANDROID, ANIMISM, ARTIFICIAL LIFE.

EURISKO

One of the earliest programs in artificial intelligence (AI), in which a computer actually became able to learn from its mistakes, was developed by Douglas Lenat in Austin, Texas. The program was called *EURISKO*.

One of the most annoying bugs with AI programs is that they seem to keep finding new ways to "screw up." Lenat ran into problems with EURISKO on an almost daily basis.

Many AI researchers think that the first true AI systems will be "psychotic." When computers can think in ways as complex as people, they argue, delusions and hallucinations will inevitably occur with machines, just as they do with people. The challenge will be to develop an AI system that is "well." See also ARTIFICIAL INTELLIGENCE.

exclusive-NOR gate

An *exclusive-NOR (XNOR) gate* is a digital logic circuit with two inputs and one output. It performs the logical "exclusive-OR" operation, followed by inversion (logic NOT or negation).

exclusive-OR gate

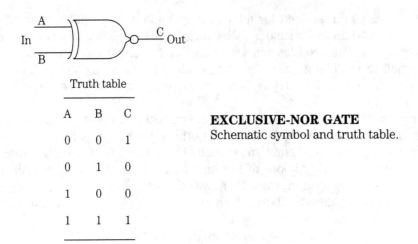

Truth table

A	B	C
0	0	1
0	1	0
1	0	0
1	1	1

EXCLUSIVE-NOR GATE
Schematic symbol and truth table.

The output of an XNOR gate is high, or logic 1 ("true") if and only if the input states are the same. If the input states are different, then the XNOR output is low, or logic 0 ("false").

Logic gates are extensively used in digital computers and other electronic devices.

The schematic symbol for an XNOR gate, along with its logical truth table, is shown in the figure. See also AND GATE, EXCLUSIVE-OR GATE, INVERTER, NOR GATE, OR GATE.

exclusive-OR gate

An *exclusive-OR (XOR) gate* is a digital logic circuit with two inputs and one output. It performs the logical "exclusive-OR" operation.

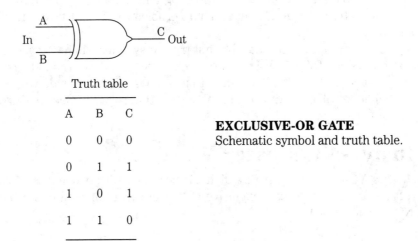

Truth table

A	B	C
0	0	0
0	1	1
1	0	1
1	1	0

EXCLUSIVE-OR GATE
Schematic symbol and truth table.

The output of an XOR gate is high, or logic 1 ("true") if and only if the input states are different. If the input states are the same, then the XOR output is low, or logic 0 ("false").

The schematic symbol for an XOR gate, along with its logical truth table, is shown in the figure. See also AND GATE, EXCLUSIVE-NOR GATE, INVERTER, NOR GATE, OR GATE.

exoskeleton

An *exoskeleton* is a type of robot that, so far, has been seen mainly in science fiction. It is like a suit of armor you can get into, and which will amplify all of your movements, making you super strong. You could, for example, lift a car over your head; the steel frame of the exoskeleton would be able to withstand the weight and pressure. Or you could kick a football a mile (if it didn't pop). Or literally knock the cover off a baseball. The armor would also protect you against blows and, perhaps, even bullets.

Actually, General Electric Company conceived a design for an exoskeleton called "Hardiman" in the 1960s. It has never really been tested. You might think that exoskeletons would be great for soldiers in war; however, experience seems to show that it's far cheaper and easier to use tanks and aircraft. See also MILITARY ROBOTS.

expert systems

The term *expert systems* refers to a method of reasoning in artificial intelligence (AI). Sometimes this scheme is called *rule-based*. Expert systems are used in the control of smart robots. They can also be employed in computers all by themselves.

The drawing is a block diagram of a typical expert system. The heart of the device is a set of facts and rules. In the case of a robotic system the facts consist of data about the robot's environment, such as a factory, an office, or a kitchen. The rules are statements of the logical form "If X, then Y," similar to many of the statements in high-level programming languages.

An inference engine decides which logical rules should be applied in various situations. Then, it instructs the robot to carry out certain tasks. But the operation of the system can only be as sophisticated as the data supplied by human programmers.

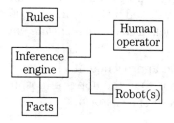

EXPERT SYSTEMS
A form of artificial intelligence using facts and rules.

Expert systems can be used in computers to help people do research, make decisions and make forecasts. A good example is a program that assists a physician in making a diagnosis. The computer asks questions, and arrives at a conclusion based on the answers given by the patient and doctor.

One of the biggest advantages of expert systems is the fact that reprogramming is easy. As the environment changes, the robot can be taught new rules, and supplied with new facts. See also ARTIFICIAL INTELLIGENCE.

extended/expanded memory

Extended memory and *expanded memory* refer to additional memory that increases the amount of data a computer can store and work with. In personal computers (PCs), extra memory is necessary in order to run some software. Extended memory is similar to expanded memory; extended memory is faster and easier for a computer to use.

Extra memory does not, all by itself, make a computer more intelligent. You might have wished for a better memory when cramming for a big test; you didn't need to be any brighter, necessarily, but only to remember more things all at once. No matter how much memory a computer has, it needs advanced software, along with high speed, to have artificial intelligence (AI). Nevertheless, a huge memory capacity does give a computer the capability to handle AI software.

A typical PC can work with 640 kilobytes of memory. The memory capacity is increased by adding more "memory chips," which are integrated circuits (ICs) with room to store digital information. There is theoretically no limit to the amount of extended memory you can add to a PC, although 32 megabytes is typical. See also ARTIFICIAL INTELLIGENCE, BYTE, COMPUTER, INTEGRATED CIRCUIT, MEMORY, SOFTWARE.

extrapolation

When data is available within a certain range, an estimate of values outside that range can be made by a technique called *extrapolation*. This can be "educated guessing," but it can also be done using a computer. The more sophisticated the computer software, the more accurately it can extrapolate. The best extrapolation is done using computers with some artificial intelligence (AI).

An example of extrapolation is shown in the drawing. A hurricane approaches the Florida coast. Knowing its path up to the present moment (solid line), a range of possible future paths is developed by the computer. Factors that can be programmed into the computer to help it make an accurate extrapolation include:

- paths of hurricanes in past years that approached in a similar way;
- weather conditions ("steering currents") in the upper atmosphere;
- weather conditions in the general path of the storm.

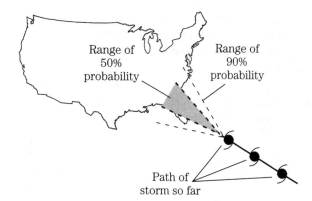

EXTRAPOLATION Computer predicts future path of hurricane.

The farther out (into the future) an extrapolation is made, the less accurate it will be. While a weather computer might do a good job of predicting the hurricane's path for the next 24 hours, no machine yet devised could tell where the storm will be in a week. See also INTERPOLATION.

eye-in-hand system

For a robot "hand" to find its way, a camera can be placed in the gripper mechanism. The camera must be equipped for work at close range, from about 1 meter (3.1 feet) down to a few millimeters (about an inch). The positioning error must be as small as possible, preferably less than 0.5 millimeter. To be sure that the camera gets a good image, a lamp is included in the gripper along with the camera (see the drawing).

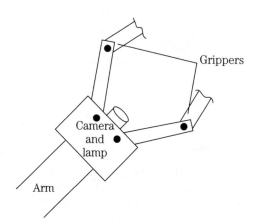

EYE-IN-HAND SYSTEM
Camera and lamp are contained in robot gripper.

eye-in-hand system

This eye-in-hand system can be used to precisely measure how close the gripper is to whatever object it is seeking. It can also make positive identification of the object, so that the gripper doesn't go after the wrong thing.

The eye-in-hand system uses a servo. The robot is equipped with, or has access to, a controller (computer) that processes the data from the camera and sends instructions back to the gripper. See also ROBOT GRIPPERS, SERVOMECHANISM, SERVO ROBOTS, SERVO SYSTEM.

factory automation

See INDUSTRIAL ROBOTS. See also DEVOL, GEORGE C., JR. and ENGELBERGER, JOSEPH F.

fantasy robots

For years, robots have played roles in movies, television, and books. Examples in the U.S. include *R2D2* from *Star Wars*, *Data* from *Star Trek: The Next Generation*, and *Robocop*. These fantasy robots are usually humanoid in form, and are almost always autonomous. The technology for them might be developed and perfected someday, but the real things will probably be much different from what you see and read about.

In Japan, fantasy robots have taken animal form, as well as humanoid form. One of the most popular was a pet robot cat called *Doraemon*, invented by two humorists in 1970 and portrayed in an illustrated children's book. The cat was sent from about 200 years in the future, as a gift to a child from his grandchildren. People became fond of Doraemon to the extent that a sort of cult developed. The Japanese don't have much trouble ascribing lifelike qualities to machines, so it is probably not stretching the truth to say that Doraemon was loved.

Probably the most famous fantasy robot in Western literature was the monster created by Dr. Frankenstein. See also ANIMISM, AUTONOMOUS ROBOTS, FRANKENSTEIN SCENARIO.

Fanuc

Fanuc is the name of a factory complex located at the foot of Mount Fuji in Japan. This corporation is one of the most innovative, disciplined, and profitable ventures in the world.

The complex is like a small city unto itself, with dormitories and training facilities. The level of efficiency and organization in this corporation is unlike anything most workers ever experience.

Fanuc manufactures motors, industrial robots, and other hardware. The system is almost entirely automated, robots having taken over all the dull, monotonous jobs. In fact, if you visit the plant, it seems as if it isn't even operating, because the machines are quiet and there are so few human workers.

Quality control at Fanuc is done rigorously. While most companies might check 10 to 20 percent of the components used in its products, Fanuc checks 100 percent. While some people might argue that this is going past the point of diminishing returns, and that it is too costly, the result is that Fanuc products hardly ever fail.

Fanuc is considered by many engineers to be a model for factory automation. See also AUTOMATED INTEGRATED MANUFACTURING SYSTEM, INDUSTRIAL ROBOTS.

farming robots

During the 20th century, farming has become largely automated. Near the end of the 19th century, about half of the U.S. workforce was involved in agriculture. Today, thanks to machines that can do many times the work of a single person, that figure is down to approximately three percent.

While modern farm machinery can be sophisticated, it is usually not considered to be robotic. This is because the machines do not have any degree of artificial intelligence (AI). A car is a largely automatic machine, but you do not think of it as a robot. There is at least one type of farming, however, in which robots might find significant use.

Fruit picking is one task that requires enough "smarts" that machines have a hard time with it. If the fruit is to be used for juice, the trees can be simply shaken and the fruits gathered up from the ground after they fall. But if such fruit (especially oranges and grapefruit) are to appear in the produce section of a supermarket, the tree-shaking scheme is not acceptable because it bruises the fruit. Engineers are therefore working on designs for a fruit-picking robot.

The first problem that a fruit-picking robot faces is how to find the fruit. This is similar to finding an object in a large bin (see BIN-PICKING PROBLEM). Fruit can be distinguished from leaves by color differences, and also because of shape differences. There is also a difference in the way fruit moves with a light wind, as compared to the leaves on the tree. Color vision systems, as well as tactile sensors, will be necessary to select the fruit. Then, too, the robot will need some way to know whether the fruit is ripe.

The second problem is to remove the fruit without damaging it, and without getting a bunch of leaves or wood along with it. The stem must be cleanly cut, not leaving too much stem and not nicking the fruit.

The third problem is gathering the fruit for transport to the packing center. The method most often suggested is to let the fruit roll down a trough into a bin (see the drawing). The trough must be steep enough so fruit will roll smoothly without jamming; but its incline must not be so steep that fruit gets bruised when it reaches the bin.

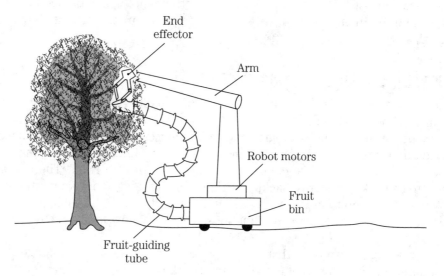

FARMING ROBOTS Fruit rolls into bin after being picked by a robot hand.

fast-food robots

Imagine walking into your favorite burger joint and telling a robot, "Two cheese-burgers, a chocolate malt, and a large fries." Further envision this robot taking your money and, without having to speak a word, pointing to the pickup counter where your order is already being set out for you. Fantasy? Today, yes; tomorrow, maybe not.

A company called AMF tried to run an automatic fast-food place in the 1960s. The restaurant was called AMFARE. It was not a success, mainly because of technical difficulties and prohibitive cost. Since the 1960s, of course, artificial intelligence (AI) has made significant advances, as have the techniques for speech recognition and speech synthesis. The technology is much better, and the cost far lower.

Fast-food work requires high speed, high efficiency, and a tolerance for boredom among its workers. Pay tends to be low, especially considering how arduous the work can be. Cleanliness is sometimes a problem. Employees are often treated badly by customers. Robots would never get tired or sick; there would be no sanitation problems; customers could be as rude as they wanted and nobody's feelings would ever get hurt (although rude customers might feel rather foolish, insulting a robot). And robots have no fear. Imagine a robber coming into a burger joint and confronting the vacant stare of an electromechanical cashier! Within seconds, half a dozen robots would have automatic weapons trained on the thug. (The law would need to be changed to accommodate robotic arrest procedures.)

Robots in fast-food preparation must have the same capabilities as those you might find in a home kitchen, with the additional requirement that they must work fast. See also FOOD-SERVICE ROBOTS.

fault-resilient system

The term *fault-resilient* can refer to either of two different characteristics of a computer/robotic system.

Suppose that all the strategic (nuclear) defenses of the United States were placed under the control of a computer. It would be imperative that this computer be sabotage-proof, and that it be impossible to turn it off. There would need to be backup systems, backup-backup systems, and so on. No matter what anyone tried to do to mess the system up, it would have to be capable of overcoming any attempt to abort its functions. This is called a *fault-resilient computer*.

Some engineers doubt that it is possible to build a totally sabotage-proof computer. As they say, "build a better mousetrap, and you'll get smarter mice." Also, any such system would have to be dreamed up by somebody, and that person (at least) ought to be able to figure out how to bust his/her own invention! And of course, no one can anticipate all of the things that can go wrong with a system. There is a favorite principle that engineers invoke, called Murphy's Law: "If something can go wrong, it will." And the corollary, less often heard but perhaps just as true, is: "If something *cannot* go wrong, it will!"

Many computers, and also computer-controlled machine systems, are designed so that if some parts fail, the system will still work at reduced efficiency and speed. This type of fault-resilience is more often called *graceful degradation*. It is achieved by means of backup systems, and also by complex design techniques that provide alternate routes for data when the main routes fail. See GRACEFUL DEGRADATION.

feedback

Feedback is a means by which a closed-loop system regulates itself. A good example of feedback can be found in a simple thermostat mechanism, connected to a heating/cooling unit. Suppose the thermostat is set for 70 degrees Fahrenheit. If the temperature rises much above 70, a signal is sent to the heating/cooling unit, telling it to cool the air in the room. If the temperature falls much below 70, a signal tells the unit to heat the room. This process is illustrated in the functional diagram.

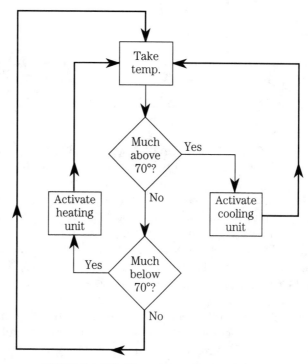

FEEDBACK Flowchart of thermostat system operation. Heavy lines show feedback paths.

There must be some leeway between heating and cooling functions. If both thresholds are set for exactly 70 degrees, the system will constantly and rapidly fluctuate back and forth between heating and cooling. A typical range might be 67 to 73 degrees. Of course, the range shouldn't be too wide, or the room temperature will vary uncomfortably.

Feedback is used extensively in robotic devices, particularly in servos. See also CLOSED-LOOP SYSTEM, SERVOMECHANISM, SERVO ROBOTS, SERVO SYSTEM.

fiberoptic data transmission

Beams of visible light, or of infrared, can be modulated to carry data. A single ray can carry thousands of signals. The data can be sent at extreme speed, because large bandwidths are possible. When the light or infrared is sent along a strand of glass or plastic called an *optical fiber*, the mode is known as *fiberoptic data transmission*.

Fiberoptic systems are used in electronic and electromechanical devices such as robots and computer-controlled cars. Fiberoptics is also replacing wire cable for data communications. The scheme is ideal for networking, or interconnecting computers.

151

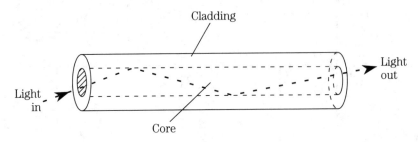

FIBEROPTIC DATA TRANSMISSION
Light or infrared energy is confined within the core.

Besides allowing control signals to be sent very fast, optical fibers are immune to electromagnetic interference. They do not produce external electromagnetic fields, either. When a signal is carried by a light beam inside an optical fiber, it's practically impossible to tap without breaking the fiber. The material from which glass fibers are made (sand) is cheap and abundant. Fiberoptic cables can be buried under the sea, and they do not corrode the way metal wires do. Therefore, fiberoptic cables are ideal for remote-controlled submarine robots.

The drawing shows a cross section of an optical fiber. The light or infrared energy is confined to the core. The core is surrounded by a cladding that has a lower index of refraction. This causes the light or infrared to stay in the core material. See also BANDWIDTH.

field-effect transistor

A *field-effect transistor (FET)* is a semiconductor device used as an amplifier or high-speed switch. In robotics and computers, FETs are generally found in large numbers in integrated circuits (ICs).

The drawing shows a cross section of an FET. The letter P represents P-type semiconductor material; the letter N represents N-type. The device shown is an N-channel FET. Some FETs have the N-type and P-type materials reversed from the drawing; that results in a P-channel FET.

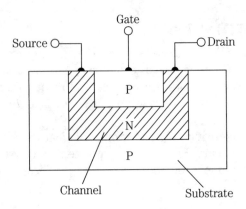

FIELD-EFFECT TRANSISTOR
Cross section of N-channel device.

The current in the FET flows from the source to the drain, along the channel. The voltage at the gate electrode causes the channel to get wider or narrower, in turn affecting the current between the source and the drain. A tiny change in gate voltage produces a large change in the channel current; this is how an FET can amplify.

Some FETs are constructed using metal oxides in addition to semiconductor materials. See COMPLEMENTARY METAL-OXIDE-SEMICONDUCTOR TECHNOLOGY, METAL-OXIDE-SEMICONDUCTOR TECHNOLOGY.

fine-motion planning

Fine-motion planning refers to the scheme used by a robot to get into exactly the right position.

Suppose a robot is told to switch on the light in a hallway. The light switch is on the wall. The robot has a computer map of the house, and this includes the location of the hallway light switch. The robot proceeds to the general location of the switch, and reaches for the wall. How does it know exactly where to find the switch, and how to position its gripper or "finger" in just the right place so that it will move the toggle on the switch?

One method would be to incorporate robot vision, such as an eye-in-hand system. Then it could actually recognize the shape of the toggle and guide itself accordingly. Another method would be to use tactile sensors, feeling along the wall in a manner similar to the way you would turn the switch with your eyes closed. When an irregularity was encountered, the robot would assume that it had found the switch, and would move its "finger" upward along the wall to toggle the switch. See also COMPUTER MAP, EYE-IN-HAND SYSTEM, GROSS-MOTION PLANNING, TACTILE SENSORS, VISION SYSTEMS.

fire-protection robots

One role that robots are especially well suited for is that of firefighter. If all firefighters were robots, there would be no risk to human life in this occupation. Robots can be built to withstand far higher temperatures than humans can tolerate. Robots do not suffer from smoke inhalation. The challenge would be to program the robots to have judgment as good as that of human beings, in a wide variety of situations.

One way to operate fire-protection robots is to have human operators at a remote point, equipped with telepresence machines. Such a machine is a "robot suit" that the operator wears, with controls attached. When the operator moves a certain way, the robot moves in exactly the same way. Television cameras in the robot transmit images to the telepresence machine. The operator can in effect go where the robot goes, without any of the attendant risk. See TELEPRESENCE.

When household robots become commonplace, one of their duties will be to ensure the safety of the human occupants. This will include escorting people from the house if it catches fire, and then putting out the fire and/or calling the fire department. It might also involve performing some first-aid tasks such as cardiopulmonary resuscitation (CPR).

firmware

Firmware is a term referring to computer programs that are permanently installed in the system. Usually this is done in read-only memory (ROM). To change the firmware, components must usually be removed and replaced. You can't just reprogram firmware like you can change software.

Firmware programming is especially common in microcomputer-controlled appliances and machinery. Simple, "dumb" control systems are well suited to the use of firmware. See also MEMORY, SOFTWARE.

first-in/first-out

A *first-in/first-out (FIFO) circuit* is a form of read-write memory. They are commonly used as buffers in machines like computers and terminals.

The operation of a FIFO circuit is as its name implies. The drawing shows a FIFO with eight characters of storage capacity. (Usually, the capacity is much larger, such as 1024 or 4096 characters.) If characters are fed in at an irregular rate of speed, they will come out at a precisely timed, regular rate, as long as the FIFO is partly full.

Not all buffers operate on the FIFO principle. Sometimes the output is reversed in sequence from the input. See MEMORY, PUSHDOWN STACK.

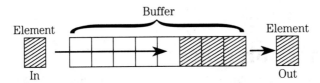

FIRST-IN/FIRST-OUT Elements leave in the same order they enter.

first-in/last-out

See PUSHDOWN STACK.

fixed-sequence robot

A *fixed-sequence robot* is any robot that does just one task or set of tasks, making exactly the same movements each time. The sequence is programmed in firmware.

Many assembly robots are of this type. For example, a robot might insert screws into a cabinet as the cabinet moves along an assembly line. Or it might place hubcaps in the wheels of cars. There is no exception or variation to the routine.

Another example of a fixed-sequence robot is a toy that goes through some routine whenever a button is pressed. These machines are especially popular in Japan. A humanoid robot might greet people entering a department store, for example. See also ASSEMBLY ROBOTS, FIRMWARE.

flatpack

The *flatpack* is a common housing for integrated circuits (ICs). A thin, flat package, usually less than a centimeter square and about a millimeter thick, is fitted with metal pins along each edge (see the drawing). The pins protrude straight outward from the package, minimizing the depth that it takes up. The number of pins can vary, depending on the complexity of the IC inside the package.

Flatpacks are mounted on circuit boards by placing them down on the surface, and then soldering each pin to the foil on the board. Installation is often done by robots. Flatpacks are rather difficult to remove and replace. When installing and replacing flatpacks, it is important that the orientation be right. It is easy, for example, to get the package rotated by 90 or 180 degrees from its proper position. See also DUAL INLINE PACKAGE, INTEGRATED CIRCUIT, SINGLE INLINE PACKAGE.

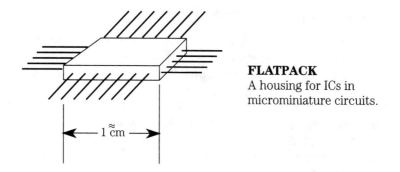

FLATPACK
A housing for ICs in microminiature circuits.

flexible automation

Flexible automation refers to the ability of a robot or system to do various different tasks. To change from one task to another, a simple software change is all that needs to be done.

A good example of flexible automation is a robot arm that can be programmed to insert screws, drill holes, sand, weld, insert rivets, and spray paint on objects in an assembly line. See ASSEMBLY ROBOTS, INDUSTRIAL ROBOTS.

When household robots become more common, they will need to be able to do many things from a single, sophisticated program. This can be thought of as the ultimate in flexible automation. The appropriate actions will result simply from verbal commands. Such robots will need a high level of artificial intelligence (AI) in order to interpret the commands and successfully carry them out. See also ARTIFICIAL INTELLIGENCE, COMMON-SENSE SUMMER PROJECT.

flexible manufacturing system

A *flexible manufacturing system (FMS)* is a manufacturing plant that can turn out a variety of different products. In particular, the term FMS refers to a robotized manufacturing plant.

A good example of an FMS is the Fanuc plant at the base of Mount Fuji, in Japan (see FANUC). Casio, the well-known maker of electronic calculators, uses an FMS to assemble many different variations of their products, ranging from simple arithmetic calculators to sophisticated machines intended for scientific experimenters.

Generally, an FMS uses one or more central computers to oversee the operation of the plant. More than one type of product might be turned out at the same time; that is, the plant might be able to work in two or more "modes" at once. See also ASSEMBLY ROBOTS, AUTOMATED INTEGRATED MANUFACTURING SYSTEM, INDUSTRIAL ROBOTS.

flight telerobotic servicer

In space missions, it is often necessary to perform repairs and general maintenance in and around the spacecraft. It's not always economical to have astronauts do this work. For this reason, various designs are under consideration for a *flight telerobotic servicer (FTS)*.

The FTS is a remote-controlled robot. The extent to which it is controlled depends on the design. The simplest FTS machines would use software from the spacecraft's main computer. More complex FTS devices might actually make use of telepresence, so that a human operator could get the feeling of "being the robot" (see TELEPRESENCE).

Because of the risk involved in sending humans into space, scientists have considered the idea of launching FTS-piloted space shuttles to deploy or repair satellites. The FTSs would be controlled from computers on the ground and in the spacecraft.

One proposed design for an FTS, that might perform routine maintenance and repairs on the exterior of a spacecraft, is shown in the drawing. It looks something like a one-legged, headless android.

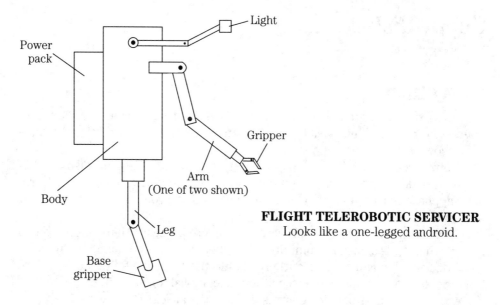

Power
pack

Light

Gripper

Arm
(One of two shown)

Body

Leg

Base
gripper

FLIGHT TELEROBOTIC SERVICER
Looks like a one-legged android.

flip-flop

A *flip-flop* is a simple electronic circuit with two stable states. The circuit changes state when it receives an input pulse or signal. The flip-flop maintains a given state indefinitely unless a pulse is received. There are several different kinds of flip-flop circuits.

A D flip-flop operates in a delayed manner, from the pulse immediately preceding the current pulse.

A J-K flip-flop has two inputs called J and K. If the J input receives a high pulse, the output is set to the high state. If the K input receives a high pulse, the output is set low. If both inputs receive high pulses, the output changes its state (high to low, or low to high).

The R-S flip-flop has two inputs called R and S. A high pulse at R sets the output low; a high pulse at S sets the output high. The circuit is not affected by pulses at both inputs.

The R-S-T flip-flop has three inputs called R, S, and T. It works just like an R-S flip-flop, except that a high pulse at the T input will cause the output to change state (high to low, or low to high).

A T flip-flop has only one input, called the T input. Each time a high pulse appears at the T input, the output state is reversed (high to low, or low to high).

Flip-flops are interconnected to form logic gates. These circuits are the workings of all digital computers, and ultimately all artificial intelligence (AI) systems. See also LOGIC GATES.

floppy disk

See DISKETTE.

flowchart

A *flowchart* is a diagram that illustrates a logical process or a computer program. It looks something like a block diagram. Boxes indicate conditions, and arrows show procedural steps.

Flowcharts are used to develop computer software. They are also used in troubleshooting of complex equipment. Flowcharts lend themselves well to robotic applications, because they indicate choices that a robot must make while it accomplishes a task.

A flowchart must always represent a complete process. There should be no places where a technician, computer, or robot will be "left hanging" without some decision being made. Also, there should not be any infinite loops, where the process goes in endless circles without accomplishing anything.

A simple example of a flowchart is given in the article FEEDBACK.

fluxgate magnetometer

A *fluxgate magnetometer* is a magnetic compass for robot guidance. The device uses coils to sense changes in the earth's magnetic field, or in an artificially generated reference field. The output of the fluxgate magnetometer is sent to the robot's artificial intelligence (AI) center.

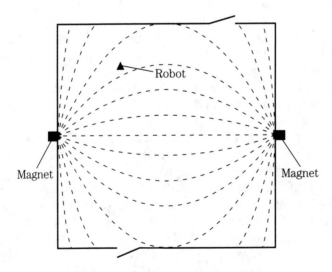

FLUXGATE MAGNETOMETER
Computer map of artificially generated magnetic flux.

Navigation within a defined area can be done by checking the orientation of magnetic lines of flux, generated by electromagnets in the walls of the room. A computer map of the flux lines might look something like the illustration.

For each point in the room, the magnetic flux would have a unique orientation and intensity. Therefore, there would be a one-to-one correspondence between magnetic flux conditions and the points inside the room. The robot's AI would know this relation in extreme detail. This would let the robot pinpoint its position, based only on the magnetic-field information. See also COMPUTER MAP.

flying eyeball

In environments hostile to humans, robots find many uses, from manufacturing to exploration. One simple undersea, or submarine, robot has sometimes been called a *flying eyeball*. This robot can see very well underwater, and can also move around. But it cannot manipulate anything; it has no arms or grippers.

A cable, containing the robot in a special launcher housing, is dropped from a boat. When the launcher gets to the desired depth, it lets out the robot, which is connected to the launcher by a tether. The tether and the drop cable convey data back to the boat (see the drawing).

The robot contains a television camera and one or more lamps to illuminate the undersea environment. It also has a set of thrusters, or propellers, that let it move around according to control commands sent through the cable and tether.

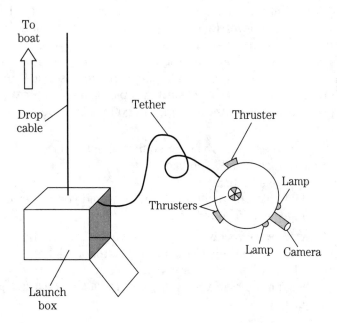

FLYING EYEBALL   A remote-controlled, submarine robot.

Human operators on board the boat can watch the images from the television camera, and guide the robot around as it examines objects on the sea floor.

In some cases, the tether can be eliminated, and radio, infrared, or visible light beams can be used to convey data from the robot to the launcher. This allows the robot to have more freedom of movement, without the concern that the tether might get tangled up in something. The range of the radio, infrared, or visible-light link is limited, as is the length of a tether.

Robots similar to that shown in the drawing have been used by the military, as well as by scientific researchers, in various underwater applications. See also SUBMARINE ROBOTS.

FMS

See FLEXIBLE MANUFACTURING SYSTEM.

food-service robots

Robots have been used to prepare and serve food. So far, the major applications have been in repetitive chores, such as placing measured portions on plates, assembly-line style, to serve a large number of people. The day probably isn't far off, when children in school lunch rooms will face a group of robots, rather than human servers, as they go through the line for their noon meal.

Robots are used in canning and bottling plants, because these jobs are simple, repetitive, mundane, and easily programmable. As a row of bottles goes by, for example, one robot fills each bottle. Then a machine checks to be sure each bottle is filled to the right level. Rejects are thrown out by another robot. Still another robot places the caps on the bottles.

Personal robots, when they are programmed to prepare or serve food, will require more autonomy than robots in large-volume food service. You might insert a disk into a home robot that tells it to prepare a meal of meat, vegetables, and beverages, and perhaps also dessert and coffee. The robot would ask you questions, such as:

- How many people will there be for supper tonight, Ma'am?
- Which type of meat would you like?
- Which type of vegetable?
- How would you like the potatoes done? Or would you rather have rice?
- What beverages would you like?

When all the answers were received, the robot would have to go through an extremely complex process to prepare the meal. The robot might also serve you as you wait at the table, and then clean up the table when you're done eating. It might do the dishes for you. See also COMMON-SENSE SUMMER PROJECT, FAST-FOOD ROBOTS, PERSONAL ROBOTS.

FORTH

FORTH is the name of a high-level computer programming language. It is especially well suited to robotic applications.

The language was originally devised for the purpose of controlling the equipment at astronomical observatories. Since its first use in the early 1970s, it has expanded into many different areas of hardware control.

In addition to being useful for robotic systems, FORTH lends itself well to factory automation, medical electronics, and electronic games. See also HIGH-LEVEL LANGUAGE, SOFTWARE.

FORTRAN

A common high-level programming language, FORTRAN has been in use since the late 1950s for scientific and mathematical problem solving. Fortran is somewhat similar in structure and syntax to BASIC. Anyone who knows BASIC can easily learn FORTRAN; commands are in English and have logical meanings. FORTRAN is somewhat more versatile than BASIC. The language was originally developed by International Business Machines (IBM).

Fortran is not especially well suited to control applications, or to business-related problems. See also BASIC, HIGH-LEVEL LANGUAGE, SOFTWARE.

frame

A *frame* is a mental symbol, a means of representing a set of things. If you're familiar with Microsoft Windows software, then you can get an excellent idea of the nature of frames by imagining them as being "windows in the mind." In artificial intelligence, objects and processes can be categorized in frames.

Suppose you tell a robot, "Go to the kitchen and pour me some water in a paper cup." The robot goes through a series of deductions concerning how to get this beverage, and how to obtain the necessary object in which you have specified that it be contained.

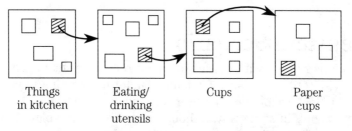

| Things in kitchen | Eating/ drinking utensils | Cups | Paper cups |

FRAME A method of depicting objects or procedures.

F.R.E.D.

First, the robot must go to the kitchen. Then it begins a search for the particular kind of object you've specified. The drawing illustrates this process. The first frame represents "all the things in the kitchen." Within this frame, a subframe is selected: "eating/drinking utensils." Within this, the appropriate frame is "cups," and within this frame, the desired category is "paper cups." Even within this subset, you might specify "six-ounce size," and then "red," and maybe even "for hot drinks" (even though you presumably want cold water).

Frames apply to procedures as well as to the selection of objects. Once the robot has the proper utensil in its grasp, what is to be done? Did you want tap water, or is there some cold water in a pitcher in the refrigerator? Or did you want some bottled mineral water? Or canned soda water? Or perhaps you wanted some of that stuff you ran out of last week, in which case the robot must either come back to you and ask for further instructions, or else make a guess as to what substitute you might accept. See also ARTIFICIAL INTELLIGENCE, COMMON-SENSE SUMMER PROJECT.

F.R.E.D.

The acronym *F.R.E.D.* stands for "friendly robotic educational device." It is a small educational robot, developed by the company Androbot, intended for children in elementary and junior-high school.

Robots like F.R.E.D. are connected to personal computers. The programming language is called LOGO. Using various LOGO commands, a child can tell the robot to move in various ways, and even to react to objects or situations in its environment. In the process of playing with the computer/robot, the child learns about programming, as well as how robots work. Also, geometric principles are taught.

For example, a child might tell the robot to "walk" around the perimeter of a right triangle having sides 30, 40, and 50 centimeters in length. Or, a command might be given to move around the perimeter of a right triangle with sides 40, 50, and 60 centimeters in length; the computer would then respond with something like "IMPOSSIBLE" because there is no right triangle with sides in these proportions. See also COMPUTER-ASSISTED INSTRUCTION, EDUCATIONAL ROBOTS, PERSONAL ROBOTS.

Frankenstein Scenario

Science-fiction is replete with stories in which robots or smart computers are characters. Sci-fi robots are usually humanoid in form, sometimes even androids (see ANDROID). Such robots or computers are invariably designed with the idea of helping humanity in some way, although sometimes they play roles in wars where one side is "helped" at the expense of the other.

A recurring theme in science fiction involves the consequences of robots, or intelligent machines, turning against their makers. This theme is often called the

Frankenstein Scenario, after the android who turned on his creator and caused a major nightmare.

A vivid example of the Frankenstein scenario is provided by the novel *2001: A Space Odyssey*, in which Hal, the smart computer on a space ship, tries to kill an astronaut. The cause for Hal's change of heart is insanity: Hal believes Dave, the astronaut, intends to destroy him. See CLARKE, ARTHUR C.

A machine might react logically to preserve its own existence when humans try to pull the plug. This could take the form of hostile behavior, in which humans become enemies that must be eliminated at all costs. Because robots are supposed to preserve themselves, this instinct can be useful, but only up to a certain point. A robot must never harm a human being (see ASIMOV'S THREE LAWS OF ROBOTICS).

Another example of the Frankenstein scenario is the computer in *Colossus: The Forbin Project*. In this case, the machine has the best interests of humanity in mind. War, the computer decides, cannot be allowed. Humans, it thinks, need structure in their lives, and must therefore have all their behavior strictly regulated. Of course, the people don't see things quite the way the computer does, and the result is a nightmare. This bears out the saying that "The road to tyranny is paved with good intentions." See ARTIFICIAL INTELLIGENCE, CARETAKERS, COLOSSUS, COMPUTERIZED DEFENSE SYSTEM.

function

A *function* is a "mapping" between sets of numbers or objects. Functions are important in mathematics, and also in logic.

The drawing shows an example of a function as a mapping between two sets, labeled x and $f(x)$. Not all of the xs necessarily have counterparts in $f(x)$, and not all $f(x)$s necessarily have counterparts in x. It is possible that more than one x might be mapped onto a single element in $f(x)$. But no element in x ever has more than one mate in $f(x)$. A function never maps a single element into more than one counterpart.

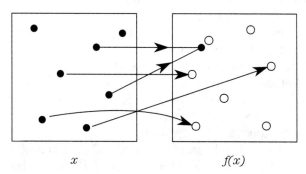

x $f(x)$

FUNCTION Mapping from set x into set $f(x)$.

The set of all xs that have mates in $f(x)$ is called the *domain* of the function f. The *range* of f is the set of all $f(x)$s with corresponding elements in the domain.

In logic, a function is an operation that takes one or more input variables, such as X, Y, and Z, and generates a specific output for each combination of inputs. Logic functions are generally simpler than mathematical ones, because the input variables can only have two values: 0 (false) or 1 (true).

An example of a logical function in three variables is shown in the table. First, the logic AND is performed on X and Y. Then the logic OR is performed between (X AND Y) and the variable Z. Some logic functions have dozens of input variables; there is only one output value, however, for each combination of inputs.

FUNCTION: Values for
$f(X,Y,Z) = (X \text{ AND } Y) \text{ OR } Z = XY + Z.$

X	Y	Z	$f(X,Y,Z)$
0	0	0	0
0	0	1	1
0	1	0	0
0	1	1	1
1	0	0	0
1	0	1	1
1	1	0	1
1	1	1	1

Logic functions are important to engineers in the design of digital circuits, including computers. Often, there are several different possible combinations of logic gates that will generate a given logic function. The engineer's job is to find the simplest and most efficient design. See also LOGIC GATES.

futurists

A *futurist* is a person who tries to predict, based on current technology and trends, what will be accomplished in a given scientific field in 5, 10, 50, or 100 years. In robotics and artificial intelligence (AI), there is plenty of work for futurists. There are many articles in this encyclopedia that deal with "probablys," "might-bes" and "doubtfuls" in robotics and AI.

Most futurists agree that robots will become more sophisticated, and more commonplace, as time goes by. There is some question as to exactly what form the robots will take. While it is fun to daydream about androids, these are often not the most practical and functional forms for robots. See, for example, ANDROID, AUTONOMOUS ROBOTS, INSECT ROBOTS.

In AI, there is theoretically unlimited potential. But in practice, things have moved more slowly than futurists of the 1950s and 1960s had hoped. Reasoning is incredibly complex.

Some futurists believe that the human mind, and all thought processes, can be broken down into interactions among atoms of matter. If this is true, then it is technically possible (although difficult) to build a computer that is as smart as a person. Other scientists are convinced that human thought involves "something more," a life force that cannot be defined or replicated in purely material terms. If this is the case, then a truly human computer might be impossible to build. See, for example, ARTIFICIAL INTELLIGENCE, COMPUTER.

Science-fiction authors have described machines and scenarios that have later come to pass. For this reason, these writers have been called futurists. Among the best known are Isaac Asimov and Arthur C. Clarke. See ASIMOV, ISAAC and CLARKE, ARTHUR C.

gardening and groundskeeping robots

There are plenty of jobs for personal robots in the yard around your house, as well as inside the house. Two obvious applications include mowing the lawn and removing snow. In addition, robots might water and weed your garden.

Riding mowers and riding snow blowers will be easy for robots to use. The robot need not be a biped, but only need the form required to sit on the chair, ride the machine around, and operate the handlebar/pedal controls. Alternatively, lawn mowers or snow blowers could be robotic devices, designed with that one task in mind.

The main challenge, once a lawn-mowing or snow-blowing robot has begun its work, is to "do its thing" everywhere it's supposed to, but nowhere else. You don't want the lawn mower in your garden, and there's no point in blowing snow from your front lawn. Therefore, these robots must be automated guided vehicles. Wires might be buried around the perimeter of your yard, and along the edges of the driveway and walkways, establishing the boundaries within which the robot must work, as shown in the illustration. See AUTOMATED GUIDED VEHICLE.

Inside the work area, edge detection can be used to follow the line between mown and unmown grass, or between cleared and uncleared pavement. This line is easily discernible because of differences in brightness and/or color. Alternatively, computer maps might be used, and the robot could sweep along controlled and programmed strips with mathematical precision. See COMPUTER MAP, EDGE DETECTION.

The hardware already exists to withstand all temperatures commonly encountered in both summer and winter, from Alaska to Death Valley. Software is more than sophisticated enough for ordinary yard-maintenance and snow-removal tasks. The only challenge remaining is to bring the cost down to where the average consumer can afford the robot. See also PERSONAL ROBOTS.

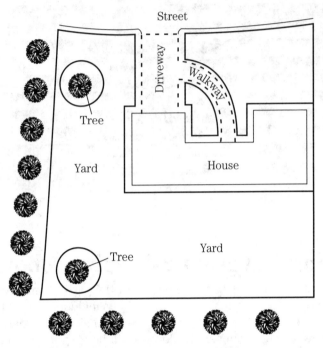

GARDENING AND GROUNDSKEEPING ROBOTS
Boundaries for lawn mowing (solid line) and snow removal (dotted line).

gas station robots

Despite the rise in popularity of self-service gas stations, there are still people who would rather sit in their cars and have someone, or something, else do the dirty work. Robots are quite capable of filling your gas tank and washing your windshield. (Changing the oil or putting air in the tires is more difficult.)

The drawing illustrates what a typical roboticized drive-through filling station might look like. When you get to the paying station, you insert your credit card. This card has information concerning the make and year of your car, as well as credit account data. This tells the robot where it can find your gas-tank fill pipe. Another method of car identification might be the use of bar coding or a passive transponder, similar to the price tags on consumer merchandise (see BAR CODING, PASSIVE TRANSPONDER).

The robot must know the position of your car to within a millimeter or so. Otherwise, the nozzle might miss the fill pipe and spill gasoline on the pavement, or worse yet, put it in the car through the window! Object recognition systems would help prevent problems like this. A biased search could be used, letting the nozzle "seek out" the fill pipe (see BIASED SEARCH, OBJECT RECOGNITION).

Roboticized filling stations let you stay in your car, without getting dirty, cold, wet, or hot. The service is generally faster than it is when done manually.

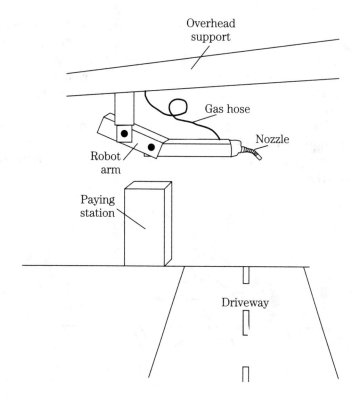

GAS-STATION ROBOTS Typical roboticized service drive-through.

Robots would be programmed not to "top off" the gas tank to get a round number for the price. (This can result in overfilling the tank, and takes unnecessary time.) The robot wouldn't forget to replace the gas cap (as you might). There would also be less chance for you to get robbed while fiddling around with nozzles, gas caps, and such.

generator

The term *generator* can refer to either of two things. A signal generator is a source of signal in an electronic circuit. An oscillator is a common example.

In a power supply, a generator is a device that produces alternating current (ac). This is done by rotating a coil within a strong magnetic field. Alternatively, a permanent magnet is rotated within a coil of wire. The rotating shaft is driven by a gasoline-powered motor, a turbine, or even human power. See POWER SUPPLY.

Small portable gasoline-powered generators, capable of delivering a few kilowatts, can be purchased in large department stores. Larger generators allow homes or buildings to keep their electrical power in the event of an interruption in the utility. The largest generators are found in power plants, and can produce in excess of a megawatt of continuous power.

Small generators are commonly used in synchro systems. These specialized generators allow remote control of robotic devices (see SYNCHRO). A generator can be used to measure the speed at which a vehicle or rolling robot is moving. The shaft of the generator is connected to one of the wheels, and the generator output voltage and frequency vary directly with the angular speed of the wheel (see the illustration).

A generator, like a motor, is an electromechanical transducer. Generators are, in fact, built in just about the same configuration as motors. Some generators can be hooked up and used as motors; these devices are called motor/generators. See also MOTOR.

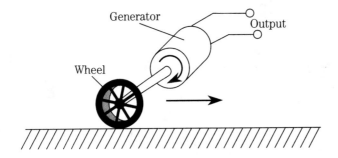

GENERATOR Connection of generator for speed measurement.

genetic engineering

The term *genetic engineering* probably makes you think of biology, not robotics or artificial intelligence. However, it has been suggested that genetic traits could be derived to create humans with computer-like brains. These beings would be automata or, perhaps more descriptively, zombies.

Genetic engineering is well known as one of the most controversial scientific pursuits. This is because of its enormous potential for abuse. See also ARTIFICIAL LIFE, BIOLOGICAL ROBOTS, CARETAKERS, CLONE, ETHICAL SLAVE.

Gennery, Don

Don Gennery is a computer expert who, while working with scientists in search of life on Mars, developed a program for determining the location of a camera, based on the view it sees of terrain below.

In order to look for signs of life, it was suggested that aerial photos be compared, taken at different times during the Martian year. The problem was that photos weren't always from the same angle. The angle difference alone made the terrain appear distorted. However, Gennery realized, the shape of the terrain features is a function of the location of the camera. There is a precise one-to-one correspondence between points in space over a given region, and the image that a video camera sees from each point.

When pictures of Mars were compared by the scientists searching for life, the images were run through Gennery's computer routine. This got rid of confusion caused by differences in camera position.

The same principle that Gennery used can be employed for robotic navigation, without the need for beacons, transponders, or other external guidance systems. Your mind uses a program similar to Gennery's all the time. It is an unconscious routine; you don't think about it. But it helps you navigate everywhere you go. See EPIPOLAR NAVIGATION.

gigabyte
See BYTE.

GMF robotics

In 1982, General Motors (GM) went into a joint venture with Fanuc, the Japanese robotic firm. American auto manufacturers were beginning to have serious trouble with Japanese competition, and GM saw the need for reliable robots in their factories. Fanuc seemed to be the ideal solution. The president of GMF (General Motors/Fanuc) Robotics was Eric Mittelstadt, who worked his way up in management at GM.

Fanuc learned from their work with the Americans, especially in the fields of artificial intelligence and vision systems.

GMF Robotics rapidly rose to the top of the American robot market. The main role of GMF was not to actually build the robots, but to handle the software and get the parts to work together in efficient systems. See also FANUC and MITTELSTADT, ERIC.

graceful degradation

When a portion of a computer system malfunctions, it is desirable to have the computer keep working even if the efficiency is impaired. If a single component causes the whole computer to fail, it is called a *catastrophic failure*. This can usually be prevented by good engineering, including the use of backup systems. In *graceful degradation*, as the number of component failures increases, the efficiency and/or speed of the system gradually declines, but doesn't instantly drop to zero (see the drawing).

In the event of a subsystem malfunction, a sophisticated computer can use other circuits to temporarily accomplish the tasks of the failed part of the system. The operator(s) are notified that something is wrong, and technicians can fix it, often with little or no downtime.

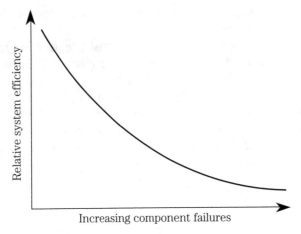

GRACEFUL DEGRADATION Efficiency
declines as the number of component failures increases.

A fiction example

An excellent and vivid example of graceful degradation is provided in the movie
2001: A Space Odyssey. The space ship's computer, Hal, goes crazy, and it becomes necessary for Dave, an astronaut, to "pull the plug."

As Dave systematically shuts off one system after another, Hal pleads, "Stop,
Dave." But Hal keeps on working, more and more slowly, with efficiency impaired to an increasing degree. Arthur C. Clarke, the author of the story, makes
it humorous and horrifying at the same time—funny because Hal acts so silly,
and scary because Dave is being forced to switch off the main control system to
the ship on which he depends for his survival.

Ultimately Hal's consciousness wanes to the point that it (he?) starts going
through test routines, identifying himself and singing songs, as if delirious. But
Hal stubbornly refuses to die completely, all the way to the bitter end.

A real-life example

I recently had an experience in which something went wrong with the diskette
drive in a personal computer (PC). This resulted in reduced system efficiency,
but did not actually make it impossible to do anything that had been possible
before.

When you use a diskette in a PC, the PC identifies it and memorizes its directory. If you change diskettes, and then read or write something from or onto
the new diskette, the PC automatically changes the directory to reflect the identity of the new diskette. But my computer developed a glitch, so that if the
diskette were changed, the old directory stayed in the computer memory.

Because of this glitch (which had not been repaired as of this writing!), any
attempt to write onto the new disk would cause the whole disk to be written over
with bits and pieces of files from either disk! Files would be cut apart with com-

plete abandon, severing paragraphs, sentences, and words without regard for their importance.

To use a new diskette, the computer had to be rebooted by pressing the reset button. To copy diskette X onto diskette Y, it became necessary to take all the files off diskette X, put them on the hard disk, reboot the computer, and then transfer all the X files on the hard disk to the new diskette Y. Although the system still worked, I got a feeling like hobbling along on a sprained ankle, or of tricking the computer into working when it didn't want to. See also CATASTROPHIC FAILURE.

grammar checking

In word processing, *grammar checking* refers to a program that goes through text and looks for things like improper word usage, sentence fragments, or illogically constructed sentences. Some grammar checkers can even look for excessive use of a word or words ("actually" is commonly abused), overly frequent use of semicolons, and other writing idiosyncrasies.

Some grammar checkers do a flash test, which is a measure of the reading level at which the text is written. If a writer is attempting to reach an audience aged 6 to 8 years, the checker can help the writer prepare the text at that level. Flash tests are especially useful to writers of young children's books.

In artificial intelligence, a grammar checker can be used to help the machine be sure it has gotten the right command from the operator. A grammatically poor sentence might cause the computer to respond with, "What?" or "Please rephrase that." Grammar checkers can be used in conjunction with speech recognition systems when spoken, rather than written, commands are used. See also ARTIFICIAL INTELLIGENCE, COMPUTER, SPEECH RECOGNITION.

grand synthesizer

Researchers in artificial intelligence (AI) sometimes talk about a hypothetical person, maybe a high-school or college student, with a mind perfectly attuned to computers. This person, the *Grand Synthesizer*, would sit down in front of a computer, and right away they would hit it off perfectly. This person would be a key figure in the history of computer technology, making gigantic contributions and revolutionizing the industry and the way people use computers. He or she would be, to the computer world, the equivalent of Bach or Mozart in the music world.

In recent years, the term *nerd* has come to mean "young genius." They're the kids who spend more time at a computer terminal than at a television set. The Grand Synthesizer might be called *Super Nerd*. He or she would be, in a sense, both a human and a computer combined into a single being.

It's quite possible that a single, ultimate Grand Synthesizer will never exist. It might just be a wild idea, dreamed up by wishful thinkers in the field of AI. If that is true, then computer advancement will continue as the result of the combined efforts of many ordinary humans, and not from the magical mind of a Super Nerd.

There have been some notable computer "whiz kids." They're sometimes called *hackers*. When they grow up, they're often the ones running the most successful computer and software companies.

graphical user interface

One of the biggest advancements in computers and artificial intelligence has been the engineering of "user friendliness" into the equipment. This includes visual and audible cues, as well as simple, straightforward verbal instructions. A *graphical user interface (GUI)* is a visual aid that makes computers easy to use.

In a GUI, symbols, called *icons*, are used in addition to simple words or word groups. The user can select functions by moving a mouse around until an arrow on the screen points to the appropriate icon; then the mouse is "clicked" to activate the function or system represented by the symbol. The illustration shows three typical icons that are commonly found in GUIs.

To make use of a GUI, a bit-mapped graphic display is needed. This means that the images are comprised of groups of dots. In addition, the computer must have a certain amount of memory, and be able to operate at a certain speed. Until the 1980s, these computers were expensive because integrated circuits (ICs) had not been miniaturized enough to make it possible to implement GUIs in personal computers. Nowadays, GUIs are commonplace. Microsoft Windows is a widely used software that employs GUIs. See also ARTIFICIAL INTELLIGENCE, COMPUTER, WINDOWS.

Set
mouse
options

Set
printer
options

Set
international
options

GRAPHICAL USER INTERFACE Examples of graphic icons.

Graphics

See COMPUTER GRAPHICS.

Grasping planning

Grasping planning refers to the scheme that a robot arm and gripper use to get hold of a chosen object.

Suppose you tell a robot to go to the kitchen and get you a spoon. The robot uses gross-motion planning to find the kitchen, and fine-motion planning to locate the correct drawer and determine which objects in the drawer are the spoons. Then the gripper must grasp a spoon, preferably by the handle, rather than by the eating end. It must not get a fork, or two spoons, or a spoon and something else like a can opener.

Hopefully the silverware is arranged logically in the drawer, so spoons aren't mixed up with forks, knives, can openers, corkscrews, and whatever else. This can be ensured by programming, as long as the robot (but only the robot!) has access to the drawer. If there are children in the household, and if they get into the silverware drawer, the robot had better be able to cope with mixed-up utensils. Then, getting a spoon becomes a bin-picking problem.

Close-up, detailed vision, such as an eye-in-hand system, can ensure that the gripper gets the right utensil in the right way. Tactile sensors might also be used, because a spoon "feels" different than any other kind of utensil. See also BIN-PICKING PROBLEM, EYE-IN-HAND SYSTEM, FINE-MOTION PLANNING, GROSS-MOTION PLANNING, ROBOT GRIPPERS, TACTILE SENSORS, VISION SYSTEMS.

Gray-scale system

The *gray-scale system* is a method of creating and displaying a digital video image. The image is made up of pixels (picture elements).

Pixels

A typical screen has an array of 256 by 256 pixels, or $2^8 \times 2^8 = 2^{16} = 65,536$ elements. This number, in computer lingo, is 64K. A high-resolution screen has an array of 512 by 512 pixels, or $2^9 \times 2^9 = 2^{18} = 262,144 = 256K$ elements.

The pixels are little squares, each with a shade of gray that is assigned a digital code. The table shows a 16-level gray scale. Four binary digits, or bits, are needed to represent each level of brightness from black = 0000 to white = 1111. This is a coarse gray scale; finer scales might use 64 or even 256 different brightnesses. These would be expressed in binary codes of six or eight bits, respectively.

GRAY-SCALE SYSTEM:
Hypothetical 16-shade binary codes.

Code	Relative shade	Percent possible brightness
0000	Black	0.00
0001		6.67
0010	Very dark gray	13.33
0011		20.00

Continued

Code	Relative shade	Percent possible brightness
0100	Dark gray	26.67
0101		33.33
0110	Medium-dark gray	40.00
0111		46.67
1000	Medium gray	53.33
1001		60.00
1010	Medium-light gray	66.67
1011		73.33
1100	Light gray	80.00
1101		86.67
1110	Off-white	93.33
1111	White	100.00

Low-resolution coarse

If a screen is of the low-resolution type, with a 16-level gray scale, then the memory needed to represent a single, nonmoving image is $256 \times 256 \times 16 = 2^8 \times 2^8 \times 2^4 = 2^{20} = 1,048,576$ bits = 1M (one meg). That's a lot! A byte consists of eight, or 2^3, bits (see BYTE); this coarse-gray, low-resolution image would therefore need $2^{20}/2^3 = 2^{17} = 128$ kilobytes. Eight such images would take up almost the entire memory of a typical computer diskette.

High-resolution fine

If the screen is of the high-resolution type, with a 256-level gray scale, then the memory required for one image to be shown once is $512 \times 512 \times 256 = 2^9 \times 2^9 \times 2^8 = 2^{26} = 67,108,864 = 64M$ bits. That's eight megabytes, which is a sizable fraction of the capacity on the hard disk of a typical personal computer (PC).

Analog versus digital

The previous examples show that analog technology still enjoys some advantages over digital technology. Remember that these massive memory figures represent a single image, sent just once; for a fast-scan television picture, at least 16 images are required per second. This is impossible with modern PC systems. Yet, even these images would represent only black-and-white pictures. Digital television programs cannot, at the present time, be effectively stored on diskettes or hard disks in your PC, simply because the required memory is too great. Yet hour-long programs are easily recorded on an analog tapes of manageable size, and played on analog equipment of reasonable cost.

Feature focus

For digital moving pictures to be possible in robot vision systems, some means will be needed to sort out relevant from irrelevant data. This will require a high level of artificial intelligence (AI), which also takes up substantial memory. When you look at your surroundings, you don't take in the whole scene all the time. Most of your attention is concentrated within a narrow zone around the fovea, or focus point of your eyes. The same will need to be done with robot vision systems. Not only that, but the programming will need to tell the robot where to look for the expected data. That is, the machine will have to anticipate the goings-on in its environment as it navigates its way around and manipulates objects. One method is called *local feature focus*, devised by Robert Bolles. See also ANALOG, DIGITAL, LOCAL FEATURE FOCUS, VISION SYSTEMS.

Grippers

See ROBOT GRIPPERS.

Gross-motion planning

Gross-motion planning is the scheme a robot employs to navigate its way around without bumping into things, falling down stairs, or tipping over.

Gross-motion planning can be done using a computer map of the environment. This tells it where tables, chairs, beds, and other objects are. Another method is to use proximity sensing or vision systems. These devices can work in environments unfamiliar to the robot, and for which it has no computer map. Still another method is the use of beacons.

Suppose you tell your personal robot, "Go to the kitchen and get me an apple from the basket on the counter." The robot would use gross motion planning to scan its computer map and locate the kitchen. Within the kitchen, it would need some way to determine where the counter is. Finding the basket, and picking an apple from it (especially if there are other types of fruit in the basket too), would make use of fine-motion planning. See also BEACON, COMPUTER MAP, FINE-MOTION PLANNING, GUIDANCE SYSTEMS, PROXIMITY SENSING, VISION SYSTEMS.

Guidance systems

A robot *guidance system* refers to the hardware and software that lets a robot find its way. In particular, it refers to gross motion, or getting around in the environment.

For detailed information see the following articles: AUTOMATED GUIDED VEHICLE, BEACON, BIASED SEARCH, COMPUTER MAP, DIRECTION FINDING, DISTANCE MEASUREMENT, EDGE DETECTION, EMBEDDED PATH, EPIPOLAR NAVIGATION, GROSS-MOTION PLANNING, GYROSCOPE, LOG-POLAR NAVIGATION, OBJECT RECOGNITION, PARALLAX, PROXIMITY SENSING, VISION SYSTEMS.

gyroscope

A *gyroscope* or *gyro* is a device useful in robot navigation. It is a form of inertial guidance system, operating from the fact that a rotating, heavy disk tends to maintain its orientation in space.

The drawing shows the principle of a gyro. The disk is mounted in a gimbal or set of bearings that allows it to turn up and down or from side to side. The disk can be driven by various different methods, such as motors.

A common use for a gyro is to keep track of a robot's direction of travel, or heading, without having to rely on external objects or on the earth's magnetic field. Gyros are effective only for a limited time, because they tend to slowly change their orientation over long periods. See also GUIDANCE SYSTEMS.

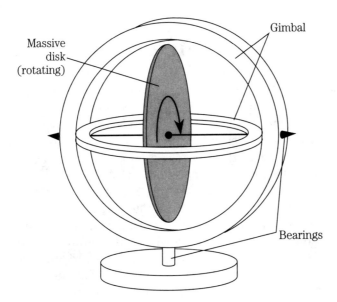

GYROSCOPE A rotating disk maintains its orientation in space.

hacker

See GRAND SYNTHESIZER.

hacker program

One of the earliest experiments with artificial intelligence (AI) was done with an imaginary robot, entirely contained within the "mind" of a computer. A student named Gerry Sussman wrote a program called *HACKER*, in a computer language known as *LISP*. The result was a little universe in which a robot could stack blocks on each other.

Sussman created laws of physics in the imaginary universe. Among them were things such as:

- Blocks X, Y, and Z each weigh five pounds.
- Blocks V and W each weigh 50 pounds.
- The robot can lift no more than 10 pounds.
- Only one object can occupy a given space at a given time.
- The robot knows how many blocks there are.
- The robot can find blocks if they're out of direct sight.

Illustration A shows the five blocks lying around, as they might appear on the computer monitor, along with the robot, a stick figure.

Sussman gave commands to the robot, such as, "Stack the blocks all up, one on top of the other." As stated, this command is impossible, because it requires the robot to lift a block weighing 50 pounds (either V or W), and the robot is capable of lifting only 10 pounds. See illustration B. What would happen? Would the robot try forever to lift a block beyond its limit of strength? Or would it tell Sussman something like "I can't do this!"? Would it go after either block V or W first, trying to get it on top of one of the light blocks, or on top of the other heavy block? Would it pick up all the lighter blocks X, Y, and Z in some sequence, stacking them vertically on top of V or W? Would it put two light blocks on V, and the remaining light block on W, and then give up? Eventually, it would run into the

A B C D

HACKER PROGRAM
In a computerized world, a robot stacks blocks. See text for discussion.

impossibility of the command—but how long would it try, and what would it try, before quitting?

Another command might be, "Stack the blocks so that lighter ones are on top of heavier ones." This, the robot can do, according to the rules written above. But there exist several different possible ways (two of these are shown in illustrations C and D). Would the robot hesitate, unable to make a decision? Or would it go ahead and accomplish the task in some way? If the experiment were repeated, would the result always be the same, or would the robot solve the problem a different way each time?

Numerous AI researchers have written programs similar to HACKER, creating "computerized universes" in an attempt to get machines to think and learn. The results have often been fascinating and unexpected. See also ARTIFICIAL INTELLIGENCE, LISP.

HAL

See CLARKE, ARTHUR C.

hallucination

In a human being, a *hallucination* occurs when the senses don't function right. This can happen in mental illness, or under the influence of certain drugs. In their most severe form, hallucinations cause a person to see, hear, or feel things that aren't really there. Hallucinations can be, and often are, combined with delusions, or misinterpretations of reality. An example is the person who thinks that spies are after him/her, and who actually sees sinister figures lurking behind trees or in dark alleyways.

Computers can have hallucinations and delusions, too. The likelihood of these malfunctions increases as the systems get more complex.

Phil Agre, a researcher in artificial intelligence (AI), has said that computer hallucinations/delusions might result from improper design and care. He believes that machines, as they evolve and become progressively "smarter," will develop hangups, just the way people do. If a child is not brought up well, (s)he will turn out to be a maladjusted adult; the same kind of thing is true, Agre implies, for computers.

I have had experiences that seem to verify Agre's theories. In Florida, without air conditioning, the indoor temperature hovers around 90 degrees Fahrenheit in the summertime. To some extent, humans can adapt to this (even learn to like it, sort of). But computers can overheat if used for long periods in temperatures exceeding about 80° F. In my case, the word-insertion functions start to go awry first. The computer misinterprets commands and does things differently than usual, although it still functions enough to be usable. But unless the machine is allowed to cool, graceful degradation of performance takes place (see GRACEFUL DEGRADATION), and eventually the problems get so bad that it becomes difficult or impossible to work on the machine. Sometimes the computer will actually crash (see CRASH).

In fiction, a vivid example of computer hallucination/delusion is portrayed by Arthur C. Clarke in *2001: A Space Odyssey*. See CLARKE, ARTHUR C. See also ARTIFICIAL INTELLIGENCE, COMPUTER.

hand

See ROBOT GRIPPER.

handshaking

In a digital communications system, accuracy can be optimized by having the receiver verify that it has received the data correctly. This is done periodically, say every three characters, by means of a process called *handshaking*.

The process goes as follows. First, the transmitter sends three characters of data. Then it pauses, and awaits a signal from the receiver that says either (a) "I recognize all three characters," or (b) "One or more characters is something I don't recognize." If the return signal is (a), the transmitter sends the next three characters. If the return signal is (b), the transmitter repeats the three characters. This process is illustrated by the flowchart.

In computer systems, the term *handshaking* refers to a method of controlling, or synchronizing, the flow of serial data between devices. The synchronization is accomplished by means of a control wire in hardware, or a control code in software. Hardware handshaking is used when direct wire or cable links are pos-

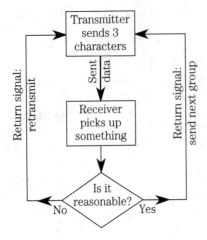

HANDSHAKING
A method of improving the
accuracy of serial data transfer.

sible, such as between a personal computer and a serial printer. Software hand-shaking is similar to the process used for communications systems (see the flow-chart). The transmitting station sends data only when the receiving station is ready for it and is "expecting" it. See also SERIAL.

hard disk

Most modern personal computer (PC) systems, whether desktop or notebook (laptop), use a hard disk (HD) to store application programs, software, and data. The term *hard disk* is technically a misnomer, because there are not one, but several disks in a typical HD (see the drawing). The individual disks are called *platters*. Most HD systems have three, four, or five platters from 3.5 to 5.25 inches in diameter. They are rigid, hence the term "hard," and are made of aluminum with a ferromagnetic coating similar to that used in diskettes.

The platters turn at 3,600 revolutions per minute (rpm). The drive mechanism uses a stepper motor or a voice-coil motor to move the read-write heads over the platters. The heads don't make contact with the surfaces, but hover very close to the magnetic material. This prevents mechanical wear on the platters. The motor, read-write heads and platters are all contained in a sealed hous-

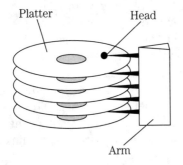

HARD DISK
Consists of several stacked
platters.

ing. This protects them from the environment, but also makes it impossible to exchange platters.

In a typical PC, the memory capacity of an HD varies from about 50 megabytes (50Mb) to more than 500Mb. In expensive business machines, the HD can have more than 1 gigabyte, or 1,000Mb, of memory. In most computers, the HD is referred to by the letter C.

The main advantages of HD systems include large memory capacity and high read-write speed. All data stored on an HD should be backed up on diskettes, because HD systems eventually malfunction. See also BYTE, COMPUTER, DISKETTE, MEMORY.

hardware

The term *hardware* refers to physical apparatus, such as motors, grippers, magnetic disks, and semiconductor devices. Hardware items are all tangible objects. They all can, and eventually do, wear out.

Hardware is contrasted with firmware and software in robotics, computers and artificial-intelligence (AI) systems. See also FIRMWARE, SOFTWARE.

hard wiring

In a computer or "smart robot," the term *hard wiring* refers to functions that are built right into the machine hardware. Hard wiring cannot generally be changed without rearranging physical components, or changing the interconnecting wires. Sometimes, the term *firmware* is used to mean hard wiring.

An "ideal computer" (that exists only in theory) would be capable of programming entirely by software. The physical components must be hooked up together somehow, but in the ideal case, functions could be changed just by reprogramming the machine. This has been realized to a large extent in recent years by the use of diskettes and hard disks.

Hard wiring does have some advantages over software control. Most significant is the fact that hard-wired functions can be done at a higher rate of speed than software processes. See also COMPUTER, DISKETTE, FIRMWARE, HARD DISK, SOFTWARE.

Hero

Hero is the name of a personal robot that was manufactured in the U.S. by Heathkit. It appeared in 1983. This robot originally cost $2,500. It might also be called a hobby robot, because the purchaser had to assemble the machine from parts supplied by Heathkit.

Hero was never really a useful robot for work around the house. A truly useful personal robot would probably cost about 10 times as much as Hero. Nevertheless, this machine will be remembered as perhaps the first true personal robot. See also HOBBY ROBOTS, PERSONAL ROBOTS.

Hertz

Hertz, abbreviated *Hz*, is the fundamental measure of alternating-current (ac) frequency. A frequency of 1 Hz is equivalent to a cycle per second. In fact, the word "Hertz" is interchangeable with the expression "cycles per second." The drawing shows a pulse train with a frequency of 3 Hz. The wave goes through three complete repetitions in one second.

Frequency is often expressed in units of kilohertz (kHz), megahertz (MHz), and gigahertz (GHz). A frequency of 1 kHz is equal to 1,000 Hz; a frequency of 1 MHz is equal to 1,000 kHz = 10^6 Hz; a frequency of 1 GHz is equal to 1,000 MHz = 10^9 Hz.

The speed at which digital computers operate is often specified in terms of frequency. The higher the frequency, the faster a microprocessor can work, and the "smarter" a computer that uses the chip can be. The reason that higher frequency translates into a "smarter" chip is simply that, as the frequency increases, more and more operations can be done within any given period of time.

An 80286 microprocessor, still common in some notebook computers, works at about 12 MHz. An 80386 microprocessor, common in desktop models, works at approximately 20 MHz. Therefore, the '386 works approximately three times as fast as the '286. The actual ratio can vary somewhat, depending on the type of processing to be done. See also COMPUTER, MICROPROCESSOR.

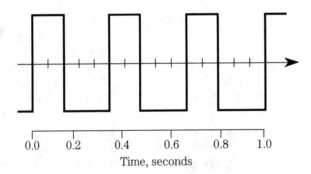

HERTZ Pulse train with frequency of 3 Hz.

heuristic knowledge

Can computers learn from their mistakes, and improve their knowledge by trial and error? Is it possible for a computer, or a network of computers, to evolve? Researchers in artificial intelligence (AI) say yes. In fact, the existence of *heuristic knowledge* (the ability of a machine to become "smarter" based on its real-world experience) is one test of AI in a system.

A fast-learning computer

Suppose a super-smart AI system is developed that can evolve to higher and higher levels of knowledge. Imagine that, one day after the machine has been put into operation, it's as smart as a 10-year-old person. After two days, it's as smart as a 20-year-old in college. After three days, it's as savvy as any 30-year-old research engineer or business fast-tracker. Suppose that more and more memory is added, so that the limit of knowledge is determined only by the speed of the microprocessor. What will this computer be like after a month? Will it have the knowledge of a 300-year-old person (if people lived that long)?

Does an increasing level of intelligence guarantee that a machine will also become *wise*?

Computers with hands

Machine knowledge becomes far more powerful when computers are given the ability to control mechanical devices—"smart robots." Knowledge alone can't build cars, bridges, aircraft, and rockets. Perhaps dolphins are as smart as people, but these marine mammals lack hands and fingers to manipulate things. A computerized robot is to a computer, as a human being is to a dolphin.

Can, and will, computers ever become smarter than, and perhaps more powerful than, their makers? These questions have yet to be answered. But AI researchers are working on it. Some scientists are genuinely scared at the possibilities for things to go wrong, or for AI to be misused, or even for AI to evolve in unexpected and unpleasant ways. See CAPEK, KAREL.

Computer evolution

If a computer is ever designed with unlimited, or practically unlimited, ability to learn from experience, will this make it possible for "evil forces" to enter and take over the machine?

At first this question seems ridiculous, the subject matter for half-baked fiction. But some well-respected scientists have suggested the possibility. Even if "evil" isn't the right word, a super-smart machine might decide that humanity would be best served by things that you and I would find impossible to live with. If the computer had control over robots and other hardware, it could force people to do things according to its will, like a totalitarian dictator.

Other researchers point to the benefits that could come from having impartial, brilliant machines make fair decisions all the time. Such computers wouldn't be any use if they didn't have some power, some way to enforce their decisions.

Imagine a courtroom in which the judge was a computer!

"It's all in the programming," say AI futurists. That's true! Every machine, no matter how advanced, had to be designed, built, and programmed by people. Thus, we can say that "It's all in the programmers." These individuals will all be rational and sane, we hope. See also ARTIFICIAL INTELLIGENCE, CARETAKERS, COLOSSUS, COMPUTER.

hexadecimal number system

The number scheme that people most commonly use, the decimal number system, has 10 symbols, or digits, arranged representing powers of the number we call 10. This is the number of fingers, including thumbs, or toes on a person. (That's probably where the term "digit" comes from.)

In computer work, other numbering systems are often preferred over the decimal scheme. The binary number system uses only zeros and ones, and is the way most digital computers "think" of numerical quantities. Another scheme, sometimes used in programming, is called the *hexadecimal number system*, so named because it has 16 symbols (according to our way of thinking), or 2^4. These digits are the usual 0 through 9 plus six more, represented by A through F, the first six letters of the alphabet.

The table shows the hexadecimal digits, along with their decimal and binary equivalents. These hexadecimal digits are used by software engineers to represent four-digit groups of binary numbers. Hexadecimal numbers can also be used just like decimal numbers; if you ever see something like 1C, 8BF it is probably a hexadecimal number. See also BINARY-CODED DECIMAL, BINARY NUMBER SYSTEM, DECIMAL NUMBER SYSTEM, OCTAL NUMBER SYSTEM.

**HEXADECIMAL NUMBER
SYSTEM: There are 16 different digits.**

Decimal	Binary	Hexadecimal
0	0000	0
1	0001	1
2	0010	2
3	0011	3
4	0100	4
5	0101	5
6	0110	6
7	0111	7
8	1000	8
9	1001	9
10	1010	A
11	1011	B
12	1100	C
13	1101	D
14	1110	E
15	1111	F

high-level language

The term *high-level language* refers to any of the programming languages used in computer work. Examples are BASIC and FORTRAN, commonly taught in junior and senior high schools; COBOL, used in business; and LISP, employed in artificial intelligence (AI).

The various different high-level languages each have advantages in some types of work, and shortcomings in others. BASIC and FORTRAN are all right for relatively simple scientific problems, but they are not very good for corporate accounting. While COBOL is good for the businessperson, it isn't much use to a student trying to solve science problems.

High-level language consists of statements in English (or some other human language). This lets humans work with computers at a sort of conversational level. Most students find high-level languages easy to get familiar with. The best way to learn these languages is to "play computer," thinking strictly by rules of logic. Because of the pure logic in programming, computers might someday be widely used to write the programs for other computers.

A computer "thinks" in low-level language. Some programs must be written in these languages, which include assembly language and machine language. It takes extensive training to learn these languages, because they are nothing like English. Translation between high-level and low-level language is done by an assembler, a compiler and/or an interpreter. See also ASSEMBLER PROGRAM, ASSEMBLY LANGUAGE, COMPILER, COMPUTER, INTERPRETER, MACHINE LANGUAGE.

hobby robots

Hobby robots are, as their name suggests, robots intended mainly for amusement and experimentation. They are toys for adults.

Hobby robots often take humanoid form (see ANDROID, HUMANOID ROBOTS). They can be programmed to give lectures, operate elevators, and even play musical instruments. Wheels are often used rather than biped designs (see BIPED ROBOTS), because wheels work better than legs, are easier to design, and cost less.

Many hobby robots are adaptations of industrial robots (see INDUSTRIAL ROBOTS). Robot arms can be attached to a main body. Vision systems can be installed in the robot's head, which can be equipped to turn to the right and left, and to nod up and down. See VISION SYSTEMS.

Speech recognition and speech synthesis allow a hobby robot to converse with its owner in plain language, rather than by means of a keyboard and monitor. This makes the machine much more human-like. See SPEECH RECOGNITION, SPEECH SYNTHESIS.

Perhaps the most important feature for a hobby robot is artificial intelligence (AI). The smarter the robot, the more fun it is to have around. It's especially interesting if a machine can learn from its mistakes, or be taught things by its owner.

Hobby robot societies have sprung up around the United States. They tend to evolve and change their names often. If you live in a large city, you might be near such an organization. Probably the best thing to do is to look in the phone book under "Robot," and/or to contact a local electronics store or manufacturer. See also AMUSEMENT ROBOTS.

homunculus

The term *homunculus* is Latin for "little person." It reflects the concept of artificial life, an idea that is probably as old as civilization. Even in ancient Rome, people wondered if it would ever be possible to create androids!

Throughout history, there have been myths, legends, and fables surrounding artificial life. Usually, these fictitious creatures can't survive in the real world. Somehow, human-made life is deemed inferior to life that has evolved naturally. People are leery of "playing God." This queasiness, with roots in religion, is apparent today among artificial-life researchers. See ANDROID, ARTIFICIAL LIFE.

household robots

See PERSONAL ROBOTS.

human engineering

Human engineering refers to the art of making machines, especially computers and robots, easy to use. This is sometimes also called *user-friendliness*.

A user-friendly computer program allows the machine to be operated by someone who knows nothing about computers. Bank automatic-teller machines (ATMs) are a good example of devices that employ user-friendly programming. Increasingly, libraries are computerizing their card catalogs, and it's important that the programs be user-friendly so that people can find the books they want. There are many other examples.

A user-friendly robot would be able to carry out orders efficiently, reliably, and reasonably fast. Ideally, you could say, "Go to the kitchen and get me a banana, please," and (assuming you had bananas in your kitchen) the robot would return in a minute or two, holding one, and bring it right to you. This ability is not easy to program into a machine, however, as researchers have found out. Even the simplest tasks are, in terms of sheer digital logic operations, unbelievably complicated. See, for example, COMMON-SENSE SUMMER PROJECT.

Artificial intelligence (AI) helps to make computers and robots user-friendly. It's much easier to communicate with a machine that's "smart," compared with one that's "stupid." It is especially enjoyable if the machine can learn from its mistakes, or show ability to reason. See ARTIFICIAL INTELLIGENCE.

Speech recognition and speech synthesis help greatly in making computers and robots user-friendly. It's much easier, and more pleasant, if one can converse with a machine, rather than typing in commands and watching data on a monitor. See SPEECH RECOGNITION, SPEECH SYNTHESIS.

humanoid robots

A *humanoid robot* is, as its name suggests, a robot that resembles a human being. Such robots have two legs, two arms, and a head. They are programmed to make motions like people, and to do things that people might do. In its most advanced form, a humanoid robot is called an android (see ANDROID).

There are some mechanical problems with design of humanoid robots. For one thing, biped robots are unstable (see BIPED ROBOTS). It's easy to push over, and knock down, a robot with only two legs. Even three-legged designs, while more stable, are two-legged whenever one of the legs is off the ground. While people are equipped with a sense of balance, this feature is extremely hard to program into a machine.

But humanoid robots have nevertheless enjoyed popularity, especially in Japan. One of the most famous was called Wasubot. It played an organ, and did so with the finesse of a professional musician. This robot became an idol at the Japanese show Expo '85, held in 1985. The demonstration showed that machines can actually have artistic beauty, as well as be functional.

Will we ever see personal robots in humanoid form, that walk along and talk with us, or that will provide companionship when we are lonely? Perhaps. But many researchers think wheels, not legs, will provide the locomotion. Wheels are cheaper than legs to design, install, program, and maintain. See also PERSONAL ROBOTS.

hung system

See CRASH.

hunting

Hunting is the result of overcompensation in a servo system. It is especially likely when there is not enough hysteresis.

Any circuit designed to lock onto something, by means of error correction, is subject to hunting. It takes the form of a back-and-forth oscillation between two conditions. If severe, it can go on indefinitely. In less serious cases, the system eventually settles on the correct level or position (see the illustration).

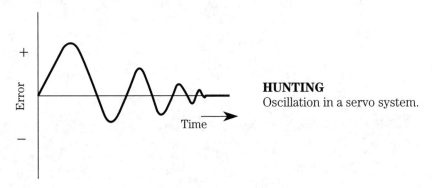

HUNTING
Oscillation in a servo system.

Hunting is eliminated by careful design of servo systems, so there is just the right amount of hysteresis. See HYSTERESIS LOOP, SERVO SYSTEM.

hysteresis loop

A *hysteresis loop* is a graph that shows the "sluggishness" of response in a servo system. The word is pronounced his-ta-REE-sis. The drawing shows a hysteresis loop for a typical thermostat, used for control of the indoor air temperature.

The horizontal scale shows the room temperature in degrees Fahrenheit. On/off conditions for heating and cooling are shown on the vertical scales. Notice that there is a small range of temperatures, from about 68 to 71 degrees, within which neither the heating nor the cooling is on. This prevents the system from rapidly oscillating back and forth between heating and cooling states. But it is a narrow enough temperature range so that the people in the room don't get too hot or cold.

All servo systems employ feedback of some kind. There must always be some "sluggishness" built into the feedback response. Otherwise, over reactions can be so severe that the system won't work. See also FEEDBACK, SERVO SYSTEM.

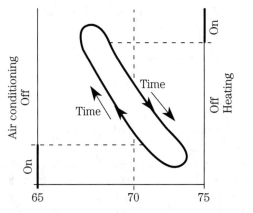

HYSTERESIS LOOP
Curve for hypothetical heating/cooling system.

IBM

See INTERNATIONAL BUSINESS MACHINES.

icon

See GRAPHICAL USER INTERFACE.

IF/THEN/ELSE

In artificial intelligence (AI) systems, choices must often be made in the execution of a program. One of the most common is called *IF/THEN/ELSE*. It can be thought of as a sentence: "If A, then B; otherwise C."

An example of an IF/THEN/ELSE process is shown in the flowchart. Suppose a computer is working with a variable x. If x is negative, then you want to multiply it by –1. If x is zero or positive, you want to leave it alone. The computer would compare the numerical value of x with zero. It would then output the absolute value of the number, either by multiplying by –1 or else by leaving it alone.

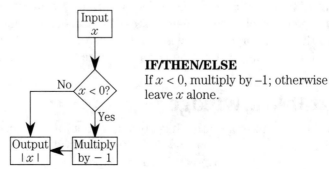

IF/THEN/ELSE
If $x < 0$, multiply by –1; otherwise leave x alone.

IF/THEN/ELSE processes are especially useful command structures for robots. You might tell a robot, "Go to the bathroom and get me a tissue." The robot would have, in its electronic memory, a command structure to follow. It would need an alternative in case there are no tissues in the bathroom. The robot's programming might take the form: "If master's command can be executed, do it. Otherwise return and say, with a tone of sincere regret, 'Your order cannot be completed as given.'"

Robotic programs can be exceedingly complicated, containing thousands of logical steps. See also ARTIFICIAL INTELLIGENCE, COMMON SENSE SUMMER PROJECT.

image orthicon

An *image orthicon* is a video camera tube, similar to a vidicon (see VIDICON).

A fine beam of electrons, emitted from an electron gun, scans a target electrode (see the drawing). Some of this beam is reflected back toward the electron gun. The amount of reflected energy depends on the emission of secondary electrons from the target electrode. This, in turn, depends on how much light is hitting the target electrode in a given place. The greatest return beam intensity corresponds to the brightest parts of the video image. The return beam is modulated as it scans the target electrode in a pattern just like the one in the TV picture tube. The return beam is picked up by a receptor electrode.

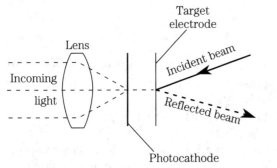

IMAGE ORTHICON
Simplified functional diagram of an image-orthicon video camera tube.

The main disadvantage of the image orthicon is that it produces significant noise in addition to the signal output. But when a fast response is needed (when there is a lot of action in a scene) and the light ranges from very dim to very bright, the image orthicon is the camera tube of choice. It is common in robot vision systems that must process images rapidly, and/or that must operate where the amount of light varies over a wide range. See also VISION SYSTEMS.

immortal knowledge

In the advanced nations, computers have brought about a transformation in human culture. This is most evident in the United States, Europe (including the United Kingdom), and Japan. Knowledge and history can be recorded, and built upon, by computers. The only role humans need to play is to put the data into the computers.

Before computers (that is, before about 1950), history was passed from generation to generation in the form of books and stories. If you read a book written 200 years ago, you will interpret things somewhat differently than the original author thought of them. This is because society is not the same as it was two centuries ago. Values have changed. We have different priorities and beliefs.

When history is put down in books, or told as stories, much information is simply lost, never to be recovered. But computers can keep data indefinitely. If artificial intelligence (AI) becomes advanced enough, computers might become able to interpret data for us, as well as store it. Then there would be less difference in thought from one generation to the next.

Computers will make information—and misinformation—more permanent. If carried to the extreme, computers could give us immortal knowledge. Memory could be backed up to prevent loss because of computer failure, sabotage, and aging of disks and tapes. Every fact, every detail, and possibly all the subtle meaning, too, would be passed along unaltered for century after century. See also ARTIFICIAL INTELLIGENCE, CARETAKERS, COMPUTER.

incompatible time-sharing system

In the early years of computers (the mid-1960s and before), writing, debugging, and running programs was a tedious and laborious process. Usually, the program was written on a terminal with no computer linked to it. The data might be stored on punched cards or paper tape. You would write your program, give it to someone, and let them run it. The results might not even come back the same day!

In the mid-1960s, the first time-sharing system was developed at Massachusetts Institute of Technology. The inventors of the system called it the "Incompatible Time-sharing System" (ITS). This name was a joke. Actually the system made several terminals compatible with the main computer, and therefore it would have more accurately been called the "Compatible Time-sharing System."

incompleteness theorem

In 1931, a young mathematician named Kurt Godel discovered something about logic that changed the way people think about reality. The *Incompleteness Theorem* tells us it's impossible to prove all true statements. In any logical system of thought, there are undecidable propositions.

In mathematical systems, certain assumptions are made. These are called *axioms*. Logical rules are employed to prove theorems based on the axioms. Hopefully there are no contradictions; then the system is consistent. If a contradiction is found, the set of axioms is inconsistent.

Generally, the "stronger" the set of axioms, the greater the chance that a contradiction can be derived from them. A logical system that is "too strong" literally falls apart, because once a contradiction is found, every statement, no matter how ridiculous, becomes provable. If a system is "too weak," then it doesn't produce much of anything meaningful. For centuries, mathematicians strove to build thought universes with elegance and substance, but free of contradictions. Pure mathematics is thus a form of art.

Godel showed that, for any consistent set of axioms, there are more true statements than provable theorems (see the drawing). In any logical system without contradictions, the "whole truth" is unknowable!

The incompleteness theorem has implications in artificial intelligence (AI). Broadly speaking, it is impossible to build a Universal Truth Machine, or a computer that can mathematically prove every true statement. There will never be a computer—or a person—that knows everything. See also ARTIFICIAL INTELLIGENCE.

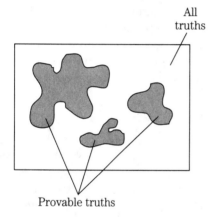

All truths

INCOMPLETENESS THEOREM
Not all true statements can be proved.

Provable truths

incremental optical encoder

See OPTICAL ENCODER.

industrial robots

An *industrial robot*, as its name implies, is a robot employed in industry. Industrial robots can work in construction, manufacturing, packing, and quality control. They are sometimes also used in laboratories.

Among the specific applications for industrial robots are the following: welding, soldering, drilling, cutting, forging, paint spraying, glass handling, heat treating, loading and unloading, plastic molding, bottling, canning, die casting, fruit picking, inspection, and stamping.

Two engineers, George Devol and Joseph Engelberger, were largely responsible for getting industries interested in robots. Businesspeople were hard to convince at first. But Devol and Engelberger translated things into language the businesspeople understood: profit! See DEVOL, GEORGE C., JR. and ENGELBERGER, JOSEPH F.

The robotization of industry has not been welcomed by everyone. Workers are concerned about being displaced by robots. But robotization seems inevitable. Competition, both within the U.S. and internationally, will make it necessary to robotize as much as possible. Companies that ignore technology will be driven out of business, because their competitors will produce better goods at lower cost.

Related terms in this encyclopedia include: ARTIFICIAL INTELLIGENCE, ASSEMBLY LINE, ASSEMBLY ROBOTS, AUTOMATED GUIDED VEHICLE, AUTOMATED INTEGRATED MANUFACTURING SYSTEM, AUTOMATION, BUILDING CONSTRUCTION ROBOTS, ECONOMIC EFFECT OF ROBOTICS, FANUC, FARMING ROBOTS, FLEXIBLE AUTOMATION, FLEXIBLE MANUFACTURING SYSTEM, LUDDITES, MANUFACTURING AUTOMATION PROTOCOL, QUALITY ASSURANCE AND CONTROL, ROBOT ARMS, ROBOT GRIPPERS, UNMANNED FACTORY.

inertial guidance

See GYROSCOPE.

inference engine

An *inference engine* is a circuit that gives instructions to a robot. It does this by applying programmed rules to commands given by a human operator. The inference engine is something like a computer that performs IF/THEN/ELSE operations on a data base of facts.

The inference engine is the functional part of an expert system. See EXPERT SYSTEM, IF/THEN/ELSE.

infinite loop

See ENDLESS LOOP.

infinite regress

An *infinite regress* is a thought pattern in which a sort of endless loop occurs (see ENDLESS LOOP). The simplest example of an infinite regress is a definition that "defines" something in terms of itself.

It is possible to define things in terms of simpler versions of themselves, without going in circles. This is a tricky kind of logic known as recursion (see RECURSION).

Arthur Samuel, one of the pioneers in artificial intelligence (AI), expressed the opinion that machines can't get smarter than their makers. Computers, Samuel wrote, never have truly original thoughts. If they could, it would imply some infinite regress. See SAMUEL, ARTHUR.

Computers process data. They cannot actually make data (unless they malfunction). This data must come from outside the machine. It might come from some other computer, but where did that computer get it? From still another computer? Ultimately, a human being must be at the root of any idea a computer gets. Either that, or the idea must come from an infinite succession of computers, one before the other before the other, without any beginning.

The drawing shows Cybo the computer getting a flash of insight from Precybo, the computer before it, which got the same data, perhaps in different form but with identical content, from Preprecybo (Pre^{2}cybo). Pre^{2}cybo got the data from Prepreprecybo, also known as Pre^{3}cybo or Prepre^{2}cybo or, if you like, Pre^{2}precybo. This gets silly in a hurry, and that's just as well. If true, it means the whole business never got started, which makes it false. The conclusion is, computers cannot generate their own ideas.

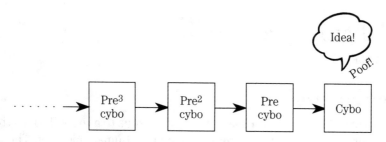

INFINITE REGRESS Cybo and its
predecessors get an idea from the infinite past (see text).

Some researchers have argued against Samuel. They ask, "How do people get their knowledge then?" This brings the debate into philosophy, biology, and even religion. When this happens, notions get rather bizarre. For example, some people believe that biological evolution takes place through mutations (malfunctions) in gene reproduction. Can computers then evolve via electronic malfunctions? See also ARTIFICIAL INTELLIGENCE, COMPUTER.

"Information processor" species

In the life sciences, the term *species* is used in reference to large classifications of plants or animals. Sometimes the term is used for nonliving things, like machines. All computers can be classified, for example, as members of the "Information Processor" species.

Are human beings members of this species? In one sense, the answer is yes; we humans do process information. Our brains and nervous systems are, in some ways, like super-sophisticated computers. But, say many scientists, humans are far more than complex biological machines. People aren't just data-processing devices. It can actually be dangerous for us humans to think of ourselves that way, because when we do, we elevate machines to our own status. Then we give the machines more power than they ought to have.

In a book called *Computer Power and Human Reason*, scientist Joseph Weizenbaum warns that it is dangerous for people to get overly dependent on computers (see WEIZENBAUM, JOSEPH). Have you ever gone to the Post Office and been unable to mail a package because the computers were down and they couldn't weigh anything? Or needed cash to get groceries and found that the automatic teller machine (ATM) didn't work? The species "Information Processor" has, in these cases, grown into something more—something with power over your life.

As a corollary to this, the danger in letting computers become too important, there is a parallel risk. When working with artificial intelligence (AI), people sometimes get a feeling of inferiority. It is as if we humans are nothing but "Information Processors," and not very efficient ones, either. Weizenbaum and many of his colleagues argue that humans are, and always will be, far more advanced than any machine. See also ARTIFICIAL INTELLIGENCE, COMPUTER.

infrared transducer

See PROXIMITY SENSING.

input/output module

An *input/output module*, usually abbreviated *I/O*, is a data link between a microprocessor and a computer's peripherals. In robotics, I/O modules transfer data from the controller to the mechanical parts, or vice-versa. Also, I/O modules can interconnect robot controllers, or link many robots to a central computer (see the drawing).

As its name suggests, an I/O circuit carries data in two directions: into and out of a microprocessor. It must do both at the same time. The incoming data might be much different from the outgoing data. See also CONTROLLER, MICROPROCESSOR.

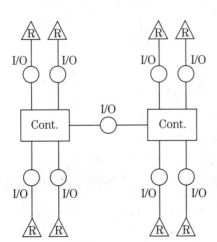

INPUT/OUTPUT MODULE
Modules (I/O) interface between controllers (CONT) and/or robots (R).

insect robots

The term *insect robot* applies to robots of a revolutionary design, that operate in large numbers under the control of a central artificial intelligence (AI) system. In particular, the term is used in reference to systems designed by engineer Rodney Brooks. He began developing his ideas at Massachusetts Institute of Technology (MIT) during the early 1990s. See BROOKS, RODNEY.

Insect robots have six legs, and often look somewhat like beetles or cockroaches. They range in size from more than a foot long to less than a millimeter across. Most significant is the fact that they work collectively, rather than as individuals.

Individual units, each with high AI, don't necessarily work in a team. People provide a good example. Professional sports teams have been assembled by buying the best players in the business, but the team won't win unless the players get along.

Insects, by contrast, are stupid in the individual. Ants and bees are like idiot robots. But an anthill or beehive is an efficient system, run by the "collective mind" of all its members.

Rodney Brooks saw this huge difference between autonomous and collective intelligence. He also saw that most of his colleagues were trying to build autonomous robots, perhaps because of the natural tendency for humans to envision robots as humanoid (see ANDROID, HUMANOID ROBOTS). To Brooks, it was obvious that a major avenue of technology was being neglected. Thus he began designing robot "colonies" consisting of many units under the control of a central AI system.

Brooks envisions microscopic insect robots that might live in your house, coming out at night to clean your floors and countertops. "Antibody robots" of even tinier proportions could be injected into a person infected with some heretofore-incurable disease. Controlled by a central microprocessor, they could seek out the disease bacteria or viruses and swallow them up. See also ARTIFICIAL INTELLIGENCE, AUTONOMOUS ROBOTS, COMPUTER.

instructional robots

See COMPUTER-ASSISTED INSTRUCTION, EDUCATIONAL ROBOTS.

integral

The term *integral* refers to the area under the curve of a mathematical function. For example, displacement is the integral of speed or velocity, which in turn is the integral of acceleration.

Drawing A shows a graph of speed as a function of time. The curve of this function is a nearly straight line. You might think of it as the steady increase in

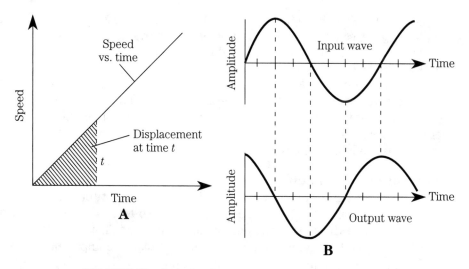

INTEGRAL At A, for drag racer; at B, for waveform.

speed of a drag racer. While the speed constantly increases, the displacement, indicated by the area under the curve, increases at an even faster rate.

In digital electronics, a circuit that continuously takes the integral of an input wave is called an *integrator*. An example of the operation of an integrator is shown in the graph of drawing B. The input is a sine wave. The output is, mathematically, a negative-cosine wave, but you can think of it as a sine wave that has been shifted by 90 degrees, or ¼ of a cycle. See also DERIVATIVE.

integrated circuit

Integrated circuits (ICs) have stimulated as much change as, and perhaps more evolution than, any other single development in the history of technology.

Most ICs look like gray or black plastic boxes with protruding pins. Common configurations are the single inline package (SIP), the dual inline package (DIP) and the flatpack. Another package looks like a transistor with too many leads. This is a metal-can package, sometimes also called a TO package.

Assets

Integrated circuits have several advantages over individual, or discrete components. These include compactness, high speed, low power consumption, reliability, ease of maintenance, and modular construction.

Compactness

An obvious asset of IC design is economy of space; ICs are far more compact than equivalent circuits made from individual transistors, diodes, capacitors, and resistors. Complex circuits can be kept to a reasonable size. Thus, you see notebook computers, also known as laptops, with capabilities more advanced than early computers that took up whole rooms. See COMPUTER.

High speed
In an IC, interconnections among components are physically tiny, making high switching speeds possible. Electric currents travel fast, but not instantaneously. The less time charge carriers need to get from component X to component Y, in general, the more computations are possible within a given span of time, and the less time is needed for complex operations.

Low power consumption
ICs need far less power than equivalent discrete-component circuits. This is important if batteries are to be used. Because ICs use so little current, they produce less heat than their discrete-component equivalents. This translates into better efficiency. It also minimizes the problems that plague equipment that gets hot with use, such as frequency drift and generation of internal noise.

Reliability
Integrated circuits fail less often, per component-hour of use, than appliances that make use of discrete components. This is mainly a result of the fact that all interconnections are sealed within the IC case, preventing corrosion or the intrusion of dust. The reduced failure rate translates into less downtime, or time during which the equipment is out of service for repairs.

Ease of maintenance
Integrated-circuit technology lowers maintenance costs, mainly because repair procedures are simplified when failures do occur. Many appliances use sockets for ICs, and replacement is simply a matter of finding the faulty IC, unplugging it, and plugging in a new one. Special desoldering equipment is used with appliances having ICs soldered directly to the circuit boards.

Modular construction
Modern IC appliances use modular construction. In this scheme, individual ICs perform defined functions within a circuit board; the circuit board or card, in turn, fits into a socket and has a specific purpose. Computers, programmed with customized software, are used by repair technicians to locate the faulty card in an appliance. The whole card can be pulled and replaced, getting the appliance back to the consumer in the shortest possible time. Then the computer can be used to troubleshoot the faulty card, getting the card ready for use in the next appliance that happens to come along with a failure in the same card.

Liabilities
No technological advancement ever comes without some sacrifice or compromise. There are some limitations in IC technology.

No inductors

While some components are easy to fabricate onto chips, other components defy the IC manufacturing process. Inductances, except for extremely low values, are one such bugaboo. Devices using ICs must generally be designed to work without inductors. Fortunately, resistance-capacitance (RC) circuits are capable of doing most things that inductance-capacitance (LC) circuits can do.

No megapower

High power necessitates a certain minimum physical bulk, because it generates large amounts of heat. This isn't a serious drawback. Power transistors are available when high power is needed.

The future

Research is constantly going on in IC technology. As time goes on, their versatility will increase. This technology has made modern computers possible. It will perhaps someday make machines nearly as smart as humans. The main emphasis is on getting more and more components into a smaller and smaller physical space. See ARTIFICIAL INTELLIGENCE, BIOCHIP, INTEGRATED-CIRCUIT FABRICATION, MICROPROCESSOR, ULTRA-LARGE-SCALE INTEGRATION, VERY-LARGE-SCALE INTEGRATION.

integration

See INTEGRAL.

intelligence

See ARTIFICIAL INTELLIGENCE, COMPUTER.

interactive program

When you work with a computer, controlling to some extent what the program does, and helping the computer to make decisions, you are using a computer with an interactive program. You interact with the machine. This is in contrast to a program that simply runs, carrying out its functions entirely independent of you.

A common example of interactive programming is found in bank automatic-teller machines (ATMs). The ATM gives instructions and/or asks questions, and you answer by pressing buttons on a keypad. Another good example is the computer card catalog in the library. See also COMPUTER, INTERFACE, SOFTWARE.

interface

An *interface* is a device that carries data between a computer and its peripherals, or between a computer and a person. An interface consists of both hardware and software. The term is also used as a verb; when you connect two devices together and make them compatible, you interface them.

Suppose you want to use a computer with a certain printer. You must ensure that they'll work together. That is, you must interface the computer with the printer. This requires the right connecting cable and connector (hardware), the use of the right data port (serial or parallel), and also the correct program (software) for the printer functions.

Computer peripherals include such things as the keyboard, modem, printer, monitor, mouse, and joystick. Each requires its own interface. In a robotic system, all the moving parts are peripherals to the controller, and must be properly interfaced with the controller. See also COMPUTER, CONTROLLER, HARDWARE, INTERACTIVE PROGRAM, JOYSTICK, KEYBOARD, MODEM, MONITOR, MOUSE, PRINTER, SOFTWARE.

International Business Machines

International Business Machines (IBM) is a large U.S. corporation, best known in recent years for its personal computers (PCs). The first IBM PC was made available in 1981. The abbreviation "PC" has become generic, like the word "kleenex." Nowadays, almost all small computers are called PCs.

You will often hear about PC clones or IBM-compatible hardware and software. Numerous imitation computers have been manufactured and distributed, similar (and sometimes practically identical) to IBM models.

The other major computer type in the U.S. today is Macintosh. Their computer line is sometimes also called Apple. The IBM and Macintosh computer types have similar memory and speed, but the software systems are somewhat different. Some people find IBM-compatible machines easier to work with; others prefer Macintosh. See also MACINTOSH.

interpolation

When there is a gap in data, but data is available on either side of the gap, an estimate of values within the gap can be made by *interpolation*. This can be "educated guessing," but it can also be done using artificial intelligence (AI).

An example is shown in the drawing. The temperature (in degrees Fahrenheit, or F) is graphed versus time for a day in April in Minneapolis. Data is taken every half hour. But the thermometer breaks between noon and 3:00 p.m. This causes a gap in the curve. The solid line represents the curve derived from actual data.

You can guess what the curve would look like if the thermometer hadn't broken. A computer would fill in the curve along the dotted line. You wouldn't expect a "spike" to +70° F or a "dip" to +20° F. This curve is easy to interpolate.

When there is insufficient data, interpolation is difficult or impossible. If temperatures are taken every Saturday at 3:00 in the morning, it is impossible to

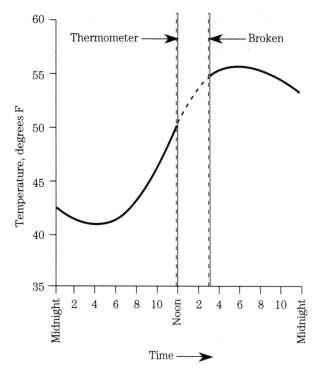

INTERPOLATION Filling in a gap in data.

interpolate the temperature at 3:00 p.m. on a Tuesday, especially in Minneapolis in April! See also EXTRAPOLATION.

interpreter

In a computer system, an *interpreter* translates the high-level language (used by the human operator) into machine language (used by the computer hardware), and vice-versa.

An interpreter does basically the same things as a compiler (see COMPILER). The difference is that, while a compiler translates everything before the program is actually run, the interpreter translates the data as it goes along.

The main advantage of an interpreter is that it can help the operator pinpoint programming errors. The main disadvantage is that an interpreter takes more time than a compiler to run a program. See also HIGH-LEVEL LANGUAGE, MACHINE LANGUAGE.

inverter

An *inverter*, also known as a *NOT GATE*, is a digital logic circuit with one input and one output. It performs the logical NOT operation, also called negation.

inverter

The output of an inverter is high, or logic 1 ("true") if and only if the input is low, or logic 0 ("false"). The output is low if and only if the input is high.

Logic gates are extensively used in digital computers and other electronic devices.

The schematic symbol for an inverter, along with its logical truth table, is shown in the figure. See also AND GATE, EXCLUSIVE-NOR GATE, EXCLUSIVE-OR GATE, NOR GATE, OR GATE.

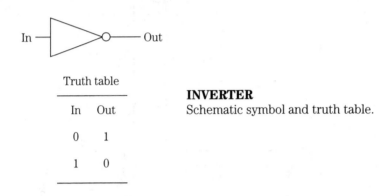

Truth table

In	Out
0	1
1	0

INVERTER
Schematic symbol and truth table.

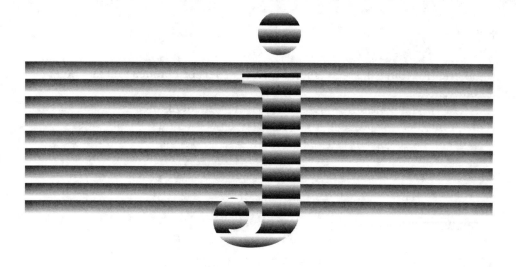

jointed geometry

See ARTICULATED GEOMETRY.

joint-force sensor

A *joint-force sensor* is a feedback device that keeps a robot joint from exerting too much force. The sensor works by detecting the resistance the robot arm encounters. The greater the resistance, the more force is being applied. The sensor can be programmed to stop the joint if this amount of resistance is exceeded. See also BACK PRESSURE SENSOR, FEEDBACK.

Joseph, Earl

Earl Joseph is a futurist, or researcher with ideas about the future of technology. He has been compared with inventor/philosopher R. Buckminster Fuller, the man who conceived the idea of the geodesic dome. See FUTURISTS.

Earl Joseph predicted the development of integrated circuits (ICs) several years before they became available. He has made other technological forecasts that later came true. Some of his predictions for artificial intelligence (AI) suggest that machines might someday evolve on their own, without human intervention. Computers, with robots to do the labor, might conduct their own research and development, run their own labs, and manufacture new and improved versions of themselves.

Joseph has suggested that homes will someday be totally automated. You won't have to spend any time doing laundry, washing dishes, cooking, mowing the lawn, or shoveling snow. You'll never have to paint the walls, vacuum the floor, clean the toilets, dust your desk, or even drive the car if you don't want to. Futurists have often observed that, even in this modern society, humans spend a lot of time doing menial, mechanical things. Total automation would free people to do creative things instead, such as write novels or compose music. See PERSONAL ROBOTS.

Some psychologists wonder, however, how mentally healthy such an existence would be. It is possible, they argue, to have too much leisure time.

Joseph has suggested that smart robots could be built to serve as soldiers. This has been written about in science fiction (see CAPEK, KAREL). Smart robots could serve as caretakers. Joseph makes mention of biological robots. And along these lines, he notes that "new species" have already been patented! See also ARTIFICIAL INTELLIGENCE, ARTIFICIAL LIFE, BIOLOGICAL ROBOTS, CARETAKERS, COMPUTER, MILITARY ROBOTS.

joystick

A *joystick* is a control device capable of movement in two or three dimensions. The drawing shows a joystick with two dimensions of movement, labeled x and y. The device consists of a movable lever or handle, and a ball-bearing within a control box. The stick is moved by hand.

The joystick gets its name from its resemblance to the joystick in an aircraft. Some joysticks can be rotated clockwise and counterclockwise, in addition to the usual two coordinates, allowing control in a third dimension.

Joysticks are used in computer games, for entering coordinates into a computer, and for the remote-control of robots. See also MOUSE, TRACKBALL.

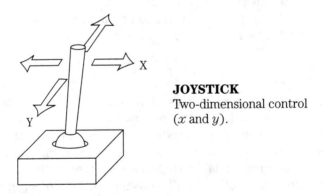

JOYSTICK
Two-dimensional control
(x and y).

Jungian-world theory

An interesting motivation for research in artificial intelligence (AI) is called the *Jungian-world theory*. According to this notion, people keep making the same mistakes in every generation. It seems that people can't learn from history. Also, humanity seems unable to foretell, or care about, the potential consequences of the things they do. It is as if the human race is time-blind. Thus, as the saying goes, history repeats itself.

This theory has been demonstrated many times. People keep fighting wars for the same reasons. Ultimate victory never comes. Wars don't solve problems. In fact, quite often they create more.

What can humanity do to stop this self-defeating vicious circle? According to researcher Charles Lecht, the answer is to develop AI to the point that machines become smarter and wiser than people (see LECHT, CHARLES). Perhaps a brilliant computer can help us humans control our destiny, so that we need not keep re-living the same old calamities. At least, this is Lecht's hope.

Many researchers doubt that machines can become smarter than people. But it is possible that computers, if they become powerful enough, might help us humans find solutions to some of our problems—solutions we might otherwise never discover. See also ARTIFICIAL INTELLIGENCE, COMPUTER, COMPUTER REASONING.

Karnaugh map

A *Karnaugh map* is a pictorial breakdown of a logical expression, showing all the individual operations. This can be done in the form of a table, or in the form of a drawing.

Consider the Boolean expression:

$$V = (W + X)(-Y + Z).$$

This is a rather complicated logical expression. Recall that, in Boolean algebra, addition refers to the logical OR operation, multiplication to the logical AND, and a minus sign to the logical NOT (see BOOLEAN ALGEBRA). The equal sign means "if and only if," which mathematicians write as "iff." The above expression might then be written:

$$V \text{ iff } (W \text{ OR } X) \text{ AND } ((\text{NOT } Y) \text{ OR } Z).$$

The table is a breakdown of this logical expression into its components, using a columnar format. It is called a truth table (see TRUTH TABLE).

KARNAUGH MAP:
Truth table for $V = (W + X)(-Y + Z)$.

W	X	Y	Z	W + X	−Y	−Y + Z	V
0	0	0	0	0	1	1	0
0	0	0	1	0	1	1	0
0	0	1	0	0	0	0	0
0	0	1	1	0	0	1	0
0	1	0	0	1	1	1	1
0	1	0	1	1	1	1	1
0	1	1	0	1	0	0	0
0	1	1	1	1	0	1	1

KARNAUGH MAP *Continued*

W X Y Z	W + X	–Y	–Y + Z	V
1 0 0 0	1	1	1	1
1 0 0 1	1	1	1	1
1 0 1 0	1	0	0	0
1 0 1 1	1	0	1	1
1 1 0 0	1	1	1	1
1 1 0 1	1	1	1	1
1 1 1 0	1	0	0	0
1 1 1 1	1	0	1	1

The drawing is a schematic diagram of the expression using digital logic symbols. In this case, there are four inputs (W, X, Y, Z) and one output (V). The individual logical operations are shown as gates (see AND GATE, INVERTER, OR GATE). Both the truth table and the logic diagram can be considered Karnaugh maps.

Often, logical expressions are exceedingly complicated. This is especially true in computer systems. A Karnaugh map of an entire computer would occupy hundreds, thousands, or millions of square miles! This gives some insight as to how sophisticated computer technology has become.

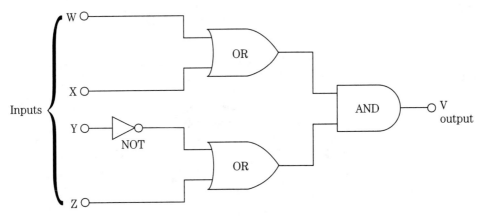

KARNAUGH MAP Logic circuit for $V = (W + X)(-Y + Z)$.

For most complex logical expressions, there are many different ways to arrange the individual functions to get the desired outputs. Some arrangements are simpler than others. The more complex the expression, the more different equivalent circuits exist. Can you find a simpler way to write the logical expression discussed here?

Kato, Ichiro

The most well-known robotics engineer and researcher in Japan is Ichiro Kato. He has developed artificial hands and arms, known as *prostheses*. His ultimate goal is to design and construct an android, or artificial human.

In the United States and Europe, robots have been built mainly for industrial use. The idea of a two-legged, or biped, robot has not been taken very seriously, because such machines are hard to engineer. In fact, most roboticists in the Western nations have dismissed the idea of androids, especially in industry, as unworkable.

But in Japan, people find humanoid robots immensely entertaining. This is, at least in part, because of the Japanese tendency to view, and respect, all things—even what Westerners call inanimate things—as living. Robots in Japan can become like good friends. Sometimes people actually get emotionally attached to them. Thus, some Japanese engineers are eagerly working towards building robots in human form.

Kato has numerous colleagues who are also trying to develop humanoid robots. One idea is to supply power to the machines through their feet by electrifying the floor. This would eliminate the need for the robots to carry batteries. One potential problem: Humans would need shoes with soles that were electrically insulated, even when wet!

Kato is known for the concept of biomechatronics, a science that is part biology, part mechanics, and part electronics. He is also involved in a theoretical pursuit known as *synalysis*, which is part synthesis and part analysis. See also ANDROID, ANIMISM, BIOMECHATRONICS, BIPED ROBOTS, HUMANOID ROBOTS, PROSTHESIS.

keyboard/keypad

A *keyboard* is a set of buttons for entering data into a computer or terminal. There are several different keyboard arrangements. Most resemble the layout of a conventional typewriter, with the addition of various function keys. Engineers have been striving, in recent years, to put the function keys in places where they are as easy as possible to use. One of the most popular keyboards today has 101 keys. This includes a calculator-like keypad in addition to alphanumeric and function keys.

In small calculators, and also in simple microcomputer-controlled machines, there is a small keypad, but not a full keyboard. You can type out English-language commands and data on a keyboard, while on a keypad you can't.

The "core" character arrangement for computer keyboards is shown in the figure for the article CHARACTER. A typical keypad is shown in the drawing for the article ELECTRONIC CALCULATOR. You will also find a standard keypad on any tone-actuated telephone set. A few telephone keypads contain keys labeled A, B, C, and D, in addition to the numerals 0 through 9, the pound symbol (#) and the star symbol (*). The drawing shows the standard layout for a 16-key keypad. See also QWERTY.

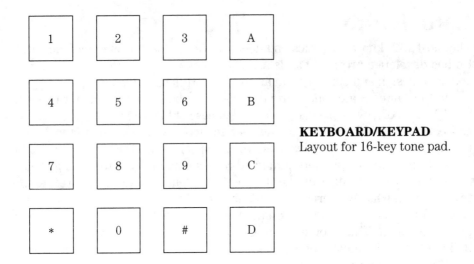

KEYBOARD/KEYPAD
Layout for 16-key tone pad.

kilobyte

See BYTE.

kilohertz

See HERTZ.

K-line programming

K-line programming is a method by which an artificially intelligent (AI) robot system can learn as it does a job, so that it will have an easier time doing the same or similar work in the future.

Suppose you have a personal robot that you use for handiwork around the house. The water heater breaks down and you instruct the robot to fix it. The robot must use certain tools to do the repair. The first time the robot repairs the water heater, it must find the tools by trial-and-error. But it encodes each tool in its memory, perhaps according to shape. It also encodes the sequence in which the tools are used to fix the water heater. Then, the next time the water heater needs repair, the robot can refer to the K line: the list of tools used before, and also the order in which they were used.

Of course, there are many different things that can go wrong with a water heater. The second time it breaks down, the K line for the first repair might not work. The robot would have to refine its knowledge, devising a second K line for the new problem. Over time, the robot might learn several different schemes for fixing a water heater, each scheme tailored to a specific problem (see the diagram). This "learning-by-experience" process is heuristic knowledge. See also ARTIFICIAL INTELLIGENCE, HEURISTIC KNOWLEDGE.

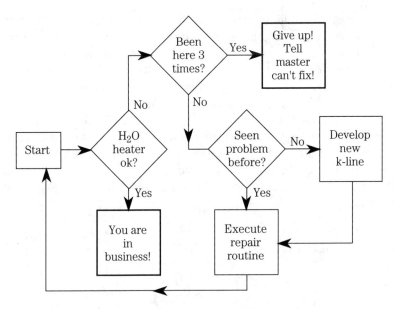

K-LINE PROGRAMMING Flowchart for repairing water heater.

kludge

A crude, useless, or grossly inefficient machine or process is called a *kludge* (pronounced "klooge"). You might say, "That contraption is a kludge robot," or "This is a kludge program."

Kludges are often useful, because they can test an idea without a lot of trouble and expense. But often, referring to a device as a "kludge" is an apology (in the case of one's own work) or a good-natured insult (in the case of someone else's work).

Perhaps you've seen, written or built kludge machines or programs. Every year, Purdue University holds a "Rube Goldberg" contest for students to build or design ridiculously inefficient machines and computer software. It is sponsored by Theta Tau, an engineering fraternity.

Eventually, almost every engineer becomes an expert at the art of kludge. But trying to sell a kludge as a finished product is an industrial faux pas—a huge no-no—unless, of course, the intent is to make people laugh. See also PROTOTYPE.

knowledge

The term *knowledge* refers to the data stored in a computer, artificial-intelligence (AI) system or human mind. Also, the term refers to how well a mind, be it electronic or biological, makes use of the data it does have.

knowledge

In expert systems, engineers speak of knowledge acquisition, the process by which machines obtain their data. Generally, all computer knowledge must come from human beings. Machines don't generate their own new knowledge.

In AI, a computer can sometimes learn from its mistakes. This is not original thought, but is derived from existing knowledge. The ability of a machine to improve the use of its data is called *heuristic knowledge*.

Humans, individually and as a culture, have knowledge that changes from generation to generation. Some researchers have suggested that AI might totally eliminate the loss of human knowledge in the future. This would give us an ever-expanding storehouse of immortal knowledge. See also ARTIFICIAL INTELLIGENCE, COMPUTER, EXPERT SYSTEMS, HEURISTIC KNOWLEDGE, IMMORTAL KNOWLEDGE, MACHINE KNOWLEDGE.

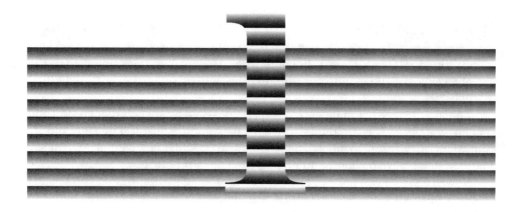

LAN

See LOCAL AREA NETWORK.

laptop computer

See COMPUTER, NOTEBOOK COMPUTER.

language

A *language* is a set of symbols, along with rules for arranging the symbols, used for controlling a computer. A language can also be used for communication among computers, or between computers and robots. Languages are classified as *low-level* or *high-level*. In addition, languages can be either audio (spoken) or visual (written).

Written languages

Low-level languages don't resemble ordinary speech or text, but instead, are more akin to long mathematical equations. Writing a program directly in a low-level language is a tedious process. Low-level languages are directly used by computers. See ASSEMBLY LANGUAGE, MACHINE LANGUAGE.

High-level languages are the ones with which you are probably most familiar. They include programming languages such as BASIC, FORTRAN, and COBOL. This is the language that a person most often uses to communicate with a computer or artificial-intelligence system. Different high-level languages are used for various applications; for example, FORTRAN is good for scientific work and COBOL is well-suited to business work. See HIGH-LEVEL LANGUAGE.

Translation between high-level and low-level languages is done either by a compiler or an interpreter. See also COMPILER, INTERPRETER.

Spoken languages

In recent years, it has become possible to literally converse with computers, just as you would talk with people. In some ways, this is more convenient than the use of a keyboard and monitor/printer. In other applications, the written method is still better.

The use of voice communication is made possible by speech recognition and speech synthesis. These circuits translate between a spoken language and its equivalent in written symbols. See SPEECH RECOGNITION, SPEECH SYNTHESIS. These systems have obvious, and increasing, use in robotics.

Nowadays, you can control a robot just by telling it what to do. The biggest problem in this technology is getting a machine to correctly choose between words that sound the same, but have different meanings, such as "way/weigh" or "do/due." To do this, a machine must be able to figure out the right word by evaluating the context in which it is used. See also ARTIFICIAL INTELLIGENCE, COMPUTER, CONTEXT, GRAMMAR CHECKING, SYNTAX.

large-scale integration

Large-scale integration (LSI) is a process of integrated-circuit (IC) manufacture, in which there are 100 to 1,000 logic gates per chip. Bipolar transistors and metal-oxide-semiconductor (MOS) devices can be used. Examples of devices using LSI include digital wristwatches and small electronic calculators. More powerful digital circuits make use of even more logic gates per chip.

There is a physical limit to the density with which components can be etched onto a microchip, simply because the atoms themselves are not infinitely small. However, technology has not yet reached the point where this limit is a factor. It's probably just a matter of time, however, before this happens. See also INTEGRATED CIRCUIT, ULTRA-LARGE-SCALE INTEGRATION, VERY-LARGE-SCALE INTEGRATION.

laser data transmission

Laser beams can be modulated to convey information, in the same way as radio waves. *Laser data transmission* allows many signals to be sent over a beam of light.

A laser-communications transmitter has a signal processor or amplifier, modulator, and laser (see drawing A). The receiver uses a photocell, an amplifier and a signal processor (see drawing B). Any form of data can be sent, including voice, television, and digital signals.

Laser communications systems are either line-of-sight or fiberoptic.

In a line-of-sight system, the beam travels in a straight line through space or clear air. Because laser beams remain narrow for long distances, long-range communication is possible. One problem is that it won't work well through clouds, fog, or other obstructions. Another problem is that the aiming must be precise.

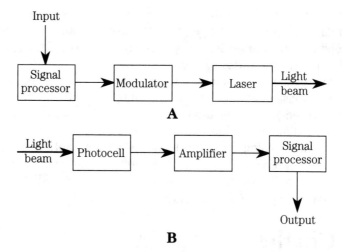

LASER DATA TRANSMISSION
Transmitter (A) and receiver (B).

In a fiberoptic system, the laser is guided through a glass or plastic filament. This is like wire or cable communications, but with far more versatility. Fiberoptic systems lend themselves well to robot control, especially in hostile environments such as the deep sea. See FIBEROPTIC DATA TRANSMISSION.

laser printer

See PRINTER.

laser ranging

See PROXIMITY SENSORS.

learning

According to Alan Turing, a pioneer in the field of artificial intelligence (AI), a machine must be able to learn things if it is truly intelligent. Turing's concept of learning involved more than just rote memory.

When a student takes courses in school, he/she gains much knowledge simply by memorizing things like names, dates, and mathematical equations. But this knowledge is useless unless the student also finds ways to put this information together into meaningful thoughts.

It's easy to store the equation:

$$E = mc^2$$

in a computer memory. This is the equation made famous by Albert Einstein, stating the relationship between energy (E), mass (m) and the speed of light (c). In a machine with complex learning capacity of the sort Turing envisioned, there would be many other facts stored.

Suppose you asked the machine, "What was the most important result of Einstein's energy/mass equation?" If the machine had learned things (and not just memorized them), it might answer, "The atomic bomb." If the machine hadn't learned anything, it wouldn't have a good answer.

As with human learning, machine learning will probably involve opinions. This fascinates AI researchers. Different people, taking the same classes in school, grow up to have vastly different outlooks on the world. Will the same thing happen with smart machines? See also ARTIFICIAL INTELLIGENCE and TURING, ALAN.

Lecht, Charles

Charles Lecht is a futurist who believes that robots and "smart computers" will eventually do all the mundane work that humans do today. Lecht worked for International Business Machines and then started his own computer software company, called Lecht Sciences, Inc.

Lecht is optimistic. He thinks the benefits of robots and artificial intelligence (AI) far outweigh the dangers.

Are robots a threat to workers? Lecht says no; in fact, the opposite is true. Robots can do monotonous jobs and set people free to do more interesting work.

Lecht points out that much of our time, even in modern society, is taken up by busy work. We mow lawns, do laundry, and take cars to the shop. We must call a repair person to fix the air conditioner or television set or telephone. The wealthiest people have servants who do these things. Lecht sees society evolving to the point where everyone can afford to have robots do these jobs. See also ECONOMIC EFFECT OF ROBOTICS, FUTURISTS, PERSONAL ROBOTS.

legs

See ROBOT LEGS.

life

Modern science, for all its advances since 1900, still has not come up with a rigorous definition of life. Yet roboticists and computer scientists, as well as biologists, talk about artificial life.

One tentative definition is this: living things increase order within and around themselves. They operate against the natural process called *entropy*, which constantly makes things degenerate toward a state of chaos.

According to this definition, a living machine must be able to "create order from chaos." See ARTIFICIAL LIFE.

limbs

See ROBOT ARMS, ROBOT GRIPPERS, ROBOT LEGS.

linear programming

Linear programming is a process of optimizing two or more variables that change in different, independent ways.

A simple example of linear programming is shown in the drawing. The two variables are x and y. They represent the positions of two robots as they move in straight lines within their work area. The path of robot A is shown by the solid line; the path of robot B is shown by the dotted line. Robot A moves at 3.5 feet per second; robot B creeps along at 1.5 feet per second. The starting points are shown by heavy dots.

Linear programming can answer these questions:

- How long after their startup will the robots be closest to each other?
- What will be the coordinates of robot A at that time?
- What will be the coordinates of robot B at that time?

Computers can be programmed to solve these problems quickly and easily. Robots with artificial intelligence (AI) would solve such problems when necessary, without having to be told by a human operator. See ARTIFICIAL INTELLIGENCE.

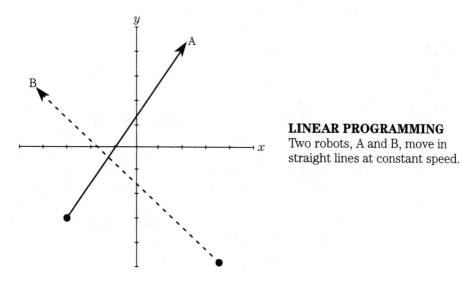

LINEAR PROGRAMMING
Two robots, A and B, move in straight lines at constant speed.

link

A *link* is a connection between two different computers, or the act of making such a connection. The term can also apply to programs or files within a single computer system.

Sometimes many computers are linked together via wire, radio, satellite, or fiberoptic networks. This makes it possible to combine the information in the computers, creating a sort of "supercomputer." The main problem with this is that the data cannot travel faster than the speed of light in free space (about 186,282 miles per second, or 299,792 kilometers per second). This limits the speed at which such networks can operate.

For example, if a link is through a geostationary satellite orbiting 22,300 miles up, the data takes about ¼ second to get from one computer to the other (see the illustration). This is fine for one-way data transfer, but it severely hampers two-way communications speed. See also NETWORK.

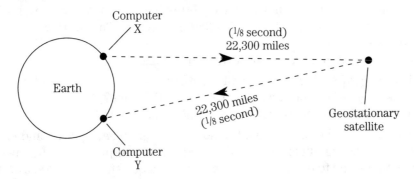

LINK With a geostationary satellite, there is always a delay.

LISP

LISP is a computer programming language that was first devised by John McCarthy at Massachusetts Institute of Technology (MIT) in 1968. The name is a contraction of the words *LISt Processing*.

LISP differs from most high-level languages in that it is written in the form of lists. Other languages usually have statements that tell the computer what to do. In LISP, a program can be interpreted directly as data, and vice-versa. The entire language is derived from a few basic functions. LISP programs are easy to debug. Although its structure is straightforward, some programmers consider LISP difficult to read.

LISP is used extensively in artificial intelligence (AI) research, particularly in the United States. The language has various "dialects" or variants. See ARTIFICIAL INTELLIGENCE, HIGH-LEVEL LANGUAGE, and PROLOG.

load/haul/dump

Load/haul/dump, abbreviated *LHD*, is a robot used in mining and construction. It does just what its name implies. With the aid of a human operator, LHDs load cargo, haul it from one place to another, and dump it in a prescribed location.

In mining, LHDs have an easier time than in general construction. Mine geometries are easily programmed into the robot controller; the layout changes very slowly. Reprogramming does not need to be done often. But in construction, the landscape is more complicated, and it changes rapidly as work progresses. Therefore, the computer maps must be revised often.

In mining, all the loads are usually the same in terms of weight and volume. Coal is coal, and taconite is taconite. But in construction, the nature of the load can vary. Small bricks differ from cinder blocks, and neither of these behave anything like sand.

LHD vehicles use various methods for navigation, including beacons, computer maps, position sensors, and vision systems. LHDs might be autonomous, although there are advantages to using a single controller for many robots. See AUTONOMOUS ROBOTS, BEACON, COMPUTER MAP, INSECT ROBOTS, POSITION SENSING, and VISION SYSTEMS.

local area network

A *local area network (LAN)* is a group of computers that are all linked together within a relatively small area. The interconnections are usually made with cables. All the computers are within a short distance of each other, so the link delays are not long.

There are two main ways in which the computers in a LAN can be connected. These are shown in the drawing. At A, there is a large central computer called a *file server*, to which all the computers are linked. This is called a *client-server LAN*. At B, every computer is simply linked to every other. This is a peer-to-peer LAN.

In any LAN, the combination of computers is far more powerful than any single computer by itself. See also LINK, NETWORK, and WIDE AREA NETWORK.

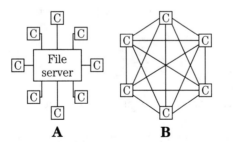

LOCAL AREA NETWORK
Client-server (A) and peer-to-peer (B). Computers are boxes labeled C.

local feature focus

In a robotic vision system, it is usually not necessary to use the whole image to perform a function. Often, only a single feature, or a small region within the image, is needed. To minimize memory space and to optimize speed, *local feature focus* can be used.

Suppose a robot needs to get a pair of pliers from a toolbox. This tool has a characteristic shape that is stored in memory. Several different images might be stored, representing the pliers as viewed from various angles. The vision system quickly scans the toolbox until it finds an image that matches one of its images for pliers. This saves time, compared with trial-and-error, in which the robot picks up tool after tool until it finds pliers.

Your mind/eye system uses local focus without thinking. If you're driving in a forest and see a sign "*Watch for deer*," you'll be on the lookout for animals on or near the roadway. A tractor parked on the shoulder won't arouse you, but a horse standing there definitely will. See also VISION SYSTEMS.

logic

Logic can refer to either of two things in electronics, computer science, and artificial intelligence (AI).

Boolean algebra is the representation of statements as symbols, along with operations, generating equations. This form of logic is important in the design of digital circuits, including computers. Boolean algebra is deductive logic, because the conclusions are derived, or deduced, in a finite number of steps. See BOOLEAN ALGEBRA.

Logic can also be inductive. In this form, a statement is proven true for an infinite sequence of cases. First, the statement is proved deductively for one case. Then it is proved that if the statement is true for some arbitrary case, it is true for the next case in the sequence. This implies truth for the whole sequence. It's "sleight-of-mind"—an infinite number of proofs is done in a finite number of steps (see drawing). Mathematicians consider this perfectly rigorous and acceptable.

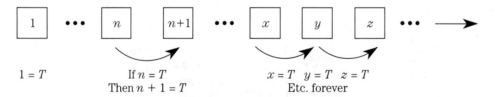

LOGIC Inductive reasoning, proving truth (T) for an infinite sequence of cases.

For a thorough discussion of deductive and inductive logic, a text on symbolic logic is recommended.

A more complex form of logic, of special interest to researchers in artificial intelligence (AI), is called *computer reasoning*. This can involve more than true/false, black/white representations. In fuzzy logic, truth value can vary over a range of several steps, or perhaps even continuously. For example, the statement "We need trees more than we need paper", is debatable, and in fact, its truth value depends on the circumstances. See ARTIFICIAL INTELLIGENCE, and COMPUTER REASONING.

logical operator

A *logical operator* or operation is a way of modifying or combining statements in logic. The most common, and most familiar, logical operators are negation, disjunction, and conjunction. These are represented in common language by the words NOT, OR, and AND. In Boolean algebra they are denoted by a minus sign, addition and multiplication, respectively. In electronics, the logical operations are carried out by circuits called *gates*. See AND GATE, BOOLEAN ALGEBRA, EXCLUSIVE-NOR GATE, EXCLUSIVE-OR GATE, INVERTER, LOGIC, NAND GATE, NOR GATE, and OR GATE.

logic gates

See AND GATE, EXCLUSIVE-NOR GATE, EXCLUSIVE-OR GATE, INVERTER, NAND GATE, NOR GATE, OR GATE.

LOGO

LOGO is a programming language especially useful for teaching children to use computers and computerized robots. In LOGO, a child can write programs to design original computer games.

Commands in LOGO take the form of simple motion descriptions. For example, LEFT 45 refers to a 45-degree turn to the left (counterclockwise). FORWARD 45 would mean "move forward 45 millimeters." By writing the commands in a sequence, the child can make a "turtle" go through these motions on a table or floor. The turtle contains tactile sensors, so that if it comes into contact with something, a signal is sent to the computer. The child might write IF TOUCH THEN RIGHT 90, meaning that if the turtle runs into anything, it should turn 90 degrees to the right (clockwise).

A child can quickly learn to write rather complex programs in LOGO. The result is that the "turtle" acquires a set of behaviors. This lets the child custom-program a robot to have artificial intelligence (AI).

Children like working with LOGO because the system is a toy. While they play, they learn. LOGO teaches concepts ranging from robotics to computer programming, and from AI to coordinate geometry, all at the same time.

log polar navigation

When mapping an image for use in a robotic vision system, it sometimes helps to transform the image from one type of coordinate system to another. In *log polar navigation*, a computer converts an image in polar coordinates to an image in rectangular coordinates.

log polar navigation

The principle of log-polar image processing is shown in the drawing. The polar system, with two object paths plotted, is shown at A. The rectangular equivalent, with the same paths shown, is at B. The polar radius (A) is mapped onto the vertical rectangular axis (B); the polar angle (A) is mapped onto the horizontal rectangular axis (B).

Radial coordinates are unevenly spaced in the polar map (A), but are uniform in the rectangular map (B). During the transformation, the logarithm of the radius is taken. This results in peripheral "squashing" of the image. The resolution is degraded for distant objects, but is improved nearby. In robotic navigation, close-in objects are usually more important than distant ones, so this is a good tradeoff (see DIRECTION RESOLUTION, DISTANCE RESOLUTION).

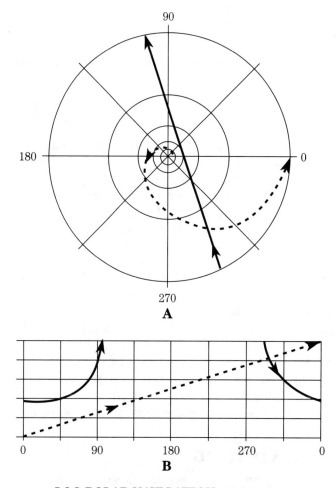

LOG POLAR NAVIGATION At A, polar
coordinates in real space. At B, coordinates after
transform. Lines show hypothetical paths in each system.

A log-polar transform greatly distorts the way a scene looks to people. A computer doesn't care about that, if each point in the image corresponds to a unique point in real space—that is, if the point mapping is a one-to-one correspondence.

What use is a log-polar transform? Robot vision systems employ of television cameras that scan in rectangular coordinates, but events in real space are of a nature better represented by polar coordinates. The transform changes real-life motions and perceptions into images that can be efficiently dealt with by a television scanning system. This is a great help in robot navigation. See also VISION SYSTEMS.

loop

A *loop* is a repeating sequence of operations in a computer program. The number of repetitions can range from two to thousands, millions, or billions. Often, the number of iterations depends on the data input.

In some programs, there are loops within loops. This is called *nesting* of loops.

Loops are useful in mathematical calculations that involve repetitious operations. Until computers were developed, many problems couldn't be solved. This was not because the problems involved esoteric principles, but because the millions and millions of steps would take a single person centuries to grind out.

Sometimes errors are made in the programming, and a computer ends up having to go through a loop without ever reaching a condition in which it can exit the loop. This is called an *infinite loop* or *endless loop*. It always results in a failure of the program to come to a satisfactory conclusion.

An example of a loop is shown in the flowchart. In this case, the program is intended to get rid of all the occurrences of a particular word in a document, and to replace that word with another word. In word processing, this is called *search-and-replace*. When all the occurrences of the word have been found and

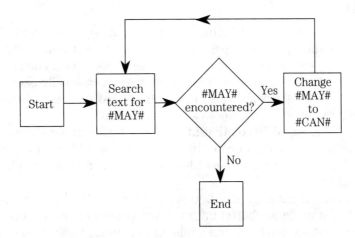

LOOP Search-and-replace function, changing all occurrences of *may* to *can*. (Symbols # denote spaces.)

changed, the program exits the loop, and is ready to continue on with whatever else the programmer might desire.

low-level language

In computer programming, the term *low-level language* refers to any language with which the computer actually works. These languages are not easy for a person to deal with, for two reasons. First, low-level programming is tedious and time-consuming. Second, low-level format is nothing at all like plain English.

Examples of low-level languages are assembly language and machine language (see ASSEMBLY LANGUAGE, MACHINE LANGUAGE).

Usually, the low-level language is converted to a high-level language. This is done via an interpreter or compiler. See also COMPILER, HIGH-LEVEL LANGUAGE, and INTERPRETER.

LSI

See LARGE-SCALE INTEGRATION.

Lucas, George

George Lucas is known for his conception of the *Star Wars* movie robots R2D2 and C3P0.

No one has built actual robots that do the things R2D2 and C3P0 did. Special effects were used in the movie. The "real" R2D2 was made from the shell of an old vacuum cleaner! Because they fulfilled people's expectations for future robots, R2D2 and C3P0 were fun to watch.

Luddites

Whenever there is a major new technological innovation, some people fear that they might lose their jobs. Job loss can occur for at least two reasons. First, greater efficiency reduces the number of people needed for a company or department to function. Second, human workers have occasionally been replaced by robots that don't get sick and don't take coffee breaks. People who fear technology for any reason are sometimes called *technophobes*.

During the Industrial Revolution in England, technophobes went on rampages and destroyed new equipment that they feared would take their jobs from them. Their leader was a man named Ned Ludd, and so these people became known as *Luddites*.

Robotization has not caused a Luddite reaction in America, Japan, or Europe. Workers nowadays are better informed than they were in Ned Ludd's time. They know that if they don't robotize while other nations do, their country won't be competitive in the global economy, and everybody's standard of living will suffer as a result. See also ECONOMIC EFFECT OF ROBOTICS.

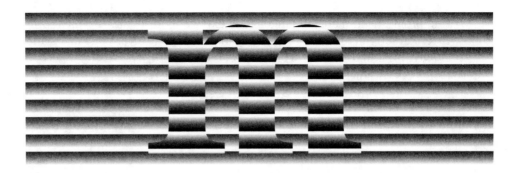

machine knowledge

Computers can store and manipulate information in ways that people find difficult or impossible. A good example is the addition of a series of five million numbers. But there are problems humans can solve that a machine cannot, and perhaps will never, be able to figure out. One example of this is the amount of medication needed to keep a hospital patient "asleep" during surgery. Machine knowledge and human knowledge are different "animals." The extent of a machine's knowledge is reflected by the types and variety of problems it can solve.

Artificial intelligence (AI) is the highest form of machine knowledge. A robot with AI can learn from its mistakes, and adjust itself to cope with its environment without having to be constantly supervised by a human operator. See ARTIFICIAL INTELLIGENCE, and KNOWLEDGE.

machine language

A computer doesn't work with words, or even with the familiar base-10 numbers. Instead, the machine uses combinations of 1's and 0's. These are the two binary states, also represented by on/off, high/low, or true/false. Data in machine language, if written down, might look like a string of numerals such as 111001010001. This can be represented pictorially by black and white blocks, as shown in the drawing. Black represents 1 and white represents 0.

When you write a program, or issue a command to a computer, you do it in high-level language. This must be converted into machine language, or the sequence of 1's and 0's, for the computer. The computer output is likewise translated from machine language into whatever high-level language you are using.

MACHINE LANGUAGE
Pictorial representation of 111001010001.

Writing programs in machine language is extremely difficult. A typical program written directly in machine language consists of page after page of 1's and 0's, seemingly random. But they are anything but random; one wrong digit can ruin the whole program.

To streamline the programming process, you write and communicate in a high-level language such as BASIC, COBOL, or FORTRAN. The compiler or interpreter takes care of the translation process. See also COMPILER, HIGH-LEVEL LANGUAGE, and INTERPRETER.

machine thought

See ARTIFICIAL INTELLIGENCE.

machining

In industrial robotics, *machining* is the modification of parts during assembly. Examples of machining are drilling, removing burrs from drilled holes, welding, sanding, and polishing. In an assembly line, many identical parts pass each workstation in rapid succession, and the worker or robot does the same tasks over and over.

There are two methods in which machining is done by robots. The robot can hold the tool while the part remains still, or the robot can hold the part while the tool stays put.

Robot holds tool

This method offers the following advantages:

- Small robots can be used if the tool is not heavy.
- Parts can be large and heavy, because they need not be moved by the robot.
- The robot can adjust easily as the tool wears down.

Robot holds part

In this method, the advantages are:

- The part can be moved to any of several different tools, without having to change the tool on the robot arm.
- Tools can be large and heavy, because they need not be moved.
- Tools can have massive, powerful motors because the robot doesn't have to hold them.

Obviously, some industrial situations lend themselves better to the first method, while some processes are done more efficiently using the second method.

There are some processes that do not lend themselves very well to robotization. These are usually tasks that require extreme accuracy, and/or that take many complicated steps to complete. Some products will probably never be made using robots, because it will not be cost-effective. An example is a custom-built automobile, put together part-by-part rather than on an assembly line.

Macintosh

In the early 1980s, Apple Computer introduced a line of computers called *Macintosh*. Since then, personal computers (PCs) in the United States have evolved along two main lines, the Macintosh and the IBM PC and its clones.

The Macintosh line became known immediately because of its versatile, innovative features. For example, the graphical user interface was first made available to consumers in the Macintosh computers.

Macintosh entered the business market in the late 1980s with desktop publishing at a reasonable price. The laser printer made desktop publishing a reality even for small companies. The main feature of the laser printer is the excellent quality of its printout, comparable to typeset.

Both Apple Computer and IBM alternately focus on various markets in attempts to gain economic advantage over the other. This has kept up the pace of progress, while minimizing cost. See also INTERNATIONAL BUSINESS MACHINES.

macroknowledge

Macroknowledge is a term used in artificial intelligence (AI). It means "knowledge in the large sense," or "knowledge about knowledge."

An example of macroknowledge is a set of definitions for different classes of living things. The two main classes are plants and animals (although some life forms share characteristics of both classes). Within the class of animals, we might focus on warm-blooded versus cold-blooded creatures.

Macroknowledge about living things might be used by a smart robot to determine, for example, whether a biped approaching it is a human or another robot—or perhaps a gorilla. See also ARTIFICIAL INTELLIGENCE, and MACHINE KNOWLEDGE.

mainframe

A *mainframe* is a large, fast computer with massive memory. Such computers can serve as a "main brain" for a set of peripheral computers and/or terminals. Mainframes are used by companies, colleges, and bureaucracies to store information in a central place, and to supply it to many different users at once.

The diagram shows the hierarchy of the computer network in a hypothetical small college. This illustration is greatly simplified; for example, the whole college has only five departments! The mainframe is the "master control center" and serves the administration and all the educational departments. This is by far the most "intelligent" computer on campus.

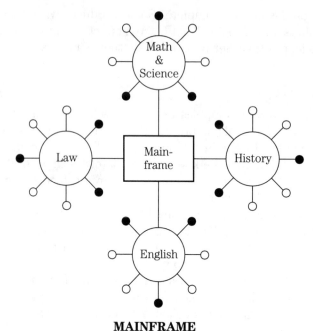

MAINFRAME
The intelligence center for an organization.

Each of the four educational departments (Mathematics/Science, History, English, and Law) has a minicomputer (large circles), connected into the mainframe. Within each department, individual employees have microcomputers (solid dots) and terminals (open small circles) that are connected into the department's minicomputer. See also LOCAL AREA NETWORK, MICROCOMPUTER, MINI-COMPUTER and TERMINAL.

manipulator

The term *manipulator* can refer to a robot arm and the gripper at the end of the arm. See ROBOT ARMS, and ROBOT GRIPPERS. The term can also refer to a remotely controlled robot. See REMOTE CONTROL SYSTEMS.

manufacturing automation protocol

In an automated factory, the different robots must all be in constant communication with the controller, or central computer. Often the robots communicate with each other, as well. This is how the operation stays synchronized, so that things proceed smoothly. *Manufacturing automation protocol (MAP)* is the set of standards used for this communication. These standards apply especially to assembly lines, such as those in automobile manufacturing plants.

magnetic disk

See DISKETTE, HARD DISK.

magnetic memory

See MEMORY.

mean time between failures

The performance of a robot, computer, or other machine can be specified in various ways. One of the most common is the *mean time between failures (MTBF)*. There are two ways in which MTBF can be defined.

For a single component

For a single component, such as an integrated circuit, the MTBF is the length of time you can expect the device to work before it fails. This is found by testing a number of components and averaging how long they keep working.

A simplified example of MTBF, calculated in hours on the basis of the performance of five identical 60-watt light bulbs, is shown in the drawing. The lifetimes are simply averaged to get the result. It is, of course, possible that five very good bulbs were chosen, or that five exceptionally bad ones were chosen for the test. Testing a large number of components, such as 1,000 or even 10,000, practically eliminates this type of coincidence.

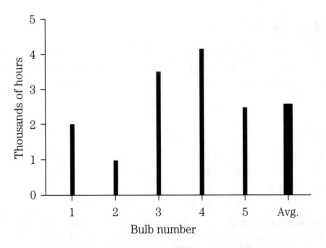

MEAN TIME BETWEEN FAILURES
Average life for five hypothetical incandescent bulbs.

For a system

In the case of a system like a robot or computer, MTBF is determined according to how often the machine breaks down. As with the component testing method, it is best to use many identical machines for this test. Conditions of the test should be as much like real life as possible.

With a large system, there are many components, any of which might malfunction. In general, the more complex the system, the shorter the MTBF will be. This does not, of course, mean that simple systems are better than sophisticated ones! Therefore, for robots and computers, the MTBF is not always a direct indicator of the quality of the machine. See also QUALITY ASSURANCE AND CONTROL.

Mechatronics

Mechatronics is a term that first appeared in Japan. It is a combination of the words *mecha*nics and elec*tronics*, and refers to the technology used in robotics. The term has the same literal meaning as "electromechanics." But "mechatronics" has a more powerful, more industrial sound. In Japan, mechatronics is synonymous with industrial and economic power.

After World War II, Japan adopted the motto "Catch and pass the West." They hoped to do this by hard work, innovation, and devotion to quality—the things that have made America prosperous. Japan exports mechatronic equipment today. Sometimes they are even called "Japan, Inc."

With the development of artificial intelligence (AI) in robotics, the emphasis on mechanics will probably dwindle, and the importance of electronics can be expected to increase. See also ARTIFICIAL INTELLIGENCE.

medical robots

Medical science has become one of the largest industries in the civilized world. There are many possible ways in which robots might be used in this industry. The most likely scenario involves a mobile robotic attendant or nurse's assistant.

As a paranurse

A robot "paranurse" (nurse's assistant) might look like the model shown in the drawing. The machine would roll along on three or four wheels in its base. Operation of the robot would be supervised from the nurses' station. This robot would work very much like proposed hobby/personal robots (see HOBBY ROBOTS, PERSONAL ROBOTS). It would deliver meals, pick up the trays, and dispense medication in pill form. A robot might also take vital signs (temperature, heart rate, blood pressure, and respiration rate).

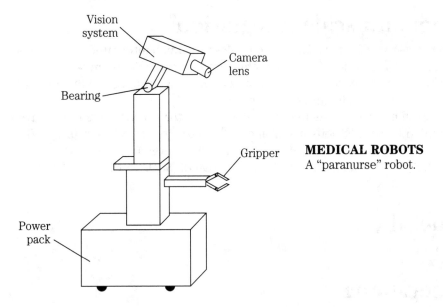

Vision
system

Camera
lens

Bearing

Gripper

MEDICAL ROBOTS
A "paranurse" robot.

Power
pack

In the body

Robotic devices can be used as artificial limbs. This has already been done, allowing amputees and paraplegics to function almost as if they had never been injured. See BIOMECHANISM, and BIOMECHATRONICS.

A more radical notion conceived by some robotics researchers involves the manufacture of robot antibodies. These microscopic creatures could be injected into a patient's bloodstream, and would go around seeking out viruses or bacteria and destroying them. The robot could be programmed to go after only a certain type of microorganism. Some researchers have suggested that organic compounds might be assembled molecule by molecule to make biological robots. See ARTIFICIAL LIFE, BIOCHIP, and BIOLOGICAL ROBOTS.

For entertainment

One of the more interesting applications of robots in medicine would be simply to entertain the patient. This is where artificial intelligence (AI) becomes important. For children, robots could play games and read stories. For adults, robots might read stories and carry on conversations. Obviously, the "smarter" the robot, the better it would be at these things.

The main argument used against medical robots is that sick people often need a human touch, which a machine can't provide. But some patients, especially children, might find robots entertaining. Anyone who has stayed in a hospital for more than a couple of days knows how boring it gets. A robot that could tell a good joke would probably brighten even the sickest patient's day. See also COMPANIONSHIP ROBOTS.

medium-scale integration

Medium-scale integration is the process by which integrated circuits, containing up to 100 individual gates per chip, are fabricated. Medium-scale integration, or MSI, allows considerable miniaturization of electronic circuits, but not to the extent of large-scale integration or LSI.

Both bipolar and metal-oxide semiconductor (MOS) technology have been adapted to MSI. Various linear and digital circuits use MSI circuitry. When extreme miniaturization is necessary, LSI is more often used. See also INTEGRATED CIRCUIT, LARGE-SCALE INTEGRATION, ULTRA-LARGE-SCALE INTEGRATION, VERY-LARGE-SCALE INTEGRATION.

megabyte

See BYTE.

megahertz

See HERTZ.

memory

Memory refers to the storage of binary data in the form of high and low levels (logic ones and zeros). There are several forms of memory.

The amount of memory storage capacity is a factor in determining how "smart" a computer can be. This is important in artificial intelligence (AI). It is also helpful in choosing the right level of computer for a robotic system. Memory is measured in bytes, kilobytes (Kb), megabytes (Mb) and gigabytes (Gb). See BYTE.

Random-access memory (RAM)

A *random-access memory (RAM)* stores data in matrices called *arrays*. The data can be addressed (selected) from anywhere in the matrix. Data is easily changed and stored back in RAM, in whole or in any part. A RAM is sometimes called a read/write memory.

An example of RAM is a word-processing computer file. This article was written in semiconductor RAM, along with all the articles for the letter M, before being stored on disk, processed, and finally printed.

There are two kinds of RAM: dynamic RAM (DRAM) and static RAM (SRAM). A DRAM employs IC transistors and capacitors; data is stored as charges on the capacitors. The charge must be replenished frequently, or it will be lost via discharge. Replenishing is done several hundred times per second. An SRAM uses a circuit called a *flip flop* to store the data. This gets rid of the need for constant replenishing of charge, but the tradeoff is that SRAM ICs require more elements to store a given amount of data.

Volatile versus nonvolatile RAM

With any RAM, the data is erased when the appliance is switched off, unless some provision is made for memory backup. The most common means of memory backup is the use of a cell or battery. Modern IC memories need so little current to store their data that a backup battery lasts as long in the circuit as it would on the shelf.

A memory that disappears when power is removed is called a *volatile memory*. If memory is retained when power is removed, it is *nonvolatile*.

Read-only memory (ROM and PROM)

By contrast to RAM, read-only memory (ROM) can be accessed, in whole or in any part, but not written over. A standard ROM is programmed at the factory. This permanent programming is known as *firmware* (see FIRMWARE). But there are also ROMs that you can program and reprogram yourself. These are called *programmable ROMs (PROMs)*.

Erasable PROM

An erasable programmable ROM (EPROM) is an IC whose memory is of the read-only type, but that can be reprogrammed by a certain procedure. It is more difficult to rewrite data in an EPROM than in a RAM; the usual process for erasure involves exposure to ultraviolet. An EPROM IC can be recognized by the presence of a transparent window with a removable cover, through which the ultraviolet is focused to erase the data. The IC must be taken from the circuit in which it is used, exposed to the ultraviolet for several minutes, and then reprogrammed via a special process.

There are EPROMs that can be erased by electrical means. Such an IC is called an *electrically erasable programmable read-only memory (EEPROM)*. These do not have to be removed from the circuit for reprogramming.

Bubble memory

Bubble memory uses magnetic fields within ICs. The scheme is especially popular in computers, because a large amount of data can be stored in a small physical volume.

A single bubble is a tiny magnetic field about 0.002 millimeters across. Logic highs and lows correspond to the existence or absence, respectively, of a bubble. The IC contains a ferromagnetic film that acts as a reprogrammable permanent magnet, on which bubbles are stored (see the illustration).

Magnetic bubbles do not disappear when power is removed from the IC. Bubbles are easily moved by electrical signals. An advantage of bubble memory is that it's a nonvolatile RAM that doesn't need a backup battery. Another asset is that data can be moved from place to place in large chunks. This process is called *block memory transfer*.

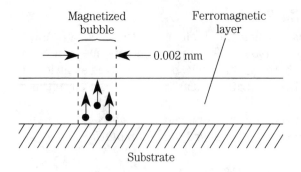

MEMORY
Principle of magnetic bubble memory.

Magnetic disks

Personal and commercial computers almost always use magnetic disks. They come in two forms: the hard disk and the floppy disk or diskette. See DISKETTE, HARD DISK.

Optical disks

Data can be stored and retrieved optically on compact disks (CDs). This medium offers several advantages over magnetic disks. See COMPACT DISK.

memory organization packets

One of the most promising aspects of artificial intelligence (AI) is its use as a tool for predicting future events, based on what has happened in the past. This process is helped by arranging the computer memory into generalizations, called *memory organization packets (MOPs)*.

Some crude examples of MOPs are the following statements:

- If the wind shifts to the east and the barometer falls, it will usually rain (or snow in the winter) within 24 hours.
- If the wind shifts to the west and the barometer rises, clearing will usually occur within a few hours.
- Light winds and a steady, high barometric pressure usually mean little weather change for at least 24 hours.
- Foul weather with a steady, low barometric pressure usually means bad weather for at least the next 24 hours.

These are broad generalizations, but they are MOPs based on the experience of meteorologists in temperate climates over the past several centuries.

In AI, the system can be programmed to find the most valid MOPs based on available data. Then it can apply these MOPs in the most effective possible way to make a forecast in a given situation.

message passing

In an artificially intelligent (AI) machine, messages are passed from one part of a computer to another, or even among two or more computers. A message might be simple data (information), or a command, or a question.

Message handling can be done in various different ways. The best method for a particular situation depends on the nature of the message. A good analogy is communication via packets. See PACKET COMMUNICATIONS, and PACKET RADIO.

When a message is passed along many times, various things can happen to change the content. You are probably familiar with this phenomenon (see the drawing). Have you ever told a friend a story, only to hear it sometime later in much different form? The same thing can happen in AI systems. Noise and distortion can alter signals, but this has largely been overcome by modern digital transmission methods. With sophisticated computers another bugaboo crops up. A super-smart machine might misinterpret a message, or even embellish it in ways the programmers did not intend and cannot predict. This is a manifestation of the fact that, when AI systems get sufficiently powerful, they can develop quirks or "neuroses." See also ARTIFICIAL INTELLIGENCE.

MESSAGE PASSING Content and meaning can be completely changed.

metal-oxide semiconductor technology

The oxides of certain metals exhibit insulating properties. So-called *metal-oxide-semiconductor (MOS)* devices have been in widespread use for years.

MOS materials include compounds like aluminum oxide and silicon dioxide. Metal-oxide semiconductor devices are noted for their low power requirements. MOS integrated circuits have high component density and high operating speed.

All MOS devices are subject to damage by the discharge of static electricity. Therefore, care must be exercised when working with MOS components. MOS integrated circuits and transistors should be stored with the leads inserted into a conducting foam, so that large potential differences cannot develop. When building, testing, and servicing electronic equipment in which MOS devices are present, the body and all test equipment should be kept at direct-current ground potential.

Metal-oxide semiconductor technology lends itself well to the fabrication of digital integrated circuits. Several MOS logic families have been developed.

Metal-oxide-semiconductor logic families are especially useful in high-density memory applications. Many microcomputer chips make use of MOS technology today. See also INTEGRATED CIRCUIT.

microchip

See INTEGRATED CIRCUIT.

microcomputer

A microcomputer is a small computer, with the central processing unit (CPU) enclosed in a single integrated-circuit package. Today, microcomputer CPU chips are available in sizes less than ¼ inch on a side. The microcomputer CPU is sometimes called a *microprocessor.*

Microcomputers vary in sophistication and memory storage capacity, depending on the intended use. Some personal microcomputers are available for less than $100. Such devices employ liquid-crystal displays and have typewriter-style keyboards. Larger microcomputers are used by more serious computer hobbyists and by small businesses. Such microcomputers typically cost from several hundred to several thousand dollars. See also COMPUTER.

Microcomputers are often used for the purpose of regulating the operation of electrical and electromechanical devices. This is known as microcomputer control. Microcomputer control makes it possible to perform complex tasks with a minimum of difficulty.

Microcomputer control is also widely used in such devices as robots, automobiles, and aircraft. For example, a microcomputer can be programmed to switch on an oven, heat the food to a prescribed temperature for a certain length of time, and then switch the oven off again. Microcomputers can be used to control automobile engines to enhance efficiency and gasoline mileage. Microcomputers can navigate and fly airplanes. It has been said that a modern jet aircraft is really a giant robot, because it can (in theory at least) complete a flight all by itself, without a single human being on board.

One of the most recent, and exciting, applications of microcomputer control is in the field of medical electronics. Microcomputers can be programmed to provide electrical impulses to control erratically functioning body organs, to move the muscles of paralyzed persons, and for various other purposes. See also BIOMECHANISM, BIOMECHATRONICS.

microknowledge

Microknowledge is detailed machine knowledge. In an artificial intelligence (AI) system, microknowledge includes logic rules, computer programs, and data stored in memory.

An example of microknowledge is the precise description of a person. If you had a personal robot in your house, its microknowledge would allow it to recognize you. It would also let the robot know if a person approaching it were someone it had never met before—say, a burglar. See also ARTIFICIAL INTELLIGENCE, and MACHINE KNOWLEDGE.

microprocessor

The central processing unit of a microcomputer is sometimes called a *microprocessor*. There is some confusion between the terms microcomputer and microprocessor, and the two are often used in place of each other.

Technically, a microcomputer consists of the microprocessor integrated circuit and perhaps one or more peripheral integrated circuits. The peripheral circuits can be integrated onto the same chip as the central processing unit, or they might be separate. The peripheral integrated circuits contain memory and programming instructions. See also COMPUTER, and MICROCOMPUTER.

Microsoft Corporation

Microsoft is a company that has become famous for its computer software. It started in the mid-1970s as a small venture run by "Grand Synthesizers" Paul Allen and Bill Gates. Since then, Microsoft has grown to take over a large share of the software market.

Some of Microsoft's software has become so common that the names are generic. Examples are DOS (Disk Operating System), Windows, and various word-processing programs.

Microsoft started out writing software for the IBM PC, the first personal computer sold by International Business Machines. Since then, Microsoft has gained a foothold in most computing markets.

In the future, Microsoft can be expected to continue to lead the way in software. For example, they might develop and sell software for "smart robots." See also DISK OPERATING SYSTEM, INTERNATIONAL BUSINESS MACHINES, and WINDOWS.

Microsoft Windows

See WINDOWS.

microwave data transmission

Microwaves are very short electromagnetic radio waves, but they have a longer wavelength than infrared energy. Microwaves travel in essentially straight lines through the atmosphere, and are not affected by the ionized layers.

Microwave frequencies are useful for short-range, high-reliability data links. Satellite communication and control is generally done at microwave frequencies. The microwave region contains a vast amount of spectrum space, and can hold many wideband signals.

Microwave radiation can cause heating of certain materials. This heating can be dangerous to human beings when the microwave radiation is intense. When working with microwave equipment, care must be exercised to avoid exposure to the rays.

A microwave repeater is a receiver/transmitter combination used for relaying signals at microwave frequencies. The signal is intercepted by a horn or dish antenna, amplified, converted to another frequency, and retransmitted (see illustration).

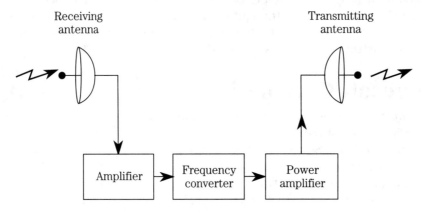

MICROWAVE DATA TRANSMISSION
Block diagram of a microwave repeater.

military robots

Whenever new technologies are developed, military experts look for ways in which they can be used in warfare. Robots have received special attention because, if machines could take the place of human beings in combat, there would be fewer human casualties. For example, androids could be used as robot soldiers, as technicians, and for many other tasks humans would otherwise have to do. Medical robots could help in hospitals for humans who are injured.

Artificial intelligence (AI) also has aroused the interest of military minds. With the aid of supercomputers, war strategy might be optimized. Computers can make decisions without being affected by emotions. In wartime, emotion is often one's own worst enemy.

The following articles concern robotic devices, computers, and AI systems that have significant potential for use in the military: ADAPTIVE SUSPENSION VEHICLE, ANDROID, AUTONOMOUS ROBOTS, BIOLOGICAL ROBOTS, COLOSSUS, COMPUTERIZED DEFENSE SYSTEM, ELECTRONIC WARFARE, FLIGHT TELEROBOTIC SERVICER, FLYING EYEBALL, INSECT ROBOTS, MEDICAL ROBOTS, POLICE ROBOTS, SECURITY ROBOTS, SENTRY ROBOTS, SUBMARINE ROBOTS, TELEOPERATION, TELEPRESENCE, and VIRUS.

million instructions per second

The speed of a computer can be specified as the number of instructions it processes in one second. The speed of modern computers is given in units of a *million instructions per second (MIPS)*. A computer at 1 MIPS can execute 1,000,000 microprocessor instructions in one second.

The MIPS specification is, in one sense, an indicator of how "powerful" a computer is. But there are other important things to consider, too, such as memory capacity and memory access time. See COMPUTER.

minicomputer

A *minicomputer* is a small digital computer. It is similar to a microcomputer, but it has larger memory capacity and higher speed of operation. The demarcation between a minicomputer and a microcomputer is not precisely defined.

Minicomputers are used by medium-size and large businesses. For small-business and personal applications, microcomputers are more often used. See also MAINFRAME, and MICROCOMPUTER.

MIPS

See MILLION INSTRUCTIONS PER SECOND.

Mittelstadt, Eric

Eric Mittelstadt was the first president of GMF Robotics, an American robotic assembly and distribution company. The abbreviation GMF stands for General Motors/Fanuc. In 1986, the Robotic Industries Association chose Mittelstadt as its president.

Mittelstadt did not get his original training in the field of robotics. Instead, he rose through the management ranks at General Motors.

Mittelstadt began a rather controversial joint venture with the Japanese. In the early 1980s, General Motors (GM) was having trouble with competition from Japanese car makers. Mittelstadt decided that robotization would help GM produce better cars. He got the robots from a Japanese firm called Fanuc. This "cooperation with the competition" worked well for both the American GM and the

modem

Japanese Fanuc. But some U.S. robot manufacturers were unhappy with the arrangement. See also FANUC, and GMF ROBOTICS.

modem

The term *modem* is a contraction of the words *mod*ulator and *dem*odulator. A modem is a two-way interfacing device that can perform either modulation or demodulation. Modems are extensively used in computer communications, for interfacing the digital signals of the computer with a transmission medium such as a radio or telephone circuit.

The illustration is a block diagram of a modem suitable for interfacing a home or business computer with an ordinary telephone. The modulator converts the digital signals from the computer into audio tones. The output is similar to that of an ordinary audio-frequency-shift keyer. The demodulator converts the incoming tones back into digital signals. The audio tones fall within the band of approximately 300 Hz to 3 kHz, so they can be efficiently transmitted over a telephone circuit or narrow-band radio transmitter.

Modems are used in computer networking. See also LOCAL AREA NETWORK, NETWORK, and WIDE AREA NETWORK.

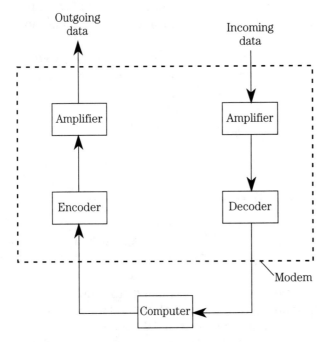

MODEM Interfaces computer with radio or telephone circuit.

modular construction

A few decades ago, electronic equipment was constructed in a much different way than it is today. Components were mounted on tie strips, and wiring was done in point-to-point fashion. This kind of wiring is still used in some high-power radio transmitters, but in recent years, *modular construction* has become the rule.

In the modular method of construction, individual circuit boards are used. Each circuit board contains the components for a certain part or parts of the system. The circuit boards are entirely removable, usually with a simple tool resembling a pair of pliers. Edge connectors facilitate easy replacement. The edge connectors are wired together for interconnection between circuit boards.

Modular construction has simplified the maintenance of complicated apparatus. In-the-field repair consists of identification, removal, and replacement of the faulty module. The faulty board is sent to a central facility, where it is fixed using sophisticated equipment. Once the board has been repaired, it can serve as a replacement module when the need arises.

modulo

The *modulus* of a number system is the number of different digital states it can attain. The scheme most often used by people is *modulo-10*, also called *decimal* or *base-10*. The digital states in this scheme are 0, 1, 2, 3, 4, 5, 6, 7, 8, and 9.

The table compares decimal values and those in modulos of 2 (binary), 4, 8 (octal), 12, and 16 (hexadecimal). In general, the larger the modulo, the smaller the numeral for a given value.

MODULO: Decimal numbers in various modulos.

Decimal	Binary	Base-4	Base-8	Base-12	Base-16
0	0	0	0	0	0
1	1	1	1	1	1
2	10	2	2	2	2
3	11	3	3	3	3
4	100	10	4	4	4
5	101	11	5	5	5
6	110	12	6	6	6
7	111	13	7	7	7
8	1000	20	10	8	8
9	1001	21	11	9	9
10	1010	22	12	A	A

Continued

Decimal	Binary	Base-4	Base-8	Base-12	Base-16
11	1011	23	13	B	B
12	1100	30	14	10	C
13	1101	31	15	11	D
14	1110	32	16	12	E
15	1111	33	17	13	F
16	10000	100	20	14	10
17	10001	101	21	15	11
18	10010	102	22	16	12
19	10011	103	23	17	13
20	10100	110	24	18	14

Computers work in the binary system (base-2), because there are only two states, represented in switching networks as on/off combinations. Sometimes computers use base-8 or base-16. Base-4 and base-12 are uncommon. See also BINARY NUMBER SYSTEM, BINARY-CODED DECIMAL, DECIMAL NUMBER SYSTEM, HEXA-DECIMAL NUMBER SYSTEM, and OCTAL NUMBER SYSTEM.

molecular computers

See DREXLER, ERIC.

molecular robots

See DREXLER, ERIC.

monitor

A *monitor* or *video display* is a visual interface between a computer and its operator. There are many types of monitors.

The simplest monitors are monochrome. They do not show color. A cathode-ray-tube (CRT) monochrome monitor is like a black-and-white TV without the tuning or volume controls. A liquid-crystal-display (LCD) monochrome monitor is lightweight and thin, and is found in less expensive notebook PCs.

Color monitors are more versatile than monochrome units. Color adds another dimension to the display. Color monitors are available in both CRT and LCD types.

The resolution of a monitor is important. This is the extent to which it can show detail; the better the resolution, the sharper the image.

With LCD monitors, speed of response is important. The older and cheaper LCD displays are sluggish. This can be an annoyance. It limits the ability of the display to show motion.

A good monitor is crucial for graphics work, with complex systems, or with interactive machines. Monitors are important in remote-control robotics and in action simulators. See also TELEOPERATION, TELEPRESENCE.

MOS

See METAL-OXIDE SEMICONDUCTOR TECHNOLOGY.

motor

A *motor* converts electrical energy into mechanical energy. Motors can operate from alternating or direct current, and can run at almost any speed. Motors range in size from the tiny devices in a wristwatch to huge, powerful machines that can pull a train at 100 miles per hour.

All motors operate by means of electromagnetic effects. Electric current flows through a set of coils, producing magnetic fields. The magnetic forces result in rotation. The greater the current in the coils, the greater the rotating force. When the motor is connected to a load, the force needed to turn the shaft increases. The more the force, the greater the current flow, and the more power is drawn from the power source.

The illustration is a cut-away diagram of a typical electric motor. One set of coils rotates with the motor shaft. This is called the *armature coil*. The other set of coils is fixed, and is called the *field coil*. The commutator reverses the current

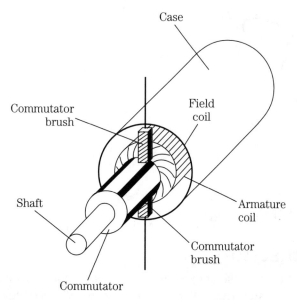

MOTOR Cutaway drawing of simple electric motor.

with each half-rotation of the motor, so that the shaft keeps turning in the same direction.

Motors are used in robots of all kinds. Many robot motors work in synchronization. See SELSYN, SERVOMECHANISM, and SERVO SYSTEM.

The electric motor operates on the same principle as an electric generator. In fact, some motors can be used as generators. See also GENERATOR.

mouse

A *mouse* is a peripheral commonly used with personal computers (PCs). By sliding the mouse around on a flat surface, a cursor or arrow is positioned on a monitor screen. A switch on the top of the unit actuates the computer to perform whatever function the cursor or arrow shows.

A mouse is user-friendly. It only takes a few minutes to get comfortable with it. The drawing shows a typical mouse. It gets its name from the way it scurries about as you push it, its wire "tail" following along. See also TRACKBALL.

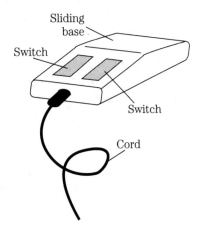

MOUSE
A user-friendly computer-control device.

MSI

See MEDIUM-SCALE INTEGRATION.

MTBF

See MEAN TIME BETWEEN FAILURES.

Mukta Institute

The *Mukta Institute* is a select group of Japanese roboticists and computer experts. It is a "think tank" that helps Japanese corporations develop new ideas in mechatronics (see MECHATRONICS). The Mukta Institute was formed in 1970.

For awhile after World War II, Japan imitated the U.S. and Europe, importing technology and business practices. But then the Japanese saw that they could invent and develop new devices and methods all their own. The Mukta philosophy, as it is called, focuses on selfless dedication to higher goals—"abandoning the ego." The Japanese have always been good at this, and it is a major reason for their extraordinary economic and technological success in recent decades.

By forgetting all about individual recognition, say the Mukta adherents, the best creative powers are tapped. The Mukta people put limitations, fears, and problems out of their minds, focusing instead on the most ambitious possible goals. They see their science, and their way of thinking, as an art form. Some of their sessions even include meditation. See also AUTOMAX.

multiplex

Multiplex is the transmission of two or more messages over the same line or channel at the same time. Multiplex transmission is done in various ways. The most common methods are frequency-division multiplex and time-division multiplex.

Frequency-division multiplex

Any channel can be broken down into subchannels. Suppose a channel is 24 kHz wide. Then it can theoretically hold eight signals 3 kHz wide (see BANDWIDTH). The frequencies of the signals must be just right, so they don't overlap. Usually there is a little extra space on either side of each subchannel to ensure that overlapping does not occur. The illustration shows six 3-kHz signals in a 24-kHz channel.

Time-division multiplex

Sometimes, data is cumbersome to transmit in parallel form. It can be converted to serial form using *time-division multiplex*. In this mode, signals are broken

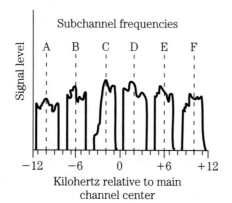

MULTIPLEX
Independent subchannels within a main channel.

into pieces "timewise," and then the pieces are sent in a rotating sequence. This slows the rate of data transfer by a factor equal to the number of signals. For example, if each of six messages is one second long sent by itself at full speed, the time-division-multiplexed signal will take six seconds. See also PARALLEL, SERIAL.

musical instrument digital interface

Computers can help composers write music. The composer plays a tune on a keyboard, and every note is entered into random-access memory (RAM). Not only is the pitch of each note memorized, but its waveform, loudness, duration, attack and decay times, and other characteristics are encoded. When the contents of the RAM are recalled and fed to an audio synthesizer, the tune is "played back." The reproduction quality is comparable to that of a compact disk (CD).

But there is a difference between the RAM and a CD! The music is digitally encoded in RAM, just like a computer program. The musician can therefore edit and debug the tune by working with the computer, literally becoming a computer programmer for awhile. The song is played on the synthesizer, edited and debugged some more, played again, and so on, until it is exactly the way the composer wants.

Some musicians swear by the musical instrument digital interface (MIDI), saying it helps them tap their creative potential in ways never before possible. It certainly gives musical composers a new approach to their craft.

Some musicians are letting computers help out with musical compositions. This is a form of artificial intelligence (AI). The problem is that computers, while seeming to have a sense of humor, don't do very well with the emotional content of music. See also ARTIFICIAL INTELLIGENCE, and COMPUTER MUSIC.

NAND gate

A *NAND gate* is a logical AND gate, followed by an inverter. The term *NAND* is a contraction of NOT AND. The illustration shows the schematic symbol for the NAND gate, along with a truth table indicating the output as a function of the input.

When both or all of the inputs of the NAND gate are high, the output is low. Otherwise, the output is high. See also AND GATE, and INVERTER.

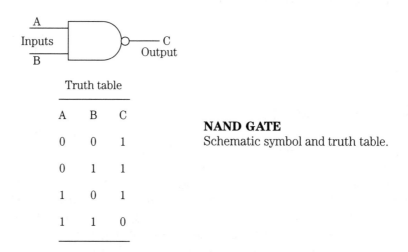

NAND GATE
Schematic symbol and truth table.

nanotechnology

Nanotechnology is the science of superminiature electronic and mechanical devices. The prefix nano-means extremely small—smaller than micro.

Nanorobots

Superminiature robots, called *nanorobots*, might find all sorts of exotic applications. Eric Drexler has suggested that such machines might serve as programmable antibodies, searching out and destroying harmful bacteria and viruses in the human body. In this way, diseases could be cured. Or the machines could repair damaged cells. See DREXLER, ERIC.

There is a dark side to nanorobotics. Anything that can be used constructively can almost always be used in some destructive way, as well. Programmable antibodies could, if they got into the hands of the wrong kinds of people, be used as biological weapons.

Nanochips

Computer researchers are always striving to get more "brain" into less space. This means superminiaturization of electronic components. This is especially important to the development of artificial intelligence (AI), especially if the intelligence is to approach that of a human being.

There is a practical limit to how many logic gates or switches can be etched onto a silicon chip of a given size. This limit depends on the precision of the manufacturing process. As methods have improved, the density of logic gates on a single chip has increased. See LARGE-SCALE INTEGRATION, VERY-LARGE-SCALE INTEGRATION, and ULTRA-LARGE-SCALE INTEGRATION. But this can only go so far.

It has been suggested that, rather than etching the logic gates into silicon to make computer chips, we might approach the problem from the opposite angle. Is it possible to build chips atom by atom? This process would result in the greatest possible number of logic gates or switches in a given volume of space. It might pave the way to building a machine that is as smart as a human—almost. See also ARTIFICIAL INTELLIGENCE, and BIOCHIP.

natural language

A *natural language* is a spoken or written language commonly used by people. Examples are English, Spanish, Russian, and Chinese.

In user-friendly artificial-intelligence (AI) systems, it is important that the machine be able to speak and/or write, and also to understand, as much natural language as possible. This is a good feature for any computer to have, in fact. The more natural language a computer can accept and generate, the more people will be able to use the machine with less time spent learning how.

Natural language will be extremely important in the future of hobby and personal robotics. If you want your robot "Susan" to get a cup of water, for example, you would like to tell it, "Susan, please get me a cup of water." You don't want to have to type a bunch of numbers, letters, and punctuation marks on a terminal, or speak in some bizarre "computerese" that is nothing like normal talk. See also SPEECH RECOGNITION, and SPEECH SYNTHESIS.

navigation

See COMPUTER MAP, GUIDANCE SYSTEMS, and VISION SYSTEMS.

nested loops

In reasoning schemes or programs, loops are often found. A loop is a logical process, or a set of program steps, that is repeated twice or more. Sometimes, loops occur inside of other loops. The loops are then said to be *nested*.

The smaller loop in a nest usually involves fewer steps per repetition than the larger loop. But the number of times the loop is followed has nothing to do with the number of steps it contains. A small, secondary loop might be repeated 1,000,000 times, while the larger loop surrounding it is repeated only 100 times.

The drawing is a flowchart that shows a simple example of nesting. Squares indicate procedural steps, such as Multiply by three and then add two (X = 3x+2). Diamonds are IF/THEN/ELSE steps, which are crucial to any loop (see IF/THEN/ELSE). The question marks inside the diamonds mean that a "question" is being asked, such as Is X greater than 587? The minus sign (–) by a diamond is like a "No" to the question, in which case the process must go back to some earlier point. The plus sign (+) is like a "Yes," telling the process to go on ahead.

Nested loops are common in computer programs, especially when there are complicated mathematical calculations. Nesting of thought loops in the human mind probably also takes place.

Reconstructing human thought processes would reveal fantastic twists and turns, more strange than the wildest science fiction. Any attempt at modeling human thought would almost certainly require the use of nested loops. This is a consideration in artificial intelligence (AI) research. See also LOOP.

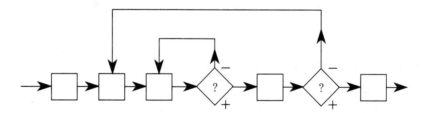

NESTED LOOPS A loop within a loop.

network

A group of computers, capable of communicating with each other or with a central mainframe, is called a *network*. There are various ways in which computers can be interconnected. The use of computers in a network is called *networking*.

Uses

Networks are commonly used by owners of personal computers (PCs). A device called a *modem* allows the PC to be hooked into the telephone lines. Sometimes radio communications is used instead of the telephone. By accessing a network, you can get all kinds of information. There are networks devoted to special interests, such as the environment, medicine, robotics, and artificial intelligence.

Businesses, educational institutions, and governments make extensive use of networks. If you want to rent an apartment, for example, the landlord might first check your credit rating through any of several different networks.

The good side

One of the big advantages of networking is that it makes a single PC enormously powerful. For a modest amount of money, anybody can have access to huge stockpiles of information. Any PC can become artificially intelligent, in the sense that it can "know" whatever data it can access.

Networks make it possible for people, companies, schools, and governments to efficiently communicate, more quickly and more cheaply than ever before possible. Bulletin board systems (BBSs) allow you to leave and retrieve messages for, and from, friends and colleagues. See also ARTIFICIAL INTELLIGENCE, BULLETIN BOARD, KNOWLEDGE, and MACHINE KNOWLEDGE.

The bad side

There are problems with networks. Sometimes, data is not updated often enough to keep it accurate. Errors can occur, and the more data you access, the greater the chance for your using a wrong piece of information without knowing it.

The most disturbing problem with networks is that the data can be used to hurt people as well as help them. It's quite easy for people to get, misuse, and sometimes even alter information that is none of their business. See also LOCAL AREA NETWORK, MODEM, PACKET COMMUNICATIONS, PACKET RADIO, and WIDE AREA NETWORK.

neurosurgery assistance robot

Robots have found a role in at least one surgical procedure. This might be surprising at first; how can a robot possibly be trusted to take human lives in its "hands"? But robotic devices are steadier, and are capable of being manipulated more precisely, than any human hand.

Drilling in the skull is one application for which robots have been used. This technique was pioneered by Dr. Yik San Kwo, an electrical engineer at Memorial Medical Center in southern California. The drilling apparatus is positioned by software derived from a computerized X-ray scan, called a *CAT scan*, of the brain. The machine was tested on nonhuman objects, such as melons, before being used on human patients. All of the robotic operation is overseen by a human surgeon.

Other possible uses for robots in surgery have been suggested. One of the more promising ideas involves using a teleoperated robot controlled by a surgeon's hands. The surgeon would watch the procedure while going through the motions, but the actual contact with the patient would be carried out entirely by the machine. Human hands always tremble a little. As a surgeon gains experience, he/she also grows older, and the trembling increases. The teleoperated robot would eliminate this problem. See also TELEOPERATION.

noise

Noise is a broadbanded alternating current or electromagnetic field. Its main characteristic is that it does not convey any information. Noise can be natural or human-made.

Noise always degrades the performance of a system. It is a major concern in any device or system in which data is sent from one place to another. The higher the noise level, the stronger a signal must be if it is to be received error-free. At a given signal power level, higher noise levels translate into more errors and reduced communications range.

The illustration is a spectral display of signals and noise, with amplitude as a function of frequency. The background noise level is called the *noise floor*. Signals A, B, C, and D are above the noise floor, and can be received. Signal D is the strongest, and will be received with the fewest errors; B is weakest, and will be subject to the most errors. Signal E is below the noise floor, and its information can't be retrieved unless a more sophisticated receiver, with a lower noise floor, is used, or else the transmitter power output is increased.

The noise level in a system can be minimized by using components that draw the least possible current. Noise can also be kept down by lowering the temperature. Some experimentation has been done at extremely cold temperatures; this is called *cryogenic* technology.

The narrower the bandwidth of the signal, in general, the better the signal-to-noise ratio will be. But this improvement takes place at the expense of data speed.

Fiberoptic systems are relatively immune to noise effects. Digital transmission methods are superior to analog methods in terms of noise immunity.

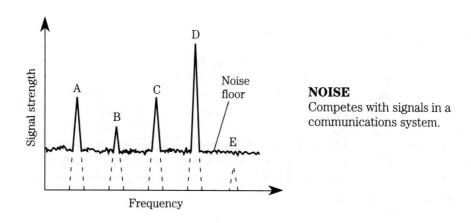

NOISE
Competes with signals in a communications system.

There is a limit to how much the noise level can be reduced; some noise will exist no matter what technology is used. See also ANALOG, BANDWIDTH, DIGITAL, and FIBEROPTIC DATA TRANSMISSION.

nonservoed system

See OPEN-LOOP SYSTEM.

NOR gate

A NOR gate is an inclusive-OR gate followed by an inverter. The expression NOR derives from NOT-OR. The NOR gate outputs are exactly reversed from those of the OR gate. That is, when both or all inputs are 0, the output is 1; otherwise, the output is 0. The illustration shows the schematic symbol for the NOR gate, along with a truth table showing the logical NOR function. See also INVERTER, and OR GATE.

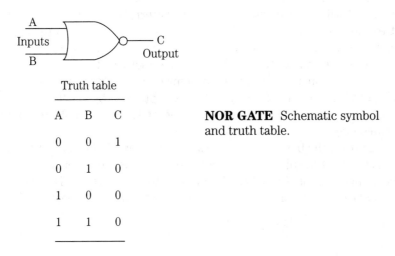

Truth table

A	B	C
0	0	1
0	1	0
1	0	0
1	1	0

NOR GATE Schematic symbol and truth table.

notebook computer

A *notebook computer*, also sometimes called a *laptop*, is a portable personal computer (PC). It gets its name from the fact that, when the fold-out monitor screen is closed up, it is about the size of a typical notebook. When the monitor screen, which uses a liquid-crystal display (LCD) or similar device, is folded out, a notebook computer looks like a little typewriter without paper (see the drawing).

In recent years, notebook computers have been improved, incorporating hard disk systems with several tens of megabytes of memory. Usually, there is one diskette drive that accepts 3.5-inch diskettes. Further improvements can be expected in coming years.

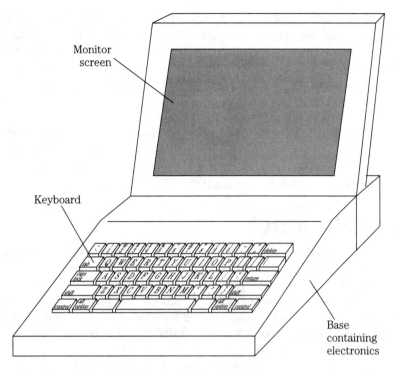

Monitor
screen

Keyboard

Base
containing
electronics

NOTEBOOK COMPUTER Looks a little like a portable typewriter.

Notebook computers operate with batteries. The battery pack must be frequently recharged. Alternatively, a power supply can be used, and the unit plugged into 117-volt utility mains.

Notebook computers can be used along with cellular telephones and modems, making it possible for you to access networks from almost anyplace. As networks become more sophisticated, you will be able to gain access to computers with artificial intelligence (AI) via notebook units. See also COMPUTER, DISKETTE, HARD DISK, MODEM, and NETWORK.

NOT gate

See INVERTER.

nuclear service robots

Robots are well suited to handling dangerous materials. This is simply because, if there is an accident, no human lives are lost. In the case of radioactive substances, robots can be used and operated by remote control, so that people will not be exposed to the radiation. The remote control is accomplished by means of teleoperation and/or telepresence (see TELEOPERATION, TELEPRESENCE).

Robots have been used for some time in the maintenance of nuclear power plants. One such machine, called ROSA, was designed and built by Westinghouse Corporation. It has been used to repair and replace heat-exchanger tubes in the boilers. The level of radiation is extremely high in this environment. It is very hard for human beings to do these tasks without endangering their health. If people spend more than a few minutes per month at such work, the accumulated radiation dose will exceed safety limits.

It is possible (but hopefully it will never be necessary) for robots to disarm nuclear warheads. If an errant missile came down without exploding the warhead, it would be better to use machines to eliminate the danger, rather than subjecting people to the risk (and mental stress) of the job.

numerical control

Numerical control (NC) is a method of programming mechanical devices, particularly machine tools. It is a primitive sort of robotic technique.

Some of the earliest NC machines were developed by George Devol (see DE-VOL, GEORGE). They used servomechanisms, and made it possible to handle dangerous substances (such as the material for nuclear reactors) from a distance. Devol also invented a programmed article transfer machine that automatically picked cartons up and loaded them onto a conveyor belt.

Numerical control was not much like the robotic devices of today. However, the idea was one of the "seeds" from which robotics has grown since Devol's inventions were tested in the early 1950s. See also SERVOMECHANISM, and SERVO SYSTEM.

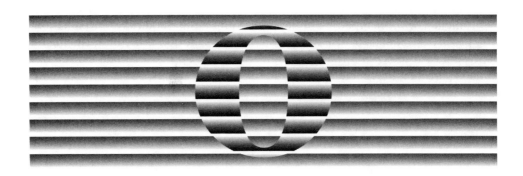

object-oriented graphics

One method by which a computer can define things is called *object-oriented graphics*. It is a powerful technique that uses analog representations, rather than digital ones.

An example of an object-oriented graphic is a circle in the Cartesian coordinate plane, defined according to its algebraic equation. Consider the circle represented by:

$$x^2 + y^2 = 1$$

This is called a *unit circle* because it has a radius of one unit, as shown in the drawing at A. The equation is easy for a computer to store in memory, and it is a precise representation.

OBJECT-ORIENTED GRAPHICS Fig. A. Unit circle in the Cartesian plane.

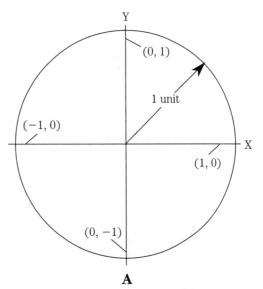

A digital, or bit-mapped, rendition of the unit circle requires approximation. The precision depends on the resolution. If the resolution is 0.1 unit, the digital representation looks like the drawing at B. Closer approximations can be obtained by making the resolution 0.01 unit or even 0.001 unit, but this will take up a large amount of computer memory.

Object-oriented graphics can be useful in robotic computer mapping. See also COMPUTER GRAPHICS, and COMPUTER MAP.

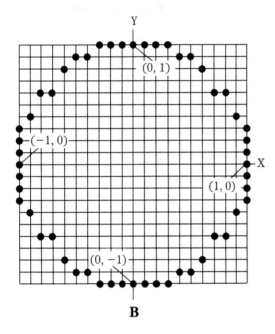

OBJECT-ORIENTED GRAPHICS
Fig. B. Unit circle in bit-mapped form with resolution 0.1 unit.

B

object-oriented language

An *object-oriented language* is a method of computer programming that uses objects to represent sets of commands. The objects are sometimes called *icons*. You can move an arrow on a monitor screen until it points to the icon you want. You can then "click" a switch and the computer will do whatever the icon represents. This is a graphical user interface.

Object-oriented languages are user-friendly. It's easy to transfer parts of an object-oriented program to some other program. But object-oriented languages take up more computer memory, and need more time to run, than conventional languages. Object-oriented languages might have advantages in communication between computers and smart robots. A robot could use its vision system to select icons on a computer screen, and thereby operate the computer. Can you imagine a smart robot at your computer, accessing a bulletin board for robot computer users? It might happen someday. See also GRAPHICAL USER INTERFACE, OBJECT RECOGNITION, and VISION SYSTEMS.

object recognition

Object recognition refers to any method that a robot uses to pick something out from among other things. An example is getting a tumbler from a cupboard. It might require that the robot choose a specific object, such as "Jane's tumbler."

Suppose you ask your personal robot to go to the kitchen and get you a tumbler full of orange juice. The first thing the robot must do is find the kitchen. Then it must locate the cupboard containing the tumblers. How will the robot pick a tumbler, and not a plate or a bowl, from the cupboard? This is a form of bin-picking problem.

One way for the robot to find a tumbler is with a vision system to identify it by shape. Another method is tactile sensing. The robot could double-check, after grabbing an object it thinks is a tumbler, whether it is cylindrical. If all the tumblers in your cupboard weigh the same, and if this weight is different from that of the plates or bowls, the robot can use weight to double-check that it has the right object.

If a particular tumbler is required, then it will be necessary to have them marked in some way, such as with bar-coding labels. See also BAR CODING, BIN-PICKING PROBLEM, TACTILE SENSORS, and VISION SYSTEMS.

octal number system

The number scheme you normally use, the decimal number system, has 10 symbols, or digits, arranged representing powers of the number we call *10*. This is the number of fingers, including thumbs, or toes on a person. (That's probably where the term *digit* comes from.)

In computers, other numbering systems can be used. The binary number system uses only zeros and ones, and is the way most digital computers "think" of numerical quantities. Another scheme, called the *octal number system*, has eight symbols (according to our way of thinking), or 2^3. These digits are the usual 0 through 7. In octal numbering, you count upwards, "1, 2, 3, 4, 5, 6, 7, 10, 11, . . ., 16, 17, 20, 21," and so on. See also BINARY-CODED DECIMAL, BINARY NUMBER SYSTEM, DECIMAL NUMBER SYSTEM, HEXADECIMAL NUMBER SYSTEM, and MODULO.

Odex

Odex is the name of a series of robots designed and built by Odetics, a company based in Anaheim, California. Odex robots have six legs, considered the ideal number by many robotics engineers. The robots are teleoperated; that is, they are run by remote control.

The six legs are arranged in a circle underneath a cylindrical main body (see the drawing). The legs can be moved and bent independently, and can extend until they are straight. Each leg is controlled by a separate microcomputer. Thus, the machine can be manipulated in a great variety of ways. It can climb stairs, squeeze through narrow doorways, and move over irregular terrain. It can

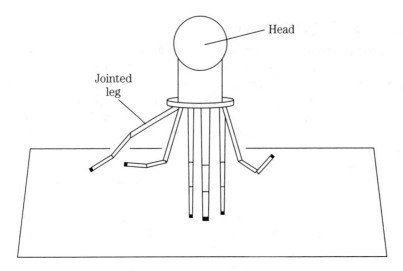

Head

Jointed
leg

ODEX Six legs move independently.

go up and down steep slopes. Odex moves slowly. Its top speed is about a meter (three feet) per second.

Odex 1 was introduced in 1983. Some people thought it was a breakthrough; others made fun of it. It was about three feet high when squatting and six feet high when standing up straight. Odex 2 was introduced a few years later, and contained some improvements over Odex 1.

Generally, Odex robots are autonomous; that is, they can work alone, independently of other robots. But they might also be used in teams, controlled by a central computer, like insect robots.

The Odex design is useful for mining and prospecting, especially in places where it is difficult, impossible or dangerous for humans to go. It has been considered for exploring other planets, particularly Mars, much of whose surface is strewn with rocks. It would be ironic to have a team of Odex robots explore Mars. Odex looks a little bit like the terrifying Martian machines that ravaged Earth in H. G. Wells' novel, *The War of the Worlds*. See also AUTONOMOUS ROBOTS, INSECT ROBOTS, ROBOTIC SPACE MISSIONS, and TELEOPERATION.

odometry

Odometry is a means of position sensing. It allows a robot to figure out where it is on the basis of two things: a starting point, and the motions it has made since leaving that point.

In one dimension

Along a straight line, or in one dimension, odometry is performed by the mileage indicator in a car. The displacement, or distance traveled, is determined by counting the number of wheel rotations, based on a certain wheel radius. (If you switch to larger or smaller size tires, you must realign the odometer in the car to get accurate readings.)

Distance traveled is equal to the integral of the speed over time. Graphically this can be represented by the area under a curve, as shown at A. The displacement changes at a rate that depends on the speed. As long as you move forward, the displacement increases. If you go backwards, the displacement decreases. Displacement can be either positive or negative with respect to the starting point.

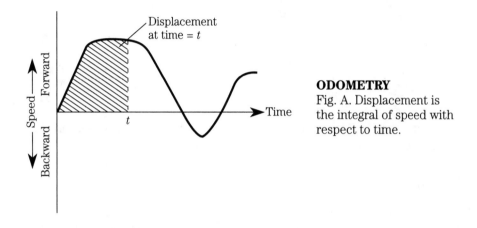

ODOMETRY
Fig. A. Displacement is the integral of speed with respect to time.

In two dimensions

In two dimensions, say in a room or over the surface of the earth, odometry is done by keeping constant track of velocity, which has components of both speed and direction.

Imagine boating in the open sea, starting from an island. You know the latitude and longitude of the island; you can measure your speed and direction constantly. You have a computer keep track of your speed and direction from moment to moment. Then, after any length of time, the computer can figure out where you are, based on past movements. It does this by integrating both components of velocity (speed and direction) simultaneously over time. Sailors know this as ded reckoning (deductive reckoning) of position.

A robot can use ded reckoning by having a microcomputer integrate its forward speed and its compass direction independently. This is called *double integration*. It is a rather sophisticated form of calculus, but a microcomputer can be programmed to do it easily. Another method is to integrate the robot speed and the wheel steering orientation independently, and then combine the data to get position as a function of time.

Illustration B shows two-dimensional odometry based on speed and compass direction. The velocity vector is constantly input to the microcomputer. The mi-

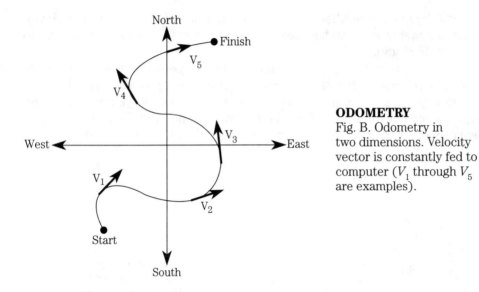

North

● Finish

V_5

V_4

West

V_3

East

V_1

V_2

● Start

South

ODOMETRY
Fig. B. Odometry in two dimensions. Velocity vector is constantly fed to computer (V_1 through V_5 are examples).

crocomputer "knows" the robot's coordinates at any moment, based on this information. See also COMPUTER MAP, and POSITION SENSING.

offloading

It has been said that a machine is something that makes people's work easier. This is especially true of robots and smart computers. These devices now do many of the tedious or dangerous jobs once done by people. As robotic technology advances, this process can be expected to continue. The replacement of human workers by robots and/or smart computers is called *robotization* or *computerization*. Robotization and computerization can take a load off our shoulders. On a personal level, the use of robots and smart computers for mundane chores is called *offloading*, a term coined by futurist Charles Lecht. (See LECHT, CHARLES.)

Some workers are afraid that robots will take their jobs and leave them without a livelihood. But robots create jobs, too. People can be trained in new lines of work that are less monotonous, or that involve less risk to their lives, than the jobs taken over by robots and computers. And the new jobs often pay better. See ECONOMIC EFFECT OF ROBOTICS.

According to Lecht, there is still much room for improvement in our lives. Even in our advanced, highly technological society, we spend time buying groceries, washing clothes, and vacuuming the floor. Bachelors, especially, often regard these things as a waste of time. But if the chores don't get done, one will go hungry, wear dirty clothes, and be tromping around in dirt all the time. Some people hire servants to do these things. But most people can't afford a butler or maid.

Robots can take over many of these chores. Lecht believes they eventually will, and that they will be affordable to almost everybody. This will free people to do more fun and creative things, Lecht says, like paint pictures, write books, compose music, or play golf! See also ECONOMIC EFFECT OF ROBOTICS, FUTURISTS, and PERSONAL ROBOTS.

Omnibot

Omnibot is the name of a series of Japanese hobby robots introduced in 1984. These robots were designed to fulfill people's fantasies about what a robot should be. The machines incorporated the latest robotic technologies, similar to the American *Hero* hobby robot.

The big difference between Omnibot and Hero was that Omnibot cost only about one-fourth the price. Both robots were designed as toys, not for practical home use. Its human-like form made Omnibot a toy to which even adults could become emotionally attached. The Japanese like to make robots in the human image. See also HERO, and HOBBY ROBOTS.

open-loop system

The term *open-loop* refers to robots without servo systems. For this reason, open-loop robots are often called *nonservoed*. This type of robot depends, for its positioning accuracy, on the alignment and precision of its parts. There is no means for correcting positioning errors. The robot operates "blind"; it can't compare its location or orientation with respect to its surroundings.

Open-loop systems can work faster than closed-loop, or servoed, robotic devices. This is because there is no feedback in an open-loop system, and therefore, no time is needed to process feedback signals and make positioning corrections.

In tasks that require extreme accuracy, open-loop systems are often not precise enough. This is especially true when a robot must make many programmed movements, one after the other. In some types of robotic systems, positioning errors can accumulate unless they are corrected from time to time. See also CLOSED-LOOP SYSTEM, SERVOMECHANISM, SERVO SYSTEM.

open systems interconnection reference model

In a network, it is important that the computers be able to communicate effectively. To ensure this, the computers' protocols must all have certain characteristics in common. A standard set of protocols is called the *open systems interconnection reference model (OSI-RM)*.

There are seven levels, or layers, in OSI-RM. These are as follows, in order from lowest to highest:

- Physical layer
- Link layer
- Network layer
- Transport layer
- Session layer
- Presentation layer
- Application layer

The physical layer moves the messages from place to place.

The link layer puts bits of data into units called *frames*, and transmits the data in this form.

The network layer finds the best routes, or paths, for the messages through the network. The first packet, or message unit, establishes the ideal route; the following packets follow the same route.

The transport layer makes sure that the source (sending station) stays in touch with the destination (receiving) station.

The session layer synchronizes the data between the source and destination stations.

The presentation layer translates between different forms of data.

The application layer interfaces messages with the application software that the system computers use. This makes it possible to program computers by remote control, and for computers to program each other.

With all seven OSI-RM layers working right, it is possible to link personal computers (PCs) together to get a "megacomputer" that is much smarter, and that has much more memory, than any of the computers alone. This is a crude form of artificial intelligence (AI). It is slow because it takes time for the signals to get from place to place, especially if the computers are far from each other. See also COMMUNICATION PROTOCOL, LINK, LOCAL AREA NETWORK, NETWORK, WIDE AREA NETWORK, PACKET COMMUNICATIONS, and PACKET RADIO.

operation envelope

See WORK ENVELOPE.

optical character recognition

Computers can translate printed matter, such as the text on this page, into digital data. The data can then be used in the same way as if you had typed it on a keyboard. This is done by means of *optical character recognition (OCR)*, also called *optical scanning*.

Reading printed matter

In OCR of printed matter, such as this page, a thin laser beam moves across the page. White paper reflects light; black ink does not. The laser beam moves in the same way as the electron beam in a television camera or picture tube. The reflected beam is modulated; that is, its intensity changes. This modulation is translated by OCR software into digital code for use by the computer. In this way, a computer can actually "read" a magazine or book.

Optical scanning is commonly used by writers, editors, and publishers to transfer printed data to computer disks. Advanced OCR software can recognize mathematical symbols and other exotic notation, as well as capital and small letters, numbers, and punctuation marks.

In robotic vision systems

Smart robots can incorporate OCR technology into their vision systems, enabling them to read labels and signs. The technology exists, for example, to build a robot with artificial intelligence (AI), along with OCR, that could get in your car and drive it anywhere. Perhaps someday, this will be commonly done. You might hand your robot a shopping list and say, "Please go get these things at the supermarket," and the robot will come back an hour later with three bags full of groceries, a couple of T shirts, and a half dozen pairs of socks.

For a robot to read something at a distance, such as a road sign, the image is observed with a television camera, rather than by reflecting a scanned laser beam off the surface. This video image is then translated by OCR software into digital data. See VISION SYSTEMS.

optical data transmission

See FIBEROPTIC DATA TRANSMISSION, LASER DATA TRANSMISSION.

optical encoder

An *optical encoder* is an electronic device that measures the extent to which a mechanical shaft has rotated. It can also measure the rate of rotation (angular speed). An optical encoder consists of a light-emitting diode (LED), a photodetector, and a chopping wheel.

The LED shines on the photodetector through the chopping wheel. The wheel has radial bands, alternately transparent and opaque (see the illustration). The wheel is attached to the shaft. As the shaft turns, the light beam is interrupted. Each interruption actuates a counting circuit. The number of pulses is a direct function of the extent to which the shaft has rotated. The frequency of pulses is a direct function of the rotational speed.

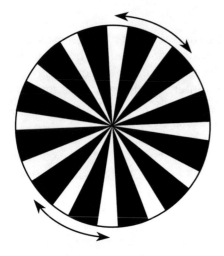

OPTICAL ENCODER
As wheel is turned by shaft at center, radial bands interrupt light beam shining through disk.

Optical encoders are used in jointed robot arms. See also ARTICULATED GEOMETRY, and ROBOT ARMS.

optic flow

See EPIPOLAR NAVIGATION.

OR gate

An *OR gate* is a digital logic gate with two or more inputs and a single output. The OR gate performs the Boolean inclusive-OR function. If all inputs are 0 (false), then the output is 0. If any of the inputs is 1 (true), then the output is 1.

The illustration shows the schematic symbol for an OR gate, along with a truth table for a two-input device. See also AND GATE, BOOLEAN ALGEBRA, INVERTER, NAND GATE, and NOR GATE.

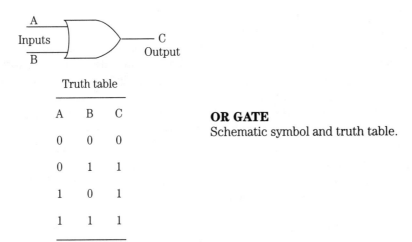

OR GATE
Schematic symbol and truth table.

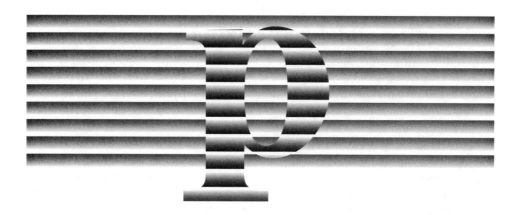

packet communications

Anybody who has a personal computer (PC) can connect it to the telephone lines using a modem. Then the PC can "talk" with other PCs. A system of computers, all communicating with each other, is a network. *Packet communications* is one way in which computers can communicate. It gets this name from the fact that the data is sent and received in units, called *packets*. Packet networks are getting larger, easier to access, and more sophisticated all the time.

Packet communications can create artificial intelligence (AI) by combining the power of many PCs. The main problem is the delay time for signals among the PCs. There is no limit to the amount of data that can be stored, but it takes time to access it from a distance. For sophisticated "thinking," the PCs must be close to each other so that the delays aren't too long. See also ARTIFICIAL INTELLIGENCE, LOCAL AREA NETWORK, MODEM, NETWORK, OPEN SYSTEMS INTERCONNECTION REFERENCE MODEL, PACKET RADIO, and WIDE AREA NETWORK.

packet radio

Packet communications can be carried on by radio, rather than over the telephone. This is commonly done by amateur radio operators ("hams"), and is known as *packet radio*. This mode can also be used by commercial networks, especially in conjunction with cellular telephone systems.

A modest ham radio station can serve as a packet station. All you need are a PC, a transceiver and a terminal node controller.

Personal-computer owners and radio hams might find a common interest in the form of "packet-radio supercomputer clubs." This will encourage PC owners to obtain amateur radio licenses. See also PACKET COMMUNICATIONS.

palletizing and depalletizing

In manufacturing processes, it is often necessary to take objects from a conveyor belt and place them on a tray designed especially to fit them. The tray is called a *pallet*, and the process of filling it is called *palletizing*. The reverse process, in which objects are removed from the pallet and placed on the conveyor, is *depalletizing*.

A complex sequence of motions is necessary to remove something from a conveyor, find an empty spot on a pallet, and place the object into the vacant spot correctly. The drawing shows a pallet with holes for eight square pegs. One hole is filled; the other seven are vacant. Suppose a robot is programmed to palletize pegs until the tray is full, and then get another tray and fill it, and so on. Its instructions might be something like this:

1. Start palletizing routine.

2. Are pegs coming along the conveyor?
   A. If not, go to step 7.
   B. If so, go to step 3.

3. Is the pallet full?
   A. If not, keep it.
   B. If so, load it on the truck, get a new pallet, and put it in place to be filled up.

4. Get the first available peg from the conveyor.

5. Place the peg in the lowest numbered empty hole in the pallet.

6. Go to step 2.

7. Await further instructions.

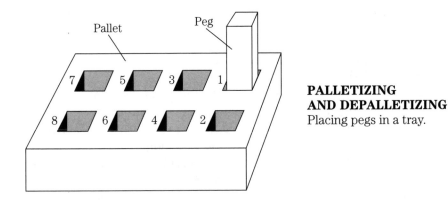

**PALLETIZING
AND DEPALLETIZING**
Placing pegs in a tray.

In practice, the robot controller would be programmed in a standard computer language, and the steps would include more detail. The robot would have to work fast enough to keep up with the conveyor, while loading pallets on the truck (or wherever they're supposed to go when they're full). The right amount of pressure would need to be used in the robot gripper. There might be several other variables that would have to be checked on and kept under control.

Depalletizing involves a similar, more or less reversed, set of steps.

paradox

In logic, a *paradox* is an unanswerable question, an unsolvable problem, or reasoning that jumps endlessly around. These "hangups" confuse artificial, as well as human, minds. An artificial intelligence (AI) system can get totally disabled by a paradox.

The simplest paradox is the sentence, "This statement is false." If you assume it's true, then it is false because that's what it says. If you assume it's false, that means it's really true, which leads to the same contradiction you got when you assumed it was true. This sort of thing can throw a computer into an infinite loop. The only escape is to assign no truth value to the sentence, and reject it as absurd.

Paradoxes sometimes arise unexpectedly. If AI is used in any kind of research, the system must recognize paradoxes, and not accept them as having truth value. A sophisticated AI research scheme must be programmed to seek logic paths that avoid paradoxes. Otherwise, you can't be certain that any of its conclusions are correct. See also ARTIFICIAL INTELLIGENCE.

parallax

Parallax is the effect that allows you to judge distances to objects and to perceive depth. Machines can use parallax for the same purpose if they have binocular robot vision. See BINOCULAR ROBOT VISION.

Parallax can be used for navigation and guidance. If you are heading toward a point, that point seems stationary while other objects seem to move away from it (see the drawing). You see this while driving down a flat, straight highway. Signs, trees, and other roadside objects appear to move radially outward from a distant point on the road. A robot vision system can use this effect to sense the direction

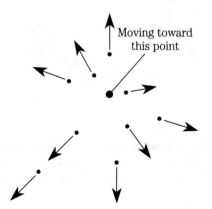

PARALLAX
Destination (large dot) seems stationary; other objects (small dots) diverge.

in which it is moving, its speed, and its location. Radio or sound beacons might be used instead of a vision system. See also BEACON, and VISION SYSTEMS.

parallel

Information can be transmitted bit by bit along one line, or it can be transmitted simultaneously along two or more lines. The latter method is called *parallel* data transfer.

In computer practice, parallel data transfer refers to the transmission of all bits in a word at the same time, over individual parallel lines, as shown in the illustration. Words are generally sent one after the other.

Parallel data transfer has the advantage of being more rapid than serial transfer. However, more lines are required, in proportion to the factor by which the speed is increased. It takes eight lines, for example, to cut the data transmission time from a serial value of 80 seconds to a parallel value of 10 seconds. See also DATA CONVERSION, and SERIAL.

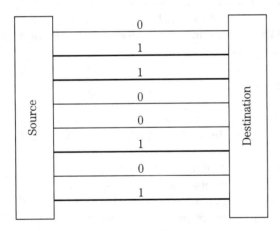

PARALLEL
Eight-bit word sent along parallel lines.

PASCAL

PASCAL is a high-level computer-programming language. It is similar to BASIC and FORTRAN in the way it is written. But it has a structure that reduces the chances for "bugs," or programming mistakes. PASCAL (with the accent on the second syllable) is used in some schools to teach computer programming.

passive transponder

A *passive transponder* is a device that allows a robot to identify an object. Bar coding is one example. Magnetic labels, such as those on credit cards, automatic-teller-machine (ATM) bank cards, and store merchandise are another example.

All passive transponders require the use of a sensor in the robot. The sensor decodes the information in the transponder. The data can be quite detailed. The transponder can be tiny, and its information can be sensed from several feet away.

Suppose a robot needs to choose a drill bit for a certain application. There might be 150 bits in a tray, each containing a magnetic label with information about its diameter, hardness, recommended operating speeds, and its position in the tray. The robot would quickly select the best bit, install it and use it. When the robot was done, the bit would be put back in its proper place. See also BAR CODING.

pattern recognition

In a robot vision system, one way to identify an object or decode data is by shape. Bar coding is a common example. Optical scanning is another. The machine recognizes combinations of shapes, and deduces their meanings using a microcomputer. See BAR CODING, and OPTICAL CHARACTER RECOGNITION.

In robots with artificial intelligence (AI), the technology of pattern recognition will become more and more important in coming years. Researchers sometimes use Bongard problems to refine pattern-recognition systems. See BONGARD PROBLEMS.

Imagine a personal robot that you keep around the house. It might identify you because of combinations of features, such as your height, hair color, eye color, voice inflections, and voice accent. Perhaps, with sophisticated enough technology, your personal robot could instantly recognize your face, just as your friends do. But this would take up a huge amount of memory, and the cost would be high. There are simpler means of identifying people.

Suppose your robot is programmed to shake hands with anyone who enters the house. It greets everyone with, "Shake hands for identification please." In this way, the robot gets the fingerprints of the person. It has a set of authorized fingerprints stored in its "brain." (These patterns are complicated, but would probably be easier to store than facial features.) If anyone refuses to shake hands, the robot might actuate a silent alarm to summon police robots. The same thing might happen if the robot didn't recognize the print of the person shaking its hand.

personal computer

A *personal computer* is a small computer with comparatively small memory capacity and low cost. Such computers are available in desktop models and notebook (also called laptop) models.

As its name implies, a personal computer (PC) is designed especially for the needs of the individual. One-person, or "Mom-and-Pop" businesses find PCs ideal for accounting and general documentation. The two major lines of PC are the IBM-compatible and the Macintosh. Most brands fall into one or the other of these categories.

The versatility of a PC is limited mainly by its clock speed and its memory. The clock speed is measured in megahertz. Memory is usually specified in megabytes.

Personal computers will become more and more important in coming decades. It will not be too long before PCs incorporate some degree of artificial intelligence (AI). A robot, or set of robots, can be controlled by a PC.

Further information can be found in the following articles: ARTIFICIAL INTELLIGENCE, BYTE, COMPUTER, HERTZ, INTERNATIONAL BUSINESS MACHINES, MACINTOSH, MEMORY, NOTEBOOK COMPUTER.

personality

Personality, in robots and smart computers, refers to characteristics that make the machine human-like. Whenever people try to make a machine in the human image, a fondness can develop. This is especially apt to happen with children, who can become "friends" even with the simplest toys. But increasingly, adults are finding that robots and computers seem to have personalities all their own.

The personality traits of a machine aren't quite the same as those of a person. As an example, consider the personal computer (PC) on which this book was written. When the temperature got above 80 degrees, and if the machine was on for an hour or two, its behavior changed. When trying to overwrite text (type new text on top of old text), the machine would insert text instead (new text would push old text out of the way, but not erase it). This was the computer's way of saying, "I'm hot." The machine still worked all right, except for that one glitch.

As PCs evolve artificial intelligence (AI), personality traits will become more sophisticated. The machine might then actually say, "I'm overheating. If you notice erratic behavior, it's probably because I'm too hot." In the event of severe overheating, it might say, "Shut me off this instant!"

Smart robots will be especially interesting as their controllers evolve AI. Some AI researchers suggest that smart machines might develop quirks, like human neuroses. There is no way to know for sure whether this will happen, unless or until it does. If a computer ever does "lose it," it might be more bizarre than anything people have ever imagined—even science-fiction authors, who have written books and movies about it. See ANTHROPOMORPHISM, ARTIFICIAL INTELLIGENCE, and CLARKE, ARTHUR C.; see also COLOSSUS, and COMPUTER EMOTION.

personal robots

For centuries, people have imagined having personal robots. Such a machine could be a sort of slave, not asking for pay (except maintenance expenses). Until the explosion of electronic technology, however, people's attempts at building robots resulted in clumsy, stupid, and often dangerous masses of metal that did little or nothing of any real use.

The sky is the limit

Today, we can buy a computer with a notebook-sized case for a reasonable price. This is largely because of integrated-circuit (IC) technology. Most scientists agree that we haven't come near the end of the electronic revolution, and that "the sky is the limit." We can expect that good, useful personal robots will someday be available. The big trick will be getting people to buy them.

Personal robots could do all kinds of mundane chores around your house. Such robots are sometimes called household robots. Personal robots might also be used in the office; these are often called service robots.

To be effective, personal robots will have to incorporate features such as speech recognition, speech synthesis, and vision systems.

Household robot duties

Examples of household-robot chores are:

- Bodyguarding
- Car washing
- Cleaning (general)
- Companionship
- Cooking
- Dishwashing
- Fire protection
- Floor cleaning
- Grocery shopping
- Intrusion detection
- Laundry
- Lawn mowing
- Maintenance
- Meal serving
- Playmate (for kids)
- Snow removal
- Toilet cleaning
- Window washing

Office robot duties

Around the office, a robot might do things like:

- Bookkeeping
- Cleaning (general)
- Coffee preparation and serving
- Delivery
- Dictation
- Equipment maintenance
- Filing documents
- Fire protection
- Floor cleaning
- Greeting visitors
- Intrusion detection
- Meal preparation
- Photocopying
- Telephone answering
- Toilet cleaning
- Typing
- Window washing

You can probably think of a lot more possible uses for personal robots. How will they look while they do these jobs? Use your imagination!

Practical robots versus toys

Some personal robots have been designed and sold. But they haven't been sophisticated enough to be of any practical benefit. They were actually hobby robots, or adult toys. A good household robot, capable of doing even a few of the above chores efficiently and reliably, would probably cost more than $100,000. In the future, as technology improves and gets less expensive, the cost (in terms of a person's real earning power) will go down.

Simpler machines make good toys for children. Interestingly, if a robot is designed and intended as a toy, it sells better than if it is advertised as a practical machine. This is probably because modern consumers are smarter than they were a few decades ago. It's hard to fool people nowadays. People don't like to spend their money on things that might not do what they claim.

Questions and concerns

Some researchers think that robot "slaves" might be abused, as people direct their hostility at machines rather than at other people. Can you imagine a society in which robots are cursed, kicked, and whipped, as slaves once were? In my opinion, this won't happen. If you break a $100,000 machine, you must pay to have it fixed, or else get along without it. People don't beat up on laundry machines, snow blowers, or personal computers. They won't vent their rage on robots, either.

Robots must be safe to be around, and not pose any hazard to their owners, especially children. This can be ensured with good design. All robots should follow Asimov's Three Laws of Robotics.

Suppose a practical personal robot were available for $30,000. Would many people buy it? It's hard to say. As boring as some of the above mentioned tasks might seem, plenty of people enjoy doing them. Lawn mowing and snow removal can be great exercise. Lots of people love to cook. Some people will never entrust a robot to do things right, no matter how efficient and sophisticated the machines get. And while $30,000 might not seem like a lot of money to some people, others might prefer to save it, or spend it on other things.

Related articles in this book include: AMUSEMENT ROBOTS, ANTHROPOMORPHISM, APPRENTICE ROBOT, ARTIFICIAL INTELLIGENCE, ASIMOV'S THREE LAWS OF ROBOTICS, AUTOMATED HOME, CHECKERS-PLAYING MACHINE, CHESS-PLAYING MACHINE, EDUCATIONAL ROBOTS, ETHICAL SLAVE, FANTASY ROBOTS, FIRE-PROTECTION ROBOTS, FOOD-SERVICE ROBOTS, GARDENING AND GROUNDSKEEPING ROBOTS, HOBBY ROBOTS, MEDICAL ROBOTS, OFFLOADING, ROBOT GENERATIONS, SECURITY ROBOTS, SPEECH RECOGNITION, SPEECH SYNTHESIS, TIN TOY ROBOTS, and VISION SYSTEMS.

phoneme

A *phoneme* is an individual sound or syllable you make when you talk. Examples are "Ssss," "Oooo," and "Ffff."

Have you ever seen your voice displayed on an oscilloscope screen? The hardware is simple: a microphone, audio amplifier, and oscilloscope (see drawing). When you speak, a jumble dances across the screen. Phonemes, like "Ssss" or "Oooo" or "Ffff" look simpler than ordinary speech. But any waveform, no matter how complex, can be generated by electronic circuits. A speech synthesizer can, in theory, be made to sound exactly like anyone's voice, saying anything. The

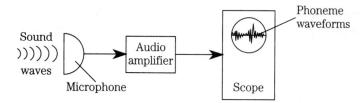

PHONEME Each sound has a unique waveform.

output of the machine then has precisely the same waveform, as seen on an oscilloscope, as your voice. See also SPEECH RECOGNITION, and SPEECH SYNTHESIS.

photoelectric proximity sensor

Reflected light can provide a way for a robot to tell if it is approaching something. A *photoelectric proximity sensor* uses a light-beam generator, a photodetector, a special amplifier, and a microprocessor. The drawing shows the principle of this device.

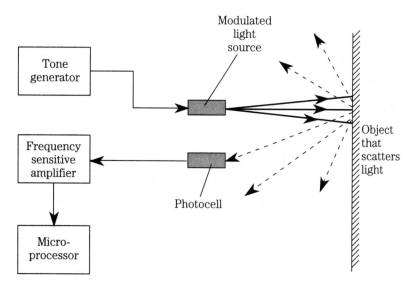

PHOTOELECTRIC PROXIMITY SENSOR Detects scattered light.

The light beam reflects from the object and is picked up by the photocell. The light beam is modulated at a certain frequency, say 1,000 Hz, and the detector has a frequency-sensitive amplifier that responds only to light modulated at that frequency. This prevents false imaging that might otherwise be caused by lamps or sunlight. If the robot is approaching an object, its microprocessor senses that the reflected beam is getting stronger. The robot can then steer clear of the object.

275

This method of proximity sensing won't work for black objects, or for things like windows or mirrors approached at a sharp angle. These would fool this system, because the light beam wouldn't be reflected back toward the photodetector. See also ELECTRIC EYE, and PROXIMITY SENSING.

picture signal

See COMPOSITE VIDEO SIGNAL.

PILOT

PILOT is an acronym that stands for *programmed inquiry learning or teaching*. It is a high-level computer language, intended for use with computer-assisted instruction (CAI). The language is simple and straightforward.

In recent years, the graphical user interface has begun to replace PILOT and other conventional programming languages. See also COMPUTER-ASSISTED INSTRUCTION, and GRAPHICAL USER INTERFACE.

pitch

Pitch is one of three types of motion that a robotic end effector can make.

Extend your arm out straight, and point at something with your index finger. Then move your wrist so that your index finger points up and down along a vertical line. This motion is pitch in your wrist. The drawing shows a robotic end effector making this motion. See also END EFFECTOR, ROLL, and YAW.

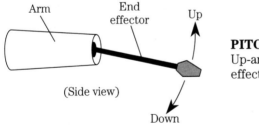

Arm

End effector

Up

(Side view)

Down

PITCH
Up-and-down movement of end effector.

pixel

Pixel is an acronym that means "picture element." A pixel is the smallest bit of information in a video image.

If you look through a magnifying glass close up at a television screen or computer monitor, you can see thousands of little dots. These are the pixels of the television or monitor screen itself. (Caution: Wear ultraviolet-protective sunglasses if you try this experiment.)

In a composite video signal, a pixel is the smallest unit that conveys information. These pixels might be, but often aren't, the same as the pixels on the screen itself.

The size of a pixel is important in robotic vision systems because this determines how much detail the robot can see. The smaller the pixels, the better the detail. Clarity of a video image is sometimes called the *resolution*. High-resolution robot vision requires better cameras, a greater signal bandwidth, and more memory than low-resolution robot vision. High-resolution systems also cost more than low-resolution ones. See also COMPOSITE VIDEO SIGNAL, RESOLUTION, and VISION SYSTEMS.

PLANNER

PLANNER is a high-level computer language sometimes used in artificial intelligence (AI). The language is "goal-oriented," in that it seeks a solution to a problem using various different schemes, if necessary.

If a computer is using PLANNER and it fails to solve a problem on the first try, it will backtrack, and try again using some other strategy. In this way, PLANNER mimics human intelligence: "If at first you don't succeed, try, try again."

Some AI researchers see backtracking as a big disadvantage. They ask, "How can you reach a goal by moving away from it?" You've probably had setbacks on the way to some goal you sought. If a strategy is wrong, it's better to acknowledge the mistake and go back, rather than waste time at a dead end. Some problems seem to require a trial-and-error approach. Mathematical proofs are often that way; so are some engineering design problems.

Ultimately, PLANNER, or some language like it, will most likely prove effective at some types of problems and not so good at others—just like things are in real life. See also ARTIFICIAL INTELLIGENCE, and PROBLEM REDUCTION.

point-to-point motion

A robot arm can move continuously, and stop at any point along its path. Or it might only be able to stop in specific places. When the device can attain only certain positions, it is using *point-to-point motion.* This requires that the coordinates of the stopping points be stored in memory.

In two dimensions, Cartesian or polar coordinates can be used. In three dimensions, the most common schemes are Cartesian, cylindrical, and spherical coordinates. The drawing shows Cartesian-plane point-to-point motion. The coordinate points are memorized in sets of ordered pairs (x, y).

A computer memory can store many sets of coordinates, so that there are hundreds or even thousands of possible stopping points. Then the motion is practically continuous. See also CARTESIAN COORDINATE GEOMETRY, CONTINUOUS-PATH MOTION, CYLINDRICAL COORDINATE GEOMETRY, POLAR COORDINATE GEOMETRY, and SPHERICAL COORDINATE GEOMETRY.

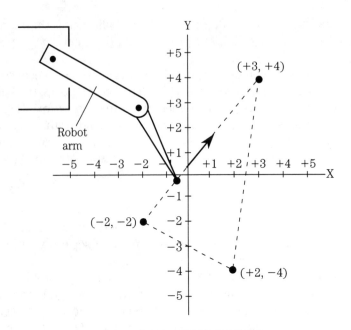

POINT-TO-POINT MOTION
Robot arm stops only at certain points.

polar coordinate geometry

Industrial robot arms can move in various different ways, depending on their intended use. One two-dimensional system is *polar coordinate geometry*. This term comes from the polar graph for mathematical functions. The drawings show standard polar coordinate systems.

The independent variable is the angle, in degrees or radians, relative to a zero line. There are two common methods of specifying the angle. If the zero line runs toward the right, or "due east," then the angle is measured counterclockwise from it (see A). If the zero line runs toward the top of the page, or "due north," then the angle is measured clockwise (as shown at B). The method at A is common for mathematical displays and robot systems. The scheme at B is used when the angle is a compass bearing or azimuth.

The dependent variable is the radius, or distance from the center of the graph. The units are usually all the same size, such as millimeters, centimeters, or inches. But sometimes, a logarithmic scale is used. This is often done when plotting transducer directional patterns. See also CARTESIAN COORDINATE GEOMETRY, CYLINDRICAL COORDINATE GEOMETRY, ROBOT ARMS, and SPHERICAL COORDINATE GEOMETRY.

police robots

Can you imagine metal-and-silicon police officers, eight feet tall, capable of lifting whole cars with one arm and, at the same time shooting 100 bullets per sec-

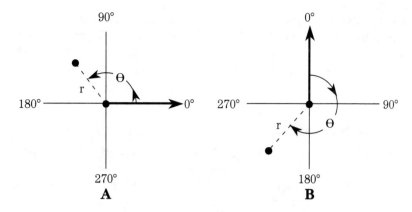

POLAR COORDINATE GEOMETRY
Zero line "east" (A) and zero line "north" (B).

ond from an end effector on the other arm? These types of police robots have been depicted in fiction. The technology to build such a monster exists right now. But a real robot cop would probably be of more modest proportions.

Advantages of robot cops

Police officers are often exposed to danger. If a remote-controlled robot could be used for any of the dangerous jobs cops are faced with, lives could be saved.

A robot police officer might work something like a robot soldier. It could be teleoperated, with a human operator stationed in a central location, not exposed to risk (see TELEOPERATION).

A mechanical cop could certainly be made far stronger than any human being. Not only this, but a machine has no fear of death, and would be able to take risks that people might back away from.

Problems with robot cops

Humans can crawl, jump, run, and climb in ways that no machine can match. A clever crook could probably elude almost any robot. The old adage applies here: "Build a better mousetrap, and you will get smarter mice." Agility will be a key concern if a robot cop is ever to apprehend anybody.

Another problem is that robot police officers might not prove cost-effective. A human operator must be paid to sit and teleoperate a robot. The robot itself would cost money to build and maintain, and if necessary, to repair or replace.

Police of the future?

People who roboticize a police force will have to weigh the saving of life against increased costs. This is not the sort of decision-making that any bureaucrat enjoys. Perhaps the cost of robotic technology will decline, and the quality will increase, until someday, part or most of our metropolitan police forces can be roboticized at a reasonable cost.

Someday, if you run a stop sign and get pulled over, don't be surprised if you find yourself making excuses to a machine. But if you think you can insult a police robot and get away with it, remember that there's a real person at the controls. See also MILITARY ROBOTS, SENTRY ROBOTS and SECURITY ROBOTS.

position sensing

Robot *position sensing* falls into either of two categories. In the larger sense, the robot can locate itself. This is important in guidance and navigation. In the smaller sense, a part of a robot can move to a spot within its work envelope, using devices that tell it exactly where it is.

Specific article titles in this book, dealing with position sensing, include: BEACON, CARTESIAN COORDINATE GEOMETRY, COMPUTER MAP, CYLINDRICAL COORDINATE GEOMETRY, DIRECTION FINDING, DIRECTION RESOLUTION, DISPLACEMENT TRANSDUCER, DISTANCE MEASUREMENT, DISTANCE RESOLUTION, EDGE DETECTION, EPIPOLAR NAVIGATION, EYE-IN-HAND SYSTEM, GUIDANCE SYSTEMS, LOG POLAR NAVIGATION, ODOMETRY, PARALLAX, PHOTOELECTRIC PROXIMITY SENSOR, POLAR COORDINATE GEOMETRY, PROXIMITY SENSING, SERVOMECHANISM, SONAR, SPHERICAL COORDINATE GEOMETRY, and VISION SYSTEMS.

power supply

A *power supply* is a circuit that provides a robot or computer with the right voltages. For robots, some motors use alternating current (ac), while others need direct current (dc). Computers use dc exclusively.

General design

An ac power supply consists of a transformer only (Fig. A). A dc power supply consists of several stages, always in the same order (Fig. B).

First, the utility ac goes through a transformer that steps the voltage either down or up, depending on the needs of the device. For an ac motor, this transformer is the only component. It must be able to handle the current that the robot motor needs.

In a dc supply, the ac must be rectified. This is done by one or more diodes. Then the voltage is filtered, or smoothed out, so that it becomes continuous with either positive or negative polarity. Finally, the dc voltage is regulated. Computers are rather finicky, insisting on the right voltage all the time.

Power supplies that provide more than a few volts must have features that protect the user from getting a shock. All power supplies need fuses and/or circuit breakers to minimize the fire hazard in case the equipment shorts out.

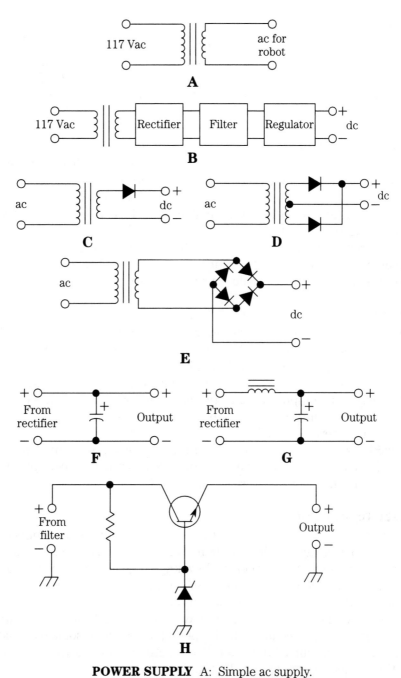

POWER SUPPLY A: Simple ac supply.
B: Block diagram of dc supply. C: Half-wave rectifier.
D: Full-wave, center-tap rectifier. E: Full-wave bridge. F:
Capacitor filter. G: Capacitor/choke filter. H: Voltage regulator.

Half-wave rectifier

The simplest rectifier circuit, called a *half-wave rectifier*, uses one diode to "chop off" half of the ac input cycle. A half-wave circuit is shown in Fig. C.

Half-wave rectification is useful in supplies that don't have to deliver much current, or that don't need to be especially well regulated. For high-current equipment, *full-wave* supplies are preferred.

Full-wave rectifier

A much better scheme for changing ac to dc is to use both halves of the ac cycle. Suppose you want to convert an ac wave to dc with positive polarity. Then you can allow the positive half of the ac cycle to pass unchanged, and flip the negative portion of the wave upside-down, making it positive instead. This is the principle behind full-wave rectification.

There are two basic circuits for the full-wave supply. One version, shown in Fig. D, uses a center tap in the transformer, and needs two diodes. The other circuit, shown in Fig. E, uses four diodes, and is known as a *bridge*.

Filtering

Electronic equipment doesn't like the pulsating dc that comes straight from a rectifier. The ripple in the waveform must be smoothed out, so that pure, battery-like dc is supplied. The filter does this.

The simplest filter is one or more large-value capacitors, connected in parallel with the rectifier output. Electrolytic capacitors are almost always used. They are polarized; they must be hooked up in the right direction. An example of this kind of filter is shown in Fig. F.

Sometimes a large-value coil, called a filter choke, is connected in series along with the capacitor in parallel. This provides a smoother dc output than the capacitor by itself. An example is shown in Fig. G.

Voltage regulation

If a Zener diode is connected across the output of a power supply, the diode will limit the output voltage of the supply by "brute force" as long as it has a high enough power rating. But a Zener-diode voltage regulator is not very efficient if the load is heavy. When a supply must deliver high current, a power transistor is used along with the Zener diode to obtain regulation (Fig. H).

In recent years, voltage regulators have become available in integrated-circuit (IC) form. You just connect the IC, perhaps along with some external components, at the output of the filter. This method provides the best possible regulation at low and moderate voltages.

Transient suppression

The ac on the utility line often has "spikes," known as *transients*, that can reach hundreds of volts.

Transients are caused by sudden changes in the load in a utility circuit. Lightning can also produce them. Unless they are suppressed, they can befuddle the operation of sensitive equipment like computers.

The simplest way to get rid of most transients is to use a commercially made transient suppressor. These are sometimes called *surge suppressors*. You insert them between the wall outlet and the input to the power supply. Then you just plug the power supply into one of the several outlets in the suppressor.

In the event of a thunderstorm locally, the best way to protect equipment is to unplug it. A nearby lightning strike can cause voltages high enough to spark right across a transient suppressor, damaging the power supply and possibly the robot or computer itself!

Fuses and breakers

If a fuse blows, it must be replaced with another of the same rating. If the replacement fuse is rated too low in current, it will probably blow out right away, or soon after it has been installed. If the replacement fuse is rated too high in current, it might not protect the equipment.

Circuit breakers do the same thing as fuses, except that a breaker can be reset by turning off the power supply, waiting a moment, and then pressing a button or flipping a switch. Some breakers reset automatically when the equipment has been shut off for a certain length of time.

Personal safety

Power supplies can be dangerous. This is especially true of high-voltage circuits, but anything over 12 volts should be treated as potentially lethal.

A power supply is not necessarily safe after it has been switched off. Filter capacitors hold a charge for a long time. In high-voltage supplies of good design, bleeder resistors are connected across each filter capacitor, so that the capacitors will discharge in a few minutes after the supply is turned off. But don't bet your life on components that might not be there, and that can and do sometimes fail.

Most manufacturers supply safety instructions and warnings with equipment that uses high voltages. But you should never assume something is harmless just because dangers aren't mentioned in the instructions. If you have any doubt about your ability to repair a power supply, *don't!*

prescription-filling robot

Much of the work done by a pharmacist (a person who fills prescriptions) is repetitive. It can get mind-numbing. A bored worker is likely to make errors. This is the sort of situation to which robots lend themselves.

In a small drugstore, it would not be cost-effective to have a robot fill prescriptions. Small shops couldn't afford robots, and the amount of work wouldn't justify it. But in a large hospital or major metropolitan discount pharmacy, the situation is different.

A robot can be told to fill a container with two dozen 500-milligram capsules of penicillin. The specific type of penicillin would be programmed in, along with the customer's preference for brand-name or generic drugs. In a pharmacy with a robot, the pharmacist-in-charge (always a human!) might have 750 prescriptions in computer memory, to be filled within 24 hours. In the evening, the list could be input to the robot. The robot would fill all 750 prescriptions during the night. In the morning, the pharmacist-in-charge would check the containers to be sure the prescriptions were filled correctly. This would take time, but not as much time as if the human pharmacist did all the work.

One concern about robotization of a pharmacy is that errors might occur. A customer might get the wrong medication, the wrong size pill, or the wrong quantity of pills. But a well-designed, well-programmed robot would probably do the job more accurately, and faster, than a human pharmacist. Prescriptions would always be double-checked by a human. Thus, there would be fewer errors than in the old days, when one overworked person did everything. See also MEDICAL ROBOTS.

pressure sensing

Touch is a useful sense for a robot. In its simplest form, this sense lets a robot detect pressure. More sophisticated pressure-sensing devices can measure force, and can tell exactly where the force is being applied.

A pressure-sensitive transducer tells a robot when it collides with something. One such device is shown in the drawing. Two metal plates are separated by a layer of nonconductive foam. This forms a capacitor. The capacitor is combined with a coil, or inductor. The coil/capacitor circuit sets the frequency of an oscillator. The transducer is coated with plastic to keep the metal from shorting out to anything. If an object hits the sensor, the plate spacing changes. This changes the capacitance, and therefore also the oscillator frequency. When the object moves away from the transducer, the foam springs back, and the plates return to their original spacing.

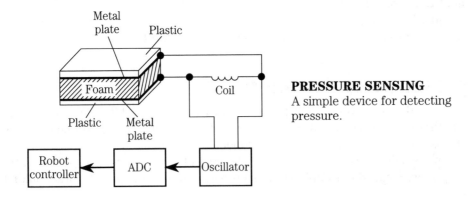

PRESSURE SENSING
A simple device for detecting pressure.

This device can be fooled by metallic objects. If a good electrical conductor comes near the transducer, the capacitance might change even if contact is not made.

Conductive foam can be placed between the plates, so that the resistance changes with pressure. A direct current is passed through the device. If something bumps the transducer, the current increases because the resistance drops. This transducer will not react to nearby conductive objects unless force is actually applied.

The output of a pressure sensor can be converted to digital data using an analog-to-digital converter (ADC). This signal can be used by the robot's controller. Pressure on a transducer in the front of a robot might cause the machine to back up; pressure on the right side might make the machine turn left. See also BACK PRESSURE SENSOR, ELASTOMER, PROXIMITY SENSING, and TACTILE SENSOR.

printed circuit

A *printed circuit* is a wiring arrangement made of foil on a circuit board. Printed circuits can be mass-produced inexpensively and efficiently. They are compact and reliable. Most electronic devices today are built using printed-circuit technology.

Printed circuits are fabricated by first drawing an etching pattern (see illustration). This is photographed and reproduced on clear plastic. The plastic is placed over a copper-coated glass-epoxy or phenolic board, and the assembly undergoes a photochemical process.

The use of printed circuits has vastly enhanced the ease with which electronic equipment can be serviced. Printed circuits allow modular construction, so that an entire board can be replaced in the field and repaired in a fully equipped laboratory. See also MODULAR CONSTRUCTION.

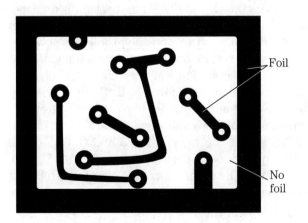

PRINTED CIRCUIT
Simple etching pattern for circuit board.

printer

A *printer* is any device that produces a permanent copy of alphanumeric data. Printers are used with computers, terminal units, word processors, and many other electronic systems.

Inexpensive printers are usually daisy-wheel type or dot-matrix type. A dot-matrix printer can produce hard copy at several hundred characters per second. The daisy-wheel printer is somewhat slower, but produces letter-quality copy.

More modern printers include laser and ink-jet printers. They are useful for graphics as well as printed matter.

problem reduction

Complex problems are often made much easier by breaking them down into small steps. This process is called *problem reduction*. It is an important part of research in artificial intelligence (AI).

Two common forms

You've used problem reduction in everyday life without being aware of it. You've probably used it consciously, too, especially in high-school geometry. Proving mathematical theorems is a good exercise in problem reduction. Another way to develop this skill is to write computer programs in a high-level language such as BASIC or FORTRAN.

When breaking a big, difficult problem down into small, easy steps, one can lose sight of the "big picture." Keeping a mental image of the goal, the progress being made, and the obstacles to come is a skill that gets better and better with practice. You can't just sit down and start to prove profound theorems in mathematics, until you've learned to prove some simple things first.

Theorem-proving machine (TPM)

Suppose you build a "theorem-proving machine" (TPM) and assign it an especially tough proof. The drawing is a rendition of this hypothetical problem. In this case, a proof exists. The desired result is true, and the problem is solvable. Often, mathematicians don't know, when setting out to prove something, if what they want to prove is true. That is, they don't know if they can ever solve the problem. In this example, there are four starting paths: A, B, C, and D. Two of these, B and C, can lead to the desired result; the other two cannot. But even if TPM starts out along B or C, there are many possible dead ends.

In this example, there is a "crossover" between paths B and C. One of the sidetracks from path B can indirectly lead to the desired result, via path C. Also, a sidetrack from path C can take TPM to the proof by moving over to path B. But these crossings-over can also lead TPM back toward the starting point, and possibly even to dead ends on the way back there!

PROBLEM REDUCTION
Breaking a big problem into little steps.

Dead ends

When TPM runs into a dead end, it can stop, turn around, and backtrack. But there's a catch. How does TPM know that it has come to a dead end? It might keep trying and trying to break through the barrier. As you know from real-life experience, sometimes persistence can get you over a tough hurdle, and in other cases the nut just doesn't want to crack. After you try for a long time to break out of a dead end, you'll give up from exasperation and turn back. But at what point should TPM get "sick and tired" of trial-and-error at a dead end?

The only answer to this question lies in the ability of TPM to learn from experience. This is one of the most advanced concepts in AI. Practice makes perfect (almost) for humans, and the same holds for truly smart machines.

A real, 100-percent reliable TPM can never be constructed. This is because there are statements in any logical system that can't be proven true or false, no matter how long one tries. There are some things that nobody—and no machine—can ever know for sure. This was proven by logician Kurt Godel, and is called the *Incompleteness Theorem*. See also ARTIFICIAL INTELLIGENCE, INCOMPLETENESS THEOREM, and PLANNER.

program

A *program* is a set of data bits, comprising instructions to perform a task. In computer practice, a program tells the machine what to do. It can help the user communicate with the machine. Programs are always written in languages. There are various languages for different kinds of programs. See also ASSEMBLER PROGRAM, ASSEMBLY LANGUAGE, COMPILER, COMPUTER, HIGH-LEVEL LANGUAGE, and INTERPRETER.

programmed article transfer machine

See NUMERICAL CONTROL.

Programmable Universal Machine for Assembly (PUMA)

One of the earliest robots in American auto manufacturing was the *Programmable Universal Machine for Assembly (PUMA)*. It was first used by General Motors.

In the 1970s, American car-building companies began suffering because of competition from Japan. General Motors (GM) decided that this was largely because Japan was using robot technology in their assembly lines, while the United States lagged behind. Thus GM began to inquire of various robot manufacturers whether they would consider developing a PUMA for use in automotive assembly lines.

A robot-manufacturing firm called Unimation, Inc., founded in 1961 by the farsighted roboticist Joseph Engelberger, already had a robot arm that fulfilled most of GM's requirements for the PUMA. (See ENGELBERGER, JOSEPH.) Unimation won the contract. There was some confusion as to who owned the rights to the acronym PUMA, and even as to what, exactly, the PUMA was. (Some people thought the robot arm was the whole machine, but other engineers included the parts-orientation devices and conveyer belts too.)

Other major auto manufacturers followed GM's lead in robotics. Eventually (and ironically), GM entered a joint venture with Fanuc, a leading Japanese robot manufacturer. This angered some American robot manufacturing companies. See also ASSEMBLY LINE, ASSEMBLY ROBOTS, and FANUC.

programming

See COMPUTER PROGRAMMING.

PROLOG

PROLOG is an acronym that stands for *pro*gramming in *log*ic. It is a high-level computer language that is of value to researchers in artificial intelligence (AI).

PROLOG is somewhat like LISP, another language often used with AI. In PROLOG, the programmer inputs knowledge along with a set of rules for the computer to follow in working with that knowledge. The program, in a sense, derives theorems from its knowledge base, by means of the rules.

One of the drawbacks of PROLOG is that it takes time to write complex programs. The number of facts in the knowledge base can become staggering. See also ARTIFICIAL INTELLIGENCE, COMPUTER, EXPERT SYSTEMS, HEURISTIC KNOWLEDGE, LISP, and MACHINE KNOWLEDGE.

proprioceptor

If you close your eyes and move your arms around, you can always tell where your arms are. You know if your arms are raised or whether they're hanging at your sides.

You know how much your elbows are bent, the way your wrists are turned, and whether your hands are open or closed. You know which of your fingers are bent and which ones are straight. You know these things because of the nerves in your arms, and the ability of your brain to interpret the signals the nerves send out.

There are advantages in a robot having some of this same sense, so that it "knows" its positioning relative to itself. A proprioceptor allows this. See also BACK PRESSURE SENSOR, ENCODING, OPTICAL ENCODER, SERVO ROBOTS, SERVOMECHANISM, SERVO SYSTEM, and TEACH BOX.

prosodic features

In human speech, meaning is conveyed by "tone of voice," as well as by the actual sounds uttered. Perhaps you've heard primitive speech synthesizers with their monotone, emotionless quality. You could understand the words perfectly, but they lacked the changes in pitch, timing, and loudness that give depth to spoken statements. These variations are called *prosodic features*.

To illustrate the importance of prosodic features, consider the sentence, "You will go to the store after midnight." Try emphasizing each word in turn:

You will go to the store after midnight.
You *will* go to the store after midnight.
You will *go* to the store after midnight.
You will go *to* the store after midnight.
You will go to *the* store after midnight.
You will go to the *store* after midnight.
You will go to the store *after* midnight.
You will go to the store after *midnight*.

Now, instead of making a statement, ask a question, again emphasizing each word in turn. Just replace the period with a question mark. You have 16 different prosodic variations on this one string of words. A few of them are rather meaningless, but the differences among most of them are striking.

Prosodic variations can also be important in speech recognition. This is because, if you say something one way, you might mean something entirely different than if you utter the very same series of words in another way. Programming a machine to pick up these subtle differences is one of the greatest challenges facing researchers in speech recognition. See also SPEECH RECOGNITION, and SPEECH SYNTHESIS.

prosthesis

A *prosthesis* is an artificial limb or part for the human body. Robotics has made it possible to build electromechanical arms, hands, and legs to replace the limbs of amputees. One of the leading researchers in this field is the Japanese roboticist Ichiro Kato. See KATO, ICHIRO. Artificial organs have also been made.

Help for the body

Mechanical legs have been developed to the point where they can let a person walk. Long pants disguise the robotic legs, so it's hard to tell that they aren't real. Artificial hands can grip; prosthetic arms can throw a ball.

Some internal organs can be replaced, at least for short periods, by machines. Kidney dialysis is one example. An artificial heart is another. Some electronic or electromechanical devices don't completely replace human body parts, but help the real organs do what they are supposed to do. An example is a heart pacemaker.

Difficulties

One of the biggest problems with prostheses is that the body sometimes rejects them as foreign objects. The human immune system, which protects against disease, thinks the machine is a deadly virus or bacteria, and attempts to destroy it. This puts life-threatening stress on the body. To keep this from happening, doctors sometimes give drugs to suppress the action of the immune system. But this can make a person more susceptible to diseases like pneumonia and various viral infections.

Prostheses have not been developed with a good sense of touch. Primitive texture sensing might be developed, but will it ever be as discerning as the real sense of touch? This depends on whether or not electronic circuits can duplicate the incredibly complex impulses that go through living nerves.

Put a penny and a dime in your pocket. Reach in and, by feel alone, figure out which is which. It's easy; the dime has a ridged edge, but the penny's edge is smooth. This data goes from your fingers to your brain as nerve impulses. Can these "nerve waves" be duplicated by electromechanical transducers? Some researchers think so; others think not. At least a few will try. See also ARTIFICIAL LIFE, BIOLOGICAL ROBOTS, BIOMECHANISM, BIOMECHATRONICS, and TEXTURE SENSING.

Prototype

In the design and manufacture of robots, a test unit is built before production begins. The test robot is perfected by experimentation, and also by trial and error. Two or more of these machines, known as *prototypes*, are often built, with each prototype debugged a little bit more than the previous one. The final prototype, when perfected, is the basis for the production units.

proximity sensing

Proximity sensing is the ability of a robot to tell when it is near an object, or when something is near it. This sense keeps a robot from running into things. It can also be used to measure the distance from a robot to some object.

Basic principle

Most proximity sensors work the same way: the output of a displacement trans-ducer varies with the distance to some object. This can take either of two forms, as shown in the illustrations. At A, the sensor output decreases as the distance gets larger. At B, the sensor output rises with increasing distance.

In theory, either type of displacement transducer will function in any appli-cation. But one type is usually easier to work with, in a given situation, than the other.

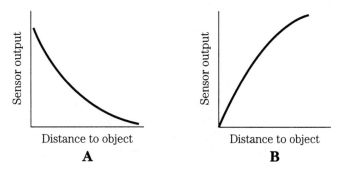

PROXIMITY SENSING At A, output diminishes with distance. At B, output increases with distance.

Bumpers and whiskers

The simplest proximity sensors don't measure distance at all. Their output is zero until they actually hit something. Then the output rises abruptly. Bumpers and whiskers work this way.

A bumper might be completely passive, just making the robot bounce away from things that it hits. More often, a bumper has a switch that closes when it makes contact, sending a signal to the controller causing the robot to back away.

When whiskers hit something, they vibrate. This can be detected, and a sig-nal sent to the robot controller. Whiskers might seem silly and primitive, but they are often a cheap and effective method to keep a machine from crashing into walls, furniture, or other obstructions.

Capacitive effects

See CAPACITIVE PROXIMITY SENSING.

Laser ranging

A laser beam can be bounced off of anything that will reflect or scatter the en-ergy. If the laser is modulated, the return-signal delay can be measured, and the distance to the object determined by the robot's computer.

Lasers can be in the visible-light range, or they can be at infrared (IR) or ultraviolet (UV). Infrared is preferred because the beam cannot be seen, and because the cost is reasonable.

One problem with laser ranging is that it works only for objects that return some of the energy. A black object would not be detectable by a visible-light laser-ranging system, for example.

Laser ranging systems are good for long-distance proximity sensing. Light beams travel fast (186,282 miles per second or 299,792 kilometers per second). It's hard to measure the return-signal delay over short distances. Sonar is better for short-range proximity sensing. See RADAR.

Parallax

See PARALLAX.

Radar

Proximity sensing can be done using radar. Radar works with ultra-high-frequency (UHF) or microwave radio signals. Pulses are transmitted and picked up after they reflect from objects. The delay time is measured, and the results sent to the robot controller. The principle is basically like that of a laser-ranging proximity sensor.

Radar won't work for objects that don't reflect UHF or microwave energy. Metallic objects reflect this energy well; salt water is fair; and trees and houses are poor. Radar, like laser ranging, is not very effective at short distances; sonar is better. See SONAR.

For further information

Related articles in this book, besides those already mentioned here, include: ARTIFICIAL STIMULUS, COMPUTER MAP, DISTANCE MEASUREMENT, GUIDANCE SYSTEMS, PHOTOELECTRIC PROXIMITY SENSOR, and VISION SYSTEMS.

PUMA

See PROGRAMMABLE UNIVERSAL MACHINE FOR ASSEMBLY.

pushdown stack

In some memory circuits, the first data in is the last to come out, and vice-versa. Such a circuit is called a *pushdown stack*. Sometimes it is called first-in/last-out. It is a read-write memory, because data can be both stored and accessed.

The drawing shows the principle of a pushdown stack. It is as if the units are stacked up, so that you must take things off the top first. Units come out reversed from the way they are put in. See also FIRST-IN/FIRST-OUT.

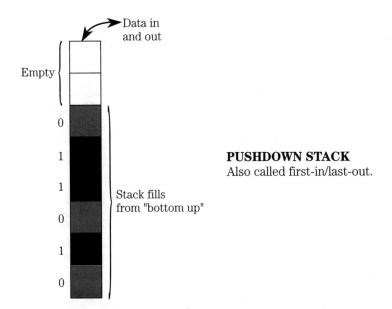

PUSHDOWN STACK
Also called first-in/last-out.

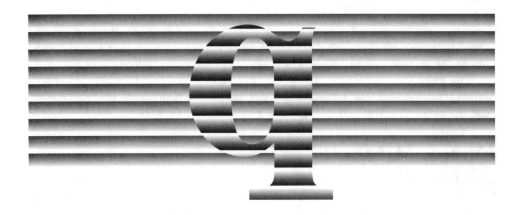

quadruped robots

Historically, people have been fond of the idea of building a robot in the human image. Such a machine would have two legs (see BIPED ROBOTS). In practice, a two-legged robot is hard to design. It tends to have a bad sense of balance; it will fall over easily. You can't build a chair that will stay upright on just two legs; a robot is the same way. As if that isn't bad enough, a two-legged robot is standing on one foot whenever it lifts one leg to move. The sense of balance, which people take for granted, is hard to build into a machine.

To guarantee stability, a robot must have at least three legs always on the surface. A four-legged machine, called a *quadruped robot*, can pick up one leg at a time while "walking," and stay stable. The only potential problem occurs when the three surface-bound legs lie on or near a common line (see drawing A). Under these conditions, even a three-legged object can fall over in either of two directions.

The best quadruped design is such that the four feet reach the ground at points nearly corresponding to the corners of a square (see drawing B). Then, when one foot is lifted for propulsion, the other three are on the surface at the vertices of a well-defined triangle. In the drawings, the solid dots represent feet on the ground; the open circle represents the foot that is lifted at the moment.

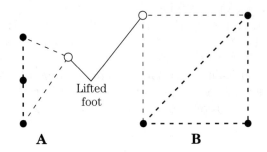

QUADRUPED ROBOTS
Poor balance (A) and solid balance (B).

Many engineers believe that six legs is the optimal number, if a robot is designed to propel itself on legs rather than by rolling on wheels. Six-legged robots can lift one or two legs at a time while walking and remain stable.

The more legs a robot has, the better its balance. But there's a practical limit. Many robotics engineers think that more than six legs represents overkill: it goes past the point of diminishing returns. See also INSECT ROBOTS.

Quain, Mitchell

In recent decades, Japan has become a serious economic competitor to the United States. Some people say this is because the Japanese people take more pride in their work, and work harder, than Americans. Others say that Americans aren't lazy at all, but just like to enjoy balanced lives, while the Japanese are workaholics to a fault. But Mitchell Quain, a factory automation specialist, has suggested another reason. Perhaps the difference is not so much in the workers themselves, but in the machinery available to them.

Quain and his colleagues point to evidence that Japan and Germany have spent more than the U.S., in relative terms, on updating their equipment. In particular, Japan has embraced robotic technology eagerly, even passionately, while the U.S. has been more cautious. During the 1980s, the U.S. stepped up its robotics research, development and manufacturing. The results are beginning to show now. See also ECONOMIC EFFECT OF ROBOTICS.

quality assurance and control

In factory work, robots can perform repetitive tasks more accurately, and faster, than human workers. Robotization has improved the quality, as well as increased the quantity, of production in many industries.

Just doing it better

An important, but often overlooked, aspect of quality assurance and control lies in the production process itself. One way to ensure perfect quality is to do a perfect job of manufacturing. Robots are ideal for this. Not all robots work faster than humans, but robots are almost always more precise. When the manufacturing process is improved, fewer "rejects" come off the assembly line. This makes the quality-assurance-and-control (QA/QC) engineer's job easier.

Some QA/QC engineers say that, ideally, their jobs shouldn't even be necessary. Flawed materials should be thrown away before they are put into anything. Assembly robots should do perfect work. This philosophy has been stated by Japanese QA/QC engineer Hajime Karatsu: *Let's do such good work that QA/QC checkers aren't needed!* He seems more interested in making a good product than in the existence of his own job. This selfless attitude is common in Japan.

Manufacturing will, of course, never be perfect. There will always be errors in assembly, or defective components that get into production units. There'll always be a need for at least one QA/QC person to keep bad units from getting to the buyers.

Robots as inspectors

Robots can sometimes work as QA/QC engineers. But they can do this only for simple inspections, because QA/QC work often requires that the inspector have a keen sense of judgment.

One simple QA/QC job is the checking of bottles for height as they move along an assembly line. A laser/robot combination can pick out bottles that are not of the right height. The principle is shown in the drawing. If a bottle is too short, both laser beams will reach the photodetectors. If a bottle is too tall, neither laser beam will reach the photodetectors. In either of these situations, a robot arm/gripper picks the faulty bottle off the line and discards it. Only when a bottle is within a very narrow range of heights (the acceptable range) will one laser reach its photodetector while the other laser is blocked. Then, the robot does nothing, and the bottle is allowed to pass.

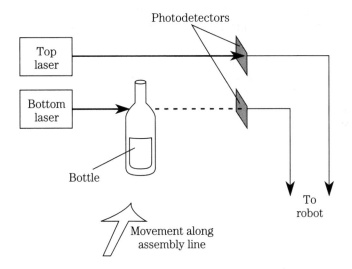

QUALITY ASSURANCE AND CONTROL
Parallel laser beams check bottle height.

Robotic QA/QC processes will get more sophisticated with the advancement of artificial intelligence (AI). Someday, robots will be able to make far more complex decisions than the simple procedure described above. But sometimes, QA/QC decisions involve intuition. This will probably always have to be done by people (see ARTIFICIAL INTELLIGENCE).

Ultimately, someone will have to oversee the manufacture of QA/QC robots, however smart they get. Hajime Karatsu will probably never quite get his wish.

QWERTY

On the traditional typewriter/computer keyboard, the letters are in a sequence that seems random. The first six letters in the top row are Q, W, E, R, T, Y. Because of this, and because these letters form an acronym when run together, the common keyboard is sometimes called *QWERTY*, or the *QWERTY keyboard*. The arrangement of this keyboard is shown in the illustration for the article CHAR-ACTER. All English-language typewriters and computers use this arrangement.

QWERTY is not the most efficient keyboard scheme. Ideally, the most-often-used keys (such as E, A, S, C) would be the easiest to strike. That would place them near the center of the keyboard. But if you look at a QWERTY keyboard, you will see that all four of these keys are near the extreme left. To hit them, you must use the small fingers on your left hand. Most people find these fingers the hardest, not the easiest, to move accurately and fast.

Other keyboard arrangements have been developed. The most well-known is called the Dvorak keyboard. But these schemes haven't caught on. The QWERTY arrangement, while not the best, has become standard. People don't want to make the switch, let alone learn to work on two completely different keyboards.

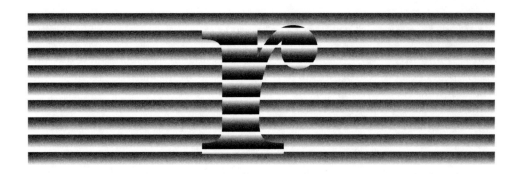

radar

Almost since the discovery of radio, it has been known that electromagnetic waves are reflected from metallic objects. The term *radar* is a contraction of the full technical description, *ra*dio *d*etection *a*nd *r*anging. Radar can be used by robots as a navigation aid.

A radar system consists of a transmitter, a directional antenna, a receiver, and a position indicator, as shown in the drawing. The transmitter produces intense pulses of microwaves. These waves strike objects. Some objects (like cars) reflect radar waves better than others (such as wood). The reflected signals, or echoes, are picked up by the antenna. The farther away the reflecting object, the longer the time before the echo is received. The transmitting antenna is rotated so that the radar "sees" in all directions.

As the radar antenna rotates, echoes are received from various directions. In a robot, these echoes are processed by a microcomputer that gives the machine a "sense of where it is." Radar will perhaps someday be used by robotic warplanes and maybe even war spacecraft.

A special form of radar, called *Doppler radar*, is used to measure the speed of an approaching or retreating target. As its name implies, Doppler radar operates by means of Doppler effect. This is how the police radar measures the speed of an oncoming vehicle.

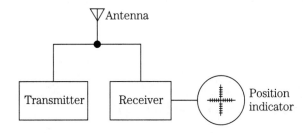

RADAR Block diagram of a radar system.

radio-frequency interference

Radio-frequency interference (RFI) is a phenomenon in which electronic devices upset each other's operation. In recent years, this problem has been getting worse because consumer electronic devices are proliferating, and they have become more susceptible to RFI.

Unfortunately, much RFI results from inferior equipment design. To some extent, faulty installation methods also contribute to the problem. Computers produce wideband radio-frequency (RF) energy that is radiated if the computer is not well shielded. Computers can malfunction because of strong RF fields, such as those from a nearby broadcast transmitter. This can, and often does, happen when the broadcast transmitter is working perfectly. In these cases, and also in cases involving cellular telephones, Citizens' Band (CB) radios and amateur ("ham") radios, the transmitting equipment is *almost never* at fault; the problem is *almost always* improper or ineffective shielding of the computer system.

RFI is often picked up on power and interconnecting cables. There are methods of bypassing or choking the RF on these cables, preventing it from getting into the computer. But you must be certain that the bypass or choke does not interfere with the transmission of data through cables. For advice, consult the dealer or manufacturer of the computer.

Power lines can cause RFI. Such interference is almost always caused by arcing. A malfunctioning transformer, or a bad street light, or a salt-encrusted insulator can all be responsible. Often, help can be obtained by calling the utility company.

A transient suppressor, also called a surge suppressor, in the power cord is essential for reliable operation of a personal computer. Line filters, consisting of capacitors between each side of the power line and ground, can help prevent RF from getting into a computer via the utility lines. Sometimes you might have to write or call the manufacturer of the PC to solve RFI problems.

As personal computers become more portable and more common, RFI problems can be expected to worsen unless manufacturers pay more strict attention to electromagnetic shielding. As computers are increasingly used as robot controllers, potential problems multiply. An errant robot might cause accidents. The danger is greatest with medical or life-support devices.

RADIX

See MODULO.

random-access memory

See MEMORY.

range of function

The *range of a function* is the set of things (usually numbers) onto which objects in the domain are mapped. Every x in the domain of a function f is mapped onto exactly one value y. There might be, and often are, y values that don't have anything mapped onto them by function f. These points are outside the range of f.

Suppose you are given the function $f(x) = x^2$. The graph of this function is shown in the drawing. This function never maps anything onto a negative number. No matter what value you pick for x in the domain, x^2 is never negative. When you square a real number, the result is always zero or greater. Therefore, the range of $f(x) = x^2$ is the set of non-negative real numbers.

Computers work extensively with functions, both analog and digital. Functions are important in robotic navigation, location and measurement systems. See also DOMAIN OF FUNCTION, and FUNCTION.

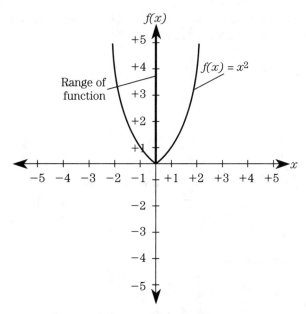

RANGE OF FUNCTION
The set of values for which a function is defined.

range sensing and plotting

Range sensing is the measurement of distances to objects in a robot's environment in a single dimension. *Range plotting* is the creation of a graph of the distance (range) to objects, as a function of the direction in two or three dimensions.

To do one-dimensional (1-D) range sensing, a signal is sent out, and the robot measures the time it takes for the echo to come back. This signal can be sound, in which case the device is sonar. Or it can be a radio wave; this is radar. Laser beams can also be used. One-dimensional range sensing is more often called *proximity sensing*. See PROXIMITY SENSING, RADAR, and SONAR.

Two-dimensional (2-D) range plotting involves mapping the distance to various objects, as a function of their direction. One method is shown in the drawing. The robot is at the center of the plot, near the entrance to a room containing a desk, a wastebasket, two chairs, a floor lamp, and a floor plant. The range is measured every 10 degrees around a complete circle, resulting in the set of points shown. A better plot would be obtained if the range were plotted every five degrees, every two degrees, or even every degree or less. But no matter how detailed the direction resolution, the 2-D range plot can show things in only one plane, such as the floor level or some horizontal plane above the floor.

Three-dimensional (3-D) range plotting requires spherical coordinates. The distance must be measured for a large number of directions at all orientations. A 3-D range plot would show ceiling fixtures, things on the floor, objects on top of the desk, and other details not visible with a 2-D plot. See also COMPUTER MAP, DIRECTION RESOLUTION, and DISTANCE RESOLUTION.

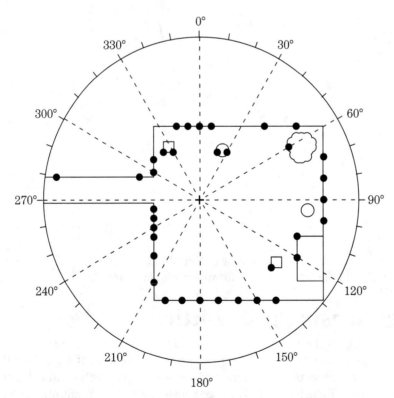

RANGE SENSING AND PLOTTING A 2-D range plot.

read-only memory

See MEMORY.

real time

In communications or data processing, operation done "live" is called *real-time operation*. The term applies especially to computers. Real-time data exchange allows a computer and the operator to converse.

Real-time operation is convenient for storing and verifying data within a short time. This is the case, for example, when making airline reservations, checking a credit card, or making a bank transaction. But real-time operation is not always necessary. It is a waste of expensive computer time to write a long program at an active terminal. Long programs are best written off-line, tested in real time, and then debugged off-line.

In a large computer system, real-time operation can be obtained from many terminals simultaneously. The most common method of achieving this is called time sharing. The computer pays attention to each terminal for a small increment of time, constantly rotating among the terminals at a high rate of speed. In a powerful system, this provides great convenience for a large number of operators. See also COMPUTER.

rectangular coordinate geometry

See CARTESIAN COORDINATE GEOMETRY.

recursion

Recursion is a special form of logical process, in which one or more tasks are put on "hold" while the main argument is being made.

Recursion is common in computer programs, where it takes the form of nested loops. Recursion is also useful in proving math theorems and in legal arguments. On the way to the final goal, or solution to a problem, it is often necessary to take little detours.

Keep the final goal in mind

Recursion can be extremely intricate, and is one of the most advanced types of human reasoning. But for recursion to work, the overall direction of progress must always be toward the final goal. Sidetracking might sometimes seem to have nothing to do with the intended result, but in recursion, there is always a reason for it. All the sub-arguments must eventually be brought out and put to some use in the main argument.

Computers are ideally suited for recursive arguments. The sub-arguments can be done and the results put into memory. Humans tend to get confused when there are too many sidetracks. Not so with computers. They will do precisely what they are programmed, without getting distracted, no matter how many little sidetracks might be going on.

In a complicated recursive argument, the sidetracks can be backed up one on top of another, like airplanes in a holding pattern, waiting to land at a large airport. The sub-arguments are held in pushdown stacks, or first-in/last-out memory registers. The sidetracked results are pulled out of the stacks when needed. The drawing shows a recursive argument with several pushdown stacks.

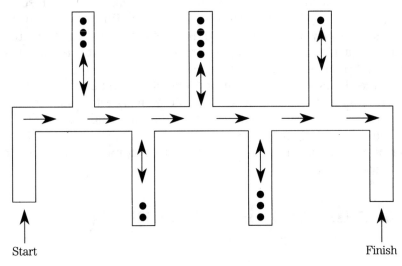

Start

Finish

RECURSION Components of an argument
(shown as dots) are stored in pushdown stacks.

Hangups in recursion

If you use recursive logic and get sidetracked too much, you might lose sight of your final objective, or just go around and around in your mind. When this happens in a computer program, it is called an *endless loop* or *infinite loop*. This is not reasoning at all, and makes it impossible to solve the problem.

There is another logical trap into which you can easily fall when making a recursive argument. This is to "prove" something by inherently assuming that it is already true. Perhaps one of your high-school or college math professors has caught you at this!

Still another potential bugaboo of recursion results in a meaningless conclusion that doesn't tell you anything—or, worse, that contradicts itself. Logicians sometimes say, "From a contradiction, anything follows." Suppose a lawyer-computer proves "beyond all reasonable doubt" that Joe is both guilty and not guilty of a crime of which he has been accused.

Artificial gurus

One of the most ambitious dreams of computer engineers is to build recursion into a machine with artificial intelligence. This might result in a computer with the reasoning power of the best lawyers and the most famous mathematicians.

Imagine a supercomputer that could outwit any spy, and could solve problems that would take even a brilliant detective centuries to unravel! Such a machine would not only be *smart*; it would be *wise*. Researchers debate whether "artificial wisdom" can ever be developed. But some computer engineers are trying, and will keep trying, to do it. See also ARTIFICIAL INTELLIGENCE, ENDLESS LOOP, NESTED LOOPS, PARADOX, PROBLEM REDUCTION, and PUSHDOWN STACK.

reduced instruction set computer

One of the most important aspects of computer "power" is the speed at which it works. Even if a computer can store massive amounts of data, its level of artificial intelligence (AI) is limited by how fast it can work with the data.

Fewer instructions, higher speed

When a computer processes information, it must work with instructions. It is always best to keep the number of instructions as small as possible. The fewer instructions needed to accomplish some task, the faster the task can be done. In recent years, computer researchers have kept this in mind, and the result has been the reduced instruction set computer (RISC). The RISC has its data organized in such a way that it's usually easier to get at specific items, compared with the older complex instruction set computer (CISC) technology.

Example: Data files

Here is an analogy that will help show how RISC works. Consider the files in which this book was written. The book abbreviation is RAI. The text file RAI.A contains all the article titles starting with A, RAI.B has all the article titles starting with B, and so on. This article, therefore, was written in file RAI.R. To find this article, you would first call up RAI.R, and then scroll down until you got to this title. This wouldn't take long. The same would be true for any other title. If a title wasn't in this book, it wouldn't take you long to find that out.

But suppose the articles had each been given a separate file, say a number from 0001.RAI (for the first article) to xxxx.RAI (for the last one). To find an article, you would have to guess at its number. Suppose you were looking for the article ROBOT. You might guess 0623.RAI, and get MICROCOMPUTER. So you'd know ROBOT was in a file with a larger number than 0623.RAI. You'd have to abort the file and call up another one. Your next guess might be 0999.RAI; suppose this came up as VIRUS. Then you'd know that ROBOT must have a number greater than 0623.RAI and less than 0999.RAI. You'd abort 0999.RAI and call up something like 0733.RAI. This would be closer to ROBOT. But ROBOT is not an article title in this book! So you'd keep calling up files until you finally deduced that ROBOT is not an article title. Clearly, it is far better to have one computer file for each letter of the alphabet, because it saves a lot of work when you want to find something.

RISC versus CISC

An RISC works something like the way you'd use the letter-of-the-alphabet file scheme; a conventional, slower computer works more like the way you'd suffer through the one-article-per-file method. The data is exactly the same in either case, but the RISC can get to it quicker than the more traditional CISC.

An RISC is not always better than a CISC. One tradeoff with RISC is that programs must be more detailed than with CISC. Sometimes the high speed of RISC is not important enough to warrant the longer programs that are needed. But in a "smart robot," high computer speed will be very important.

In AI, the high speed of RISC technology holds promise. In graphics, it seems to work very well. This is especially true of graphics that must depict fast motion with fine detail. Robot vision systems might prove a major market for RISC technology. See also ARTIFICIAL INTELLIGENCE, COMPUTER, and VISION SYSTEMS.

reductionism

Reductionism is the belief, or the theory, that all human thought can be duplicated by machines. Can everything you think and feel ultimately be reduced to logic ones and zeros? A *reductionist* would say yes.

The human brain is far more complicated than any computer yet devised. But your brain is made of a finite number of individual cells. For any finite number, no matter how large, there exists a larger number. If your brain has a zillion little logic gates, then there can be, at least in theory, a computer chip with ten zillion logic gates. The reductionist will tell you that all your deepest feelings are really nothing more than the sum total of a zillion logic gates working in many ways. Very many ways—but only a finite number.

If the reductionists are right, then computers can be literally made into living entities. Some researchers are enthusiastic about this, and others are afraid of the possible results. One of those who fears reductionism is Joseph Weizenbaum. He sees a darker side to the idea. Reductionism could be used to reduce human beings to the level of machines. Or worse, humans could be reduced to mere servants of the machines they have made. Some people joke that this is already true. See also ARTIFICIAL LIFE, and WEIZENBAUM, JOSEPH.

redundancy

Redundancy is the use of backup parts or systems to guard against possible malfunction of a machine. *Hardware redundancy* is used in computers so that operation can continue even if some components fail. See GRACEFUL DEGRADATION.

In data storage, several copies of the information are usually made to guard against damage or loss. For example, two or three backup copies might be made for each disk in a robot's software repertoire. This is *software redundancy*.

reinitialization

Sometimes a microcomputer will operate improperly because of stray voltages. When this happens, the microcomputer becomes useless until it is reinitialized.

Reinitialization consists of setting all of the microcomputer lines to low or zero. Most microcomputers are automatically reinitialized every time power is removed and reapplied. But not all microcomputers have this feature; a specific procedure must be followed to reinitialize such devices.

The deliberate removal and reapplication of power, for the purpose of reinitialization, is called a "cold boot." The reinitialization of a microcomputer without total removal of power is called a "warm boot." See also COMPUTER, MICROCOMPUTER.

reliability

Reliability is an expression of how well, and for how long, machines keep working.

Reliability is the proportion of units that still work after they have been used for a certain length of time. Suppose 1,000,000 units are placed in operation on January 1, 2010. If 920,000 units are operating properly on January 1, 2011, then the reliability is 0.92, or 92 percent per year. On January 1, 2012, you can expect that $920,000 \times 0.92 = 846,400$ units will be working. The number of working units declines according to the reliability factor, year after year.

The better the reliability, the flatter the decay curve in a graph of working units versus time. This is shown in the drawing. (The terms excellent, good, fair,

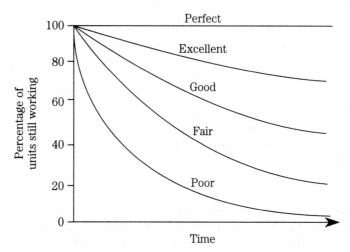

RELIABILITY The better the reliability, the slower the drop in the number of working units.

and poor are relative, and depend on many different factors.) A perfect reliability curve (100 percent) is always a horizontal line on such a graph.

Reliability is a function of design, as well as of the quality of manufacture and parts. Even if a machine is well made, and the components are of good quality, failure is more likely with poor design than with good design. Reliability can be optimized by a quality-assurance-and-control (QA/QC) procedure. See QUALITY ASSURANCE AND CONTROL.

remote control systems

Robots can be operated from a distance by human beings. Computers can also be controlled from places far removed from the machines themselves. This is done by means of *remote control systems*.

A simple example of a remote-control system is the control box for a television (TV) set. Another example is a transmitter you might have used to fly a model airplane. The TV control employs infrared to carry the data. The model airplane gets its commands via radio signals. In this sense, both the TV set and the model airplane are robots.

Remote control can be done by wire, or by optical fiber links. Undersea robots have been operated in this way. A person sits at a terminal in the comfort of a boat or submarine bubble, and operates the robot, watching a screen that shows what the robot "sees."

The range of remote control is limited when wires or optical fibers are used. It's impractical to have a cable longer than a few miles. A special problem exists for long-distance undersea remote control. Radio waves won't penetrate the oceans, but extremely long cables present mechanical problems.

When the control station and the robot are very far away from each other, even radio signals take a long time to cover the distance. A remotely controlled robot on the moon is about 1.3 light seconds away. It would be 2.6 seconds from the time a command was sent to a robot on the moon, before the operator would see the results of the command. This delay is much greater for space probes; the fringes of the Solar System are light hours away.

One of the most dramatic examples of radio remote control is the telemetry that sends commands to space probes as they fly through the Solar System. In these cases, the separation distance is on the order of millions, or even billions, of miles. As Voyager passed Neptune, and a command was sent to the probe, the results were not observed for hours. Remote control of this type is a special challenge.

There is an absolute limit to the practical distance that can exist between a remotely controlled robot and its operator. There is (as yet) no known way to transmit data faster than the speed of radio waves or visible light. The nearest stars are several light years away; remote control over that distance would require us humans to become far more patient than we now are!

Related articles in this book include FIBEROPTIC DATA TRANSMISSION, FLIGHT TELEROBOTIC SERVICER, FLYING EYEBALL, GUIDANCE SYSTEMS, JOYSTICK, LASER DATA TRANSMISSION, MICROWAVE DATA TRANSMISSION, NEUROSURGERY ASSISTANCE ROBOT, POLICE ROBOTS, ROBOTIC SPACE MISSIONS, SECURITY ROBOTS, SELSYN, SYNCHRO, TELEOPERATION, and TELEPRESENCE.

resolution

Resolution is the ability of a system to distinguish between things that are close together. Within objects, resolution is the extent to which the system can bring out details about the object. It is a precise measure of image quality. It is sometimes also called *definition*.

In a robot vision system, the resolution is the "sharpness" of the image. Poor resolution can be the result of poor focus, too few pixels in the image, or a signal bandwidth that is not wide enough. The drawing shows two objects that are far away and close together, as they might appear to a robot vision system with poor resolution, fair resolution, good resolution, and excellent resolution.

When an analog image is converted into digital form, the resolution is the number of different digital levels that are possible. An analog signal has infinitely many different levels; it can vary over a continuous range. The greater the digital resolution, the more accurate the digital representation of the signal.

In position sensing, you might hear about *direction resolution* or *distance resolution*. These terms refer to the ability of the robot sensor to differentiate between two objects that are separated by a very small angle, or that are almost the same distance away. See also ANALOG, BANDWIDTH, DATA CONVERSION, DIGITAL, DIRECTION RESOLUTION, DISTANCE RESOLUTION, PIXEL, POSITION SENSING, and VISION SYSTEMS.

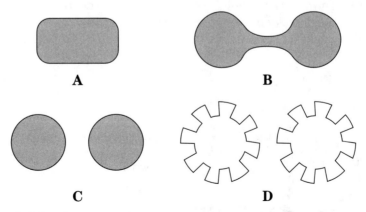

RESOLUTION Poor (A), fair (B), good (C), and excellent (D).

reverse engineering

It is often possible to build a machine that does the same things as some other machine, using a different design. When this is done with computers, it is called *cloning*. The more complex a device, the more equivalent designs there are. *Reverse engineering* is a process by which a given machine is "copied" functionally, but not literally.

If you can duplicate the things a patented machine will do, but use a new and different approach that you thought of independently, have you infringed on the patent? No. If you invent something like a smart robot and then get it patented, does this give you a patent for *what it does*? No. For example, you cannot design a car-waxing robot, and then get a patent that will keep anyone else from building a robot that can wax cars.

But suppose some people reverse-engineered your product by dismantling it and then rebuilding it almost, but not quite, the same way. They wouldn't have invented a new design. The people would have used your work in slightly altered form to make their own "new" product. If you could prove that they used your design and did not make substantial changes, then you could sue them for patent infringement. But proving that sort of thing can be extremely difficult, time consuming, and expensive. See also CLONE.

revolute geometry

Industrial robot arms can move in various different ways, depending on their intended use. One mode of movement is known as *revolute geometry*.

The drawing shows a robot arm capable of moving in three dimensions using revolute geometry. The whole arm can rotate through a full circle (360 degrees) at the base, or "shoulder." There is also an elevation joint at the base that can move the arm through 90 degrees, from horizontal to vertical. A joint in the middle of the robot arm, at the "elbow," moves through 180 degrees, from a straight position to doubled back on itself. There might be, but is not always, a "wrist" joint, too, that might flex like the elbow or twist around and around.

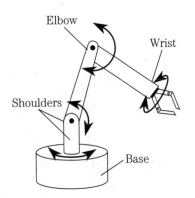

REVOLUTE GEOMETRY
Movement in shoulder, elbow, and wrist.

A well-designed revolute robot arm can reach any point within a half sphere. The radius of the half sphere is the length of the arm when its elbow and wrist (if any) are straightened out. See also CARTESIAN COORDINATE GEOMETRY, CYLINDRICAL COORDINATE GEOMETRY, POLAR COORDINATE GEOMETRY, ROBOT ARMS, and SPHERICAL COORDINATE GEOMETRY.

robot arms

There are several different ways in which robot arms can be made. Different configurations are used for different purposes. Robot arms are sometimes called *manipulators*. Some robots, especially industrial robots, are really just sophisticated robot arms.

A robot arm can be categorized according to its geometry. Some can work in two dimensions only (within a flat plane). Most can work within a defined region of three-dimensional space. This region is the work envelope.

Many robot arms resemble human arms, more or less. The joints in these machines can be given names like "shoulder," "elbow," and "wrist." Some robot arms are so much different from human arms that these names don't make sense. An arm that employs revolute geometry is similar to a human arm, with a shoulder, elbow, and wrist. An arm with Cartesian geometry is far different from a human arm, and moves along axes (x,y,z) that are best described as "up-and-down," "right-and-left," and "front-to-back."

Robot arm designs are discussed in these articles: ARTICULATED GEOMETRY, CARTESIAN COORDINATE GEOMETRY, CYLINDRICAL COORDINATE GEOMETRY, POLAR COORDINATE GEOMETRY, REVOLUTE GEOMETRY, SPHERICAL COORDINATE GEOMETRY, and WORK ENVELOPE.

Robot arms must have an end effector or gripper to perform the task for which they are designed. These devices can often move independently of the rest of the robot arm. See END EFFECTOR, and ROBOT GRIPPERS.

robot assemblers

See AUTOMATED INTEGRATED MANUFACTURING SYSTEM.

robot generations

Some researchers have analyzed the evolution of robots, marking progress according to *robot generations*. This has been done with computers and integrated circuits, so it only seems natural to do it with robots too. One of the first engineers to make formal mention of robot generations was the Japanese engineer Eiji Nakano.

First generation

According to Nakano, first generation robots are simple mechanical arms, with no artificial intelligence (AI) at all. These machines have the ability to make precise motions at high speed, many times, for a long time. Such robots still find widespread industrial use today.

First-generation robots can work in groups, such as in an automated integrated manufacturing system, if their actions are synchronized. The operation of these machines must be constantly watched, because if they get out of alignment and are allowed to keep working, the result can be a series of bad production units. See AUTOMATED INTEGRATED MANUFACTURING SYSTEM.

Second generation

A second-generation robot has some level of AI. Such a machine is equipped with sensors that tell it things about the outside world. These might include pressure sensors, proximity sensors, tactile sensors, and vision systems. A controller processes the data from these sensors and adjusts the operation of the robot accordingly. These devices came into common use around 1980.

Second-generation robots can stay synchronized with each other, without having to be overseen constantly by a human operator. Of course, periodic checking is needed with any machine, because things can always go wrong—and the more complex the system, the more ways it can malfunction. (That is part of a set of jokes called *Murphy's Law*.) See CONTROLLER, PRESSURE SENSING, PROXIMITY SENSING, TACTILE SENSORS, and VISION SYSTEMS.

Third generation

Third generation robots were mentioned by Nakano, but since the publication of his paper, some things have changed. Two major avenues are developing for advanced, "smart" robot technology. These are the autonomous robot and the insect robot. Both of these technologies hold promise for the future.

An autonomous robot can work on its own. It contains a controller and can do things largely without supervision, either by an outside computer or by a human being. A good example of this type of third-generation robot is the personal robot about which some people dream. See PERSONAL ROBOTS.

There are some situations in which autonomous robots just don't do very well. In these cases, many simple robots, all under the control of one central computer, might be used. They would work like ants in an anthill, or like bees in a hive. While the individual machines are "stupid," the group is intelligent and efficient. See also AUTONOMOUS ROBOTS, and INSECT ROBOTS.

Fourth generation and beyond

Nakano did not write about anything past the third generation of robots; however, a fourth generation of robots are a sort yet to be seriously worked on. An example of this might be robots that reproduce and evolve, or that are partly or wholly "alive." See ARTIFICIAL LIFE, and ROBOTIC REPRODUCTION AND EVOLUTION.

Past that, a fifth-generation robot is something no one has yet dreamed of at all—or if someone has, that person hasn't said anything about it.

The table is a summary of robot generations, their times of development, and their capabilities.

ROBOT GENERATIONS:
Robots are getting smarter and more nimble.

Generation	Time First Used	Capabilities
First	Before 1980	Mainly mechanical Stationary Good precision High speed Physical ruggedness Use of servomechanisms No external sensors No artificial intelligence
Second	1980–1990	Tactile sensors Vision systems Position sensors Pressure sensors Microcomputer control Programmable
Third	mid-1990s and after	Mobile Autonomous Insect-like Artificial intelligence Speech recognition Speech synthesis Navigation systems Teleoperated
Fourth	Future	Design not yet begun Able to reproduce? Able to evolve? Artificially alive? As smart as a human? True sense of humor?
Fifth	?	Not yet discussed Capabilities unknown

robot grippers

Robot grippers can take either of two general forms: hand-like, and nonhand-like. These two main designs arise from different philosophies.

Some researchers say that the human hand is an advanced device, having evolved by natural selection. Therefore, they say, robotics engineers would do well to follow Nature's lead and imitate human hands when designing and building robot grippers.

Other roboticists argue that specialized grippers should be used, because robots ordinarily must do only a few specific tasks. Human hands are used for many things, but such versatility might be unnecessary, and even detrimental, in a robot made for just one job.

Most likely, neither of these arguments will be the final word on the subject of robot grippers. "The proof of the pudding is in the eating." The best gripper for a given robot is the one that works the best, lasts the longest, and costs the least.

For more information about robot grippers and related subjects, please refer to the following articles: ACTIVE CHORD MECHANISM, BACK PRESSURE SENSOR, END EFFECTOR, EYE-IN-HAND SYSTEM, FINE-MOTION PLANNING, GRASPING PLANNING, JOINT-FORCE SENSOR, NEUROSURGERY ASSISTANCE ROBOT, PITCH, PRESSURE SENSING, PROPRIOCEPTOR, PROSTHESIS, ROLL, SERVOMECHANISM, SERVO SYSTEM, TACTILE SENSORS, TEXTURE SENSING, TWO-PINCHER GRIPPER, WRIST-FORCE SENSOR, and YAW.

robot hearing

See BINAURAL ROBOT HEARING, SOUND TRANSDUCER.

robotic reproduction and evolution

One of the most "far-out" ideas in robotics is the notion that machines might build other machines, all by themselves, without any help from people. Such a system would need advanced robots with a high degree of artificial intelligence (AI).

Robots already build radios, calculators, camcorders, and computers. Robots can build other robots. The complexity of tasks that an automated factory can perform depends on two things: the sophistication of the robots, and the sophistication of the controller or controllers. It is quite practical to design a robot factory, whose workers are all robots that build duplicates of themselves. This is *robotic reproduction*.

Take this a step further. Imagine that the robots thus produced are put to work in more factories, building still more robots just like themselves. Suppose also that the robots build the new factories in which they are to work. If this goes on long enough, the robot population will increase geometrically. The only limiting factors will be the availability of materials, the energy supply, and whether or not any human being decides to "pull the plug."

No system works perfectly; computers sometimes make mistakes. This would cause failures in some of the robots, and perhaps factory shutdowns. Would that keep robots from reproducing indefinitely, causing the system to eventually peter out? Or might improvements sometimes result? Maybe the

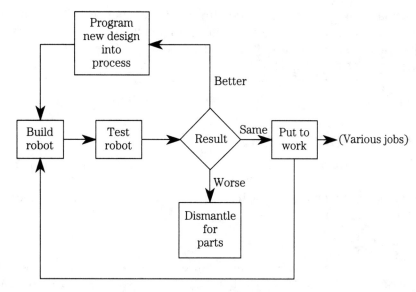

ROBOTIC REPRODUCTION AND EVOLUTION
Could a system take advantage of its mistakes?

whole process could be programmed to make use of improvements that resulted from errors in the process (see the flowchart). Could this go on for generation upon generation—*robotic evolution*—making better and smarter robots after thousands or millions of years? Could such machines end up being better suited to the Earth's environment than humans?

Will we ever visit a planet on which this has happened, and which, therefore, has a whole civilization of robots? For now, it's science fiction. But truth can be, and often is, stranger than fantasy.

Perhaps this idea seems ridiculous. But just 10 human generations ago, a "horseless carriage" was considered impractical, and anyone who believed humans would ever set foot on the moon was a candidate for an insane asylum. See also ARTIFICIAL INTELLIGENCE, and ARTIFICIAL LIFE.

robotic ship

A modern passenger jet can be, and to a large extent is, flown by a computer. It has been said that a jumbo jet could take off from Chicago, Illinois, fly to Sydney, Australia, and land perfectly without a single human being on board. Such an aircraft would actually be a robot. In a similar way, oceangoing vessels could be controlled by computers.

A robotic ship might be designed for combat, and built solely for the purpose of winning battles at sea. With no humans on board, there would be no risk to human lives. The ship would need no facilities for people. The only necessity would be to protect the robot controller from damage as much as possible.

Imagine being the captain of a destroyer, and going up against another ship that you knew had absolutely no one on board! Such an enemy would have no fear and, therefore, would be extremely dangerous.

Ultimately, wars might be fought entirely by robots. After a few such wars, people might see how ridiculous (and expensive) war is, and put their energy into making human life better, rather than worse. See also MILITARY ROBOTS.

robotic space missions

The U.S. space program climaxed when Apollo 11 landed on the moon and, for the first time, a creature from Earth walked on another world.

Some people think the visitor from Earth could just as well have been, and should have been, a robot.

Space probes

Spacecraft have been remotely controlled for decades. Communications satellites use radio commands to adjust their circuits, and sometimes even to change their orbits. Space probes, such as the *Voyager* that photographed Uranus and Neptune in the late 1980s, are controlled by radio. Satellites and space probes are crude robots.

Space probes work something like other hostile-environment machines. Robots are used inside nuclear reactors, in dangerous mines, and in the deep sea. All such robots work by means of remote control. The remote-control systems are getting more and more sophisticated as technology improves.

Almost like being there

Some people say that robots should be used to explore outer space, while people stay safely back on Earth and work the robots via remote control.

A human being can don a "control suit" and have a robot mimic all movements. The robot might be some distance away. The remote-control operation of a robot is called *teleoperation*. When the remote control has feedback that gives the operator a sense of being where the robot is, the system is called *telepresence* (see TELEOPERATION, and TELEPRESENCE).

Ultimately, with technology called *virtual reality*, it might be possible to duplicate the feeling of being in a place, to such an extent that the person can imagine he/she is really there (see VIRTUAL REALITY). Stereoscopic vision, binaural hearing, and a crude sense of touch could be duplicated. Imagine stepping into a gossamer-thin suit, walking into a chamber, and existing, in effect, on the moon or Mars—free of danger from extreme temperatures or deadly radiation!

But you wouldn't *really* be there.

The biggest problem

If robots are used in space travel, with the intention of having the machines replace astronauts, then the distance between the robot and its operator cannot be very great. The reason is that the control signals can't go faster than 186,282 miles per second (299,792 kilometers per second), the speed of light.

The moon is 1.3 light seconds from Earth. (A light second is 186,282 miles or 299,792 kilometers.) If a robot, not Neil Armstrong, had stepped onto the moon on that summer day in 1969, its operator would have had to deal with a delay of 2.6 seconds between command and response. It would take each command 1.3 seconds to get to the moon, and each response 1.3 seconds to get back to Earth. True telepresence is impossible with a delay like that.

Experts say that the maximum delay for true telepresence is 0.1 second. The distance between the robot and its controller can't be more than 0.5, or 1/20, light second. That's about 9,300 miles or 15,000 kilometers—a little more than the diameter of Earth.

A possible scenario

Suppose astronauts are in orbit around a planet whose environment is too hostile to allow an in-person visit. Then a robot might be sent down. An example of such a planet is Venus, whose crushing surface pressures would kill an astronaut clad in even the most advanced pressure suit. It would be easy to sustain an orbit of less than 9,300 miles above Venus, so telepresence would be feasible.

For further information

It is impossible to cover the subject of "space robots" in depth here. A good college library will have references. Just ask the librarian for articles in professional science journals. The subject: robotic exploration of space.

Will robots make good astronauts? In some situations, maybe. But robots will always lack one quality that's essential for great space missions: a romantic sense of adventure.

robot legs

Some robots can do all their work in one place, and can therefore be stationary, or anchored down. This is true of most robots on assembly lines, for example. But others must be mobile to get around within a certain environment.

Wheels are almost always better than legs, when it is possible for something to roll around its environment. But rolling isn't always possible. If the terrain is rough, or if the machine must climb stairs or move around in a jumble of pipes or other hardware, then legs are necessary.

Humans have long dreamed of building machines in their own image. In fact, this idea is so ingrained that many people, when they think of a "robot," imagine something like C3P0 in *Star Wars*, a machine so human-like that it might be mistaken for a person from a great distance or in a dark room. In reality, humanoid robots almost always are built for amusement. Biped robots aren't usually practical. See BIPED ROBOTS, and HUMANOID ROBOTS.

When robots must have legs, stability is a major concern. A robot can fall over if it must stand on one or two legs, or if all its legs are lined up. A toppled robot would be useless, and might not be able to get up again without human help. Legged robots usually have four or six legs. The legs might be independently maneuverable, or they might move in groups. One robot, called *Odex*, has six legs, each with a separate microcomputer for control. See INSECT ROBOTS, ODEX, and QUADRUPED ROBOTS.

Robots with more than six legs haven't often been conceived. Such designs would be expensive, and would almost always be cumbersome and hard to coordinate. It is not impossible, though, that someday someone will think of a need for a "centipede robot."

robot vision

See VISION SYSTEMS.

roll

Roll is one of three types of motion that a robotic end effector can make.

Extend your arm out straight, and point at something with your index finger. Then twist your wrist. Your index finger will keep pointing in the same direction. But it will rotate along with your wrist. If your index finger were the head of a screwdriver, it would be able to turn a screw. This is roll. The drawing shows a robotic end effector making this motion. See also END EFFECTOR, PITCH, and YAW.

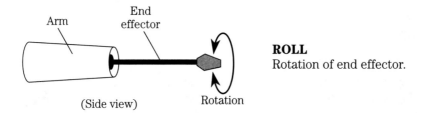

ROLL
Rotation of end effector.

rule-based systems

See EXPERT SYSTEMS.

R2D2

The name *R2D2* is familiar to anyone who saw the movie *Star Wars*. This robot and its companion, C3P0, were the creations of filmmaker George Lucas. The robots had a friendly, benevolent attitude that made them seem human. They were not only "smart," but they had a sense of humor, too.

Actually there were several different "robots" used for R2D2 in the film. It's easier to portray a robot in a movie than to build a robot that can really do the things R2D2 seemed to do. In fact, no real robot yet designed has come anywhere near the fictional capabilities of R2D2.

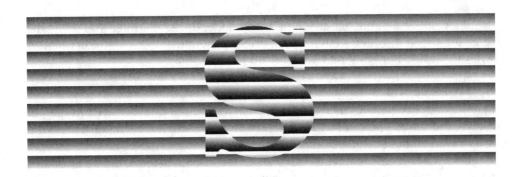

Samuel, Arthur

A pioneer in the field of artificial intelligence (AI), Arthur Samuel is known for his checkers-playing machine, as well as for a classical argument against the ability of machines to have wills of their own.

Samuel's basic argument

According to Samuel, a computer can't "know" anything that hasn't been programmed into it. The ultimate programmer must be a human being.

For anything original to happen as a result of machines programming themselves, Samuel said, there must be an *infinite regress* of machines programming machines programming machines, etc., forever back in time. This is, of course, absurd and impossible. Therefore, said Samuel, a computer cannot think original thoughts.

Samuel's argument is a little like St. Thomas Aquinas's prime-mover argument for the existence of God, with a human programmer being the computer's "God." Some outside intelligence had to start up all the causes and effects in a computer thought universe.

Perhaps.

Whirling atoms?

Some people have argued against Samuel. "Oh, yes," they say, "computers sure can think original thoughts! If a computer can't think original thoughts, then neither can we." Their argument proceeds as follows.

Suppose people's thoughts are all just the products of the motions of atoms. According to this view, a person is just a bunch of whirling particles. This is the "material world" view of human thought.

According to the "material world" view, nobody else's atoms whiz, have ever whizzed, or ever will whiz around in quite the same way as yours. This is because there are so many, many atoms in a human brain. Thus your thoughts are unique to you, and seem original.

If a complicated enough machine were made, the same thing would be true of its atoms, so it would seem to think original thoughts, too. It wouldn't matter whether or not a programmer had started the process.

Spirit forces?

Maybe thoughts come to us from some cosmic medium of which we are not yet aware, and the existence of which scientists deny or will not consider. This is the "spirit world" that occultists talk about.

If spirits can enter humans, can they enter machines too? Might a spirit see a supercomputer and decide to take control of it? This makes great science fiction, but some people really believe it. The idea elicits scorn from most scientists. But if it is true, then a machine can think original thoughts—"programmed" by ghosts!

So who is right?

The question remains: From where does human thought originally come?

Whatever the truth turns out to be, researchers will keep trying to build a machine that can think like people. Someday, perhaps AI will be so much like human reasoning that nobody will worry about whether the machine's thoughts are "original." Some computers nowadays are so powerful that they seem almost human anyway.

It is possible that, as AI becomes advanced, computers will develop thought modes completely foreign to us. Artificial life, built up from silicon chips, servomechanisms, and other electromechanical apparatus, might evolve its own new, strange, and fascinating ways of thinking. We humans like to believe that our logic is the only kind of logic that can exist. But we can't prove it.

See also ARTIFICIAL INTELLIGENCE, ARTIFICIAL LIFE, CHECKERS-PLAYING MACHINE, COMPUTER, INFINITE REGRESS, KNOWLEDGE, and MACHINE KNOWLEDGE.

satellite data transmission

A computer network can use any medium of communications. *Satellite data transmission* is one of the most interesting. It uses microwaves, but the repeaters are in space, not on the ground. Signals are sent up to the satellite, received, and retransmitted on another frequency at the same time (see the block diagram). The ground-to-satellite data is called the *uplink*; the satellite-to-ground data is the *downlink*.

The biggest problem with any wide area network is the fact that it takes time for signals to get between computers. This is true no matter how the data is sent. Nothing can go faster than the speed of light (186,282 miles, or 299,792 kilometers, per second).

Many satellites are in geostationary orbits, at fixed spots 22,300 miles above Earth's equator. When such a satellite is used, the total path length is always at least 44,600 miles (up and back), usually a little more. The smallest possible de-

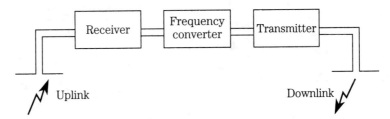

SATELLITE DATA TRANSMISSION
Block diagram of satellite transponder.

lay is therefore 44,600/186,282 = 0.24 second. High-speed, two-way data communication is impossible with a path delay that long.

One of the biggest challenges facing researchers in artificial intelligence is how to link computers that are far away from each other. There is no way to overcome the fact that the speed of light is, in computer terms, very slow! It appears that the only solution to this problem is to live with it, at least until/unless "warp speed" data transmission is developed. Most scientists don't think that will ever happen, except in science fiction. See also ARTIFICIAL INTELLIGENCE, MICROWAVE DATA TRANSMISSION, NETWORK, and WIDE AREA NETWORK.

scaling

Scaling is a principle familiar to structural engineers and physicists. In proportion to their size, small things are stronger than big things.

When things get bigger, but stay in the same relative proportions, *strength* increases as to the *square* (second power) of linear dimension—height, width, or depth. But *weight* increases according to the *cube* (third power) of the linear dimension, as shown in the drawing. Therefore, structures get weaker as they are made larger. The weight goes up much faster than the strength increases. Eventually, if an object gets big enough, it will collapse under its own weight.

In the drawing, height = h, width = w, and depth = d for the smaller cube. The cube doubles in linear dimension, so that height = $2h$, width = $2w$, and depth = $2d$. The base area of the smaller cube, A, quadruples to $4A$. The volume, V, goes up by a factor of 8, to $8V$. If the cubes are made of the same stuff, doubling the linear dimension also doubles the weight per unit surface area at the base. As the cube keeps getting bigger, eventually it will fall through the floor or sink into the ground.

If your height suddenly increased by a factor of 10, your structural strength would go up by a factor of $10^2 = 100$. But your weight would become $10^3 = 1,000$ times as great. That would be like the force of gravity getting 10 times as strong! You would fall down and wouldn't be able to get back up again.

This is why giant robots are unwieldy and impractical, while small ones are hardy and durable.

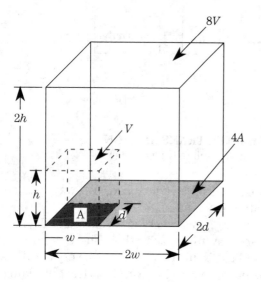

SCALING
Volume increases faster than
surface area.

SCARA

SCARA is an acronym that stands for *selective compliance assembly robot arm*. As this name suggests, it is a special type of robot arm. The SCARA was invented by Hiroshi Makino of Japan.

The idea behind the SCARA was to design a robot arm especially for assembly lines. Dr. Makino noticed that, in these situations, a robot arm must move precisely in the horizontal plane, but not so much in the vertical plane. Thus he designed the machine along the lines shown in the drawing. This is a flexible form of cylindrical coordinate geometry.

The SCARA can easily take a part from a tray or pallet, move it over to a place on a piece of equipment being assembled, and set it down there. The SCARA's end effector might then install the part, or another robot arm might install it, depending on the complexity of the task. See also CYLINDRICAL COORDINATE GEOMETRY, PALLETIZING AND DEPALLETIZING, and ROBOT ARMS.

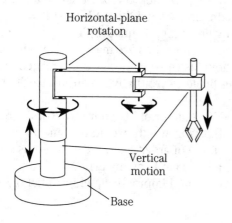

SCARA
Simplified drawing of SCARA
geometry.

security robots

A *security robot* is any robot that assists in the protection of people or property, particularly against crime.

Simple security robots

Security robots have actually been around for decades. A simple version is the electronic garage-door opener. If you lock yourself out of your house, you can still get in via the garage door if you have taken the control box out with you. And, of course, so can anyone with a control box that operates on the same frequency.

More advanced security robots include intrusion alarm systems, and electronic door/gate openers that employ digital codes. These devices can prevent unauthorized people from entering a property. Or they can detect the presence of an intruder (usually by means of ultrasound, microwaves, or lasers), and notify police through a telephone link.

The ultimate security robot

None of these things really qualify as "robots" to those accustomed to science fiction. The ultimate security robot would be like a maid or butler some of the time, and like an attack dog at other times. It would be a machine capable of preventing intrusion most of the time, perhaps by sheer intimidation. If an intruder actually did enter the property, the security robot could presumably either drive the offender away, or detain the offender until police arrived.

Robots like this have been depicted in movies, and because of these movies, many people think such machines will someday become commonplace. But there are numerous problems, some of which might be impossible to overcome.

Problems to solve

Here are some examples of the challenges facing designers of the ultimate security robot.

- Can a robot be made quick enough, and with good enough vision, to chase down an intruder, or win a fight with a human being who is in good physical condition?
- Can such robots be designed to detect any intruder, any time? Or will it be easy for a burglar to sneak by undetected?
- Can such robots be tamper-proof?
- Can such a machine be designed to withstand an assault with practically any weapon?
- If the above questions can all be answered "Yes," will the cost of such a machine ever be affordable to the average family?
- Will property owners ever be able to trust their security robots to work all the time?

- What if the robot malfunctions and thinks the owner is an intruder?
- Can a machine lawfully use deadly force, seeing as it is not "alive" and thus cannot have its "life" endangered?
- If a robot cannot legally use deadly force, of what real use is it?
- What if a security robot injures or kills an intruder? Or damages the mail-delivery robot? Or dispatches the electric-meter-reader robot?

As you can see, an effective, rugged, and affordable security robot will not be easy to design, produce, and deploy! See also POLICE ROBOTS, and SENTRY ROBOTS.

seeing-eye robot

There are as many possible applications for robots as for people and animals combined. One intriguing idea is to use robots as "seeing-eye dogs" for the blind.

A *seeing-eye robot* would need excellent vision. It would have to guide the human user as effectively as a trained seeing-eye dog would. The machine would need artificial intelligence (AI) equivalent to that of a dog. It would also need to get around well, doing such diverse things as crossing a street, going through a crowded room, or climbing stairs.

The Japanese, with their enthusiasm for robots that resemble living creatures, have designed a seeing-eye robot called *Meldog*. It is about the size of a live seeing-eye dog.

Even if seeing-eye robots become affordable, some blind people would rather have a real dog. There's nothing like being awakened in the morning by a friendly lick on the face. No robot will ever be able to do that as well as a live dog. See also COMPANIONSHIP ROBOTS, and PERSONAL ROBOTS.

selsyn

A *selsyn* is an indicating device that shows the direction in which an object is pointing. It is a synchro with the transmitting unit at the movable device, and the receiving unit in a convenient place. A common application of the selsyn is as a direction indicator for a rotatable sensor (see the drawing).

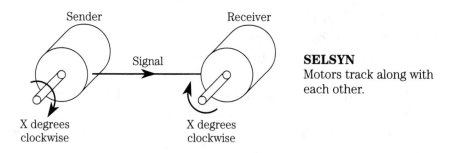

SELSYN
Motors track along with each other.

In a selsyn, the indicator rotates the same number of degrees as the moving device. A selsyn for azimuth bearings will turn 360 degrees; a selsyn for elevation bearings will turn 90 degrees. See also SYNCHRO.

semantic network

A *semantic network* is a reasoning scheme that can be used in artificial intelligence (AI). Semantic networks might someday be employed in "thinking machines" that rival human intelligence.

In a semantic network, the objects, locations, actions, and tasks are called *nodes*. The nodes are interconnected by relationships. This breaks reasoning down in a manner very much like the ways you diagram sentences in English grammar class. The main difference is that a semantic network is not limited to just one sentence; it can build on itself indefinitely, so that it represents more and more complicated situations.

An example of a sematic network is shown in the drawing. The nodes are circles, and the relationships are lines connecting the rectangles. You can tell what's happening by examining the drawing. After looking at the way this network is put together, try adding things on to it.

Some researchers think semantic networks are more versatile than another common reasoning device, known as *expert systems*. Usually, an expert system works using only one type of relationship, known as *IF/THEN/ELSE*. See also ARTIFICIAL INTELLIGENCE, EXPERT SYSTEMS, and IF/THEN/ELSE.

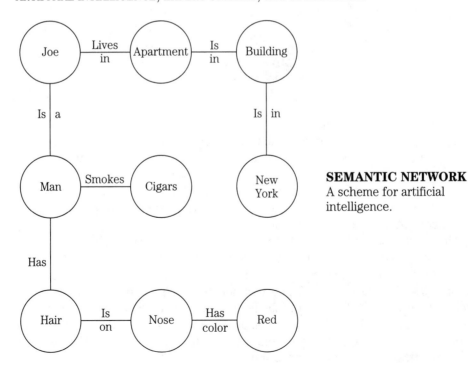

SEMANTIC NETWORK A scheme for artificial intelligence.

semiconductor

A *semiconductor* is a chemical element or compound with medium to large re-sistance. The resistance of a semiconductor is not as low as that of an electrical conductor, but it is much lower than that of an insulator. The resistance of a semiconductor depends on impurities that have been deliberately added.

Semiconductors conduct electricity in two ways: electrons and holes. Elec-trons have a negative charge and are the particles you think of as "orbiting" around and around in atoms. Holes are electron shortages, and carry a positive charge. All semiconductors have some holes and some electrons, both of which contribute to the conduction in the material. Electrons move from the negative charge pole to the positive pole, while holes go the other way, from plus to minus (see the drawing).

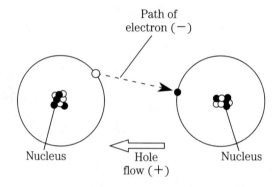

SEMICONDUCTOR
Holes flow the opposite way from electrons.

Semiconductor elements include germanium, selenium, silicon, arsenic, an-timony, boron, carbon, sulfur, and tellurium. Many compounds are semiconduc-tors; the most often-used of these include gallium arsenide, indium antimonide, and various metal oxides.

Semiconductors have revolutionized electronics. A semiconductor device can do the work of thousands of vacuum tubes, that would take up millions or even billions of times as much physical space.

The most modern semiconductor device is the integrated circuit (IC). All computers employ ICs, and they are common in other electronic devices too. Their light-weight and low-power requirements make them ideal for portable computers and "smart robots." The technology of ICs changes very fast. The newest thing today will seem old-fashioned and out of date in less than five years. New semiconductor technologies promise to keep things interesting, and eventually might make possible robots with artificial intelligence (AI). See also ARTIFICIAL INTELLIGENCE, COMPUTER, and INTEGRATED CIRCUIT.

semiconductor memory

See MEMORY.

sensors

See BACK PRESSURE SENSOR, BAR CODING, BINAURAL ROBOT HEARING, CAPACITIVE PROXIMITY SENSING, POSITION SENSING, PRESSURE SENSING, PROXIMITY SENSING, TACTILE SENSOR, TEXTURE SENSING, and VISION SYSTEMS.

sentry robots

A *sentry robot* is a security robot that simply alerts people to abnormal conditions. Such a robot might be designed to detect fire, or burglars, or water in places it shouldn't be. A sentry robot might detect abnormal temperature, barometric pressure, wind speed, humidity, or air pollution.

In industry, sentry robots can alert personnel to the fact that something is wrong. The robot might not specifically pinpoint and identify the problem, but it can let people know that something isn't right. A fire, for example, would generate smoke, as well as unnaturally high temperatures, either or both of which could be detected by a roving sentry.

A sentry robot might include features such as:

- Air-pressure sensing
- Artificial intelligence
- Autonomy
- Beacon navigation
- Computer map(s) of the environment
- Guidance systems
- Homing devices
- Intrusion detection
- Position sensing
- Radar
- Radio link to controller and central station
- Smoke detection
- Sonar
- Speech recognition
- Tactile sensors
- Temperature sensing
- Vision systems

Some of these features are described in the following articles: ARTIFICIAL INTELLIGENCE, AUTONOMOUS ROBOTS, BEACON, COMPUTER MAP, FIRE-PROTECTION ROBOTS, GUIDANCE SYSTEMS, POLICE ROBOTS, POSITION SENSING, RADAR, SECURITY ROBOTS, SMOKE DETECTION, SONAR, SPEECH RECOGNITION, TACTILE SENSORS, TEMPERATURE SENSING, and VISION SYSTEMS.

sequential access memory

In a *sequential-access memory*, data can be recalled, or addressed, only in a certain order. Any form of memory can be designed to work this way. The most common forms of sequential-access memory are the *first-in/first-out* (FIFO) and the *pushdown stack*.

If the addresses in a memory are numbered, such as 1, 2, 3, . . ., *n*, then the most common sequence for recall is upward from *1* to *n*. A downward sequence, from *n* to *1*, is less common. Still less common are "odd" sequences in which the addresses "jump around," and don't ascend or descend in any particular order.

The data from a sequential-access memory might be addressed in two or more different sequences. This is typical of read-only memory (ROM). If data can be written in and recalled in any order, then the memory is a random-access memory (RAM). See also FIRST-IN/FIRST-OUT, MEMORY, and PUSHDOWN STACK.

serial

Information can be transmitted simultaneously along two or more lines, or it can be sent bit by bit along a single line. The first method is known as *parallel data transfer*. The latter method is called *serial data transfer*. The term *serial* is used because the data is sent according to a predetermined sequence.

In computers, serial data transfer requires that a word be split up into its bits, then sent over the line, and reassembled at the other end of the line in the same order as originally (see drawing).

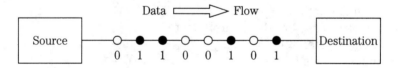

SERIAL Eight-bit word sent along a single line.

Serial data transfer requires only one transmission line, while parallel data transfer requires several lines. However, serial data transfer is somewhat slower than parallel transfer. See also PARALLEL.

servoed system

See CLOSED-LOOP SYSTEM.

servomechanism

A *servomechanism* is a form of feedback-control device.

Servomechanisms are used to control mechanical things like motors, steering mechanisms, radio controls, and robots.

An example of the use of a servomechanism is shown in the diagram. This device is an automatically tuned radio transmitting antenna. A rotary inductor is adjusted by means of an electric motor, which is in turn moved by the output signal from a sensing circuit. The sensing circuit and motor form the servomechanism; the complete antenna-tuning unit is a servo system. It is also a very simple servo robot.

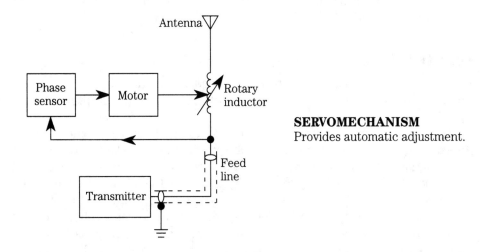

SERVOMECHANISM
Provides automatic adjustment.

When the transmitter is actuated at some frequency, the antenna might be tuned to some other frequency. If that is the case, there will be reactance at the feed point. The sensing circuit detects any reactance, and causes the motor to turn the coil in the required direction, and to the necessary extent, to get rid of the reactance. This is the optimum tuning for the antenna. The inductor stays in that position until the frequency is changed. If the frequency changes, the servomechanism automatically senses the resulting reactance. It then readjusts the inductor until there is zero reactance at the new frequency. The servomechanism constantly *seeks* a condition in which the antenna system has no reactance.

Servomechanisms are extensively used in robotics. The robot controller might tell a servomechanism to move in certain ways that depend on the inputs from sensors. Many servomechanisms, when interconnected and controlled by a sophisticated computer, can do complex chores like cook a meal. The much-sought-after personal robot will work in this way. See also CLOSED-LOOP SYSTEM, OPEN-LOOP SYSTEM, PERSONAL ROBOTS, SERVO ROBOTS, and SERVO SYSTEM.

servo robots

A *servo robot* is a robot whose movement is programmed into a computer. The robot follows the instructions given by the program, and carries out precise motions on that basis. Usually, the term is applied to industrial robots.

Servo robots can be categorized according to the way they move. In *continuous-path motion*, the robot mechanism can stop anywhere along its path. In *point-to-point motion*, it can stop only at specific points in its path.

Servo robots can be easily programmed and reprogrammed. All it takes is a change in the computer software. This might be done by exchanging diskettes, by manual programming, or by more exotic methods such as a teach box or teach pendant. See also CONTINUOUS-PATH MOTION, POINT-TO-POINT MOTION, TEACH BOX, and TEACH PENDANT.

servo system

A set of servomechanisms, including associated circuits and hardware, and intended for a specific task, is a *servo system*. Servo systems do precise, often repetitive, mechanical chores.

A computer can control a servo system made up of many servomechanisms. For example, an unmanned warplane can be programmed to take off, fly a mission, return, and land. Such an aircraft is called a *drone*, and is a type of robot. Servo systems can be programmed to do assembly-line work and other tasks that involve repetitive movement, precision, and endurance. These are industrial robots.

At what point does a complex servo system truly become a robot? This is a matter of some disagreement. Some people say that all servo systems are robots; others imagine robots as having human form and being able to do things like cook meals, mow the lawn, and do the laundry. As technology has advanced, people have raised their expectations about what robots can do. See INDUSTRIAL ROBOTS, PERSONAL ROBOTS, ROBOT GENERATIONS, and SERVOMECHANISM.

simulation

Simulation is the use of computers to mimic real-life situations. There are various kinds of simulators. Some simulators involve teaching of skills for the operation of machinery. Other simulators are programs that predict (or try to predict) events in the real world.

Interactive simulators

An interactive simulator is very much like a video game. In fact, computerized video games nowadays are more sophisticated than some simulations!

There is usually a video monitor, a set of controls, and a set of indicators. There might also be audio devices and crude motion-imitation machines. The controls depend on the scenario.

Suppose you get into a simulator intended to mimic the experience of a driver in the Indy 500. The controls include an accelerator, brakes, and a steering wheel. There is probably a speedometer and a tachometer. There might be speakers that emit noises similar to those a real driver would hear. Perhaps the seat is made to vibrate, or to rock back and forth as the "car" goes around "curves." And of course, there must be a screen that shows a realistic, perspective-enhanced view of the road, other cars, and the surroundings as they whiz by.

Interactive simulation is often used as a teaching/training aid for complex skills, such as flying an aircraft. This technique is especially useful in the military, for training in a wide variety of skills. See also COMPUTER-AIDED DESIGN, and COMPUTER-ASSISTED INSTRUCTION.

Event simulators

An event simulator is a computer program that imitates, or models, the behavior of a system. For example, you might want to start a business. How well will it operate? Will you go bankrupt? Will you make a million dollars in your first year? The event simulator, if it is sophisticated enough and if it is given enough data, can help provide answers to questions like these.

One of the most important event simulators is the hurricane forecasting model employed by the National Hurricane Center in Miami. As Hurricane Andrew approached in August 1992, the computers predicted the most likely places for landfall. Andrew took an unusual east-to-west path. Hurricanes often curve northward before they strike land, but the Hurricane Center model predicted that Andrew would keep going due west. The event simulator in this case proved remarkably accurate. The storm followed a track almost exactly the same as the one predicted by the computer.

As event simulators get more advanced, they will begin to incorporate artificial intelligence (AI) to draw their conclusions. There will always be, however, an element of uncertainty that will limit the effectiveness of event simulators. See ARTIFICIAL INTELLIGENCE.

single inline package

The *single inline package (SIP)* is a housing for integrated circuits (ICs). A flat, rectangular box containing the IC is fitted with lugs along one side, as shown in the illustration. There might be just a few pins, or as many as 12 or even 15 pins.

The SIP is easy to install in, and to remove from, a circuit board. It can be soldered in, or sockets can be used to allow easy component removal and replacement. See also DUAL INLINE PACKAGE, and INTEGRATED CIRCUIT.

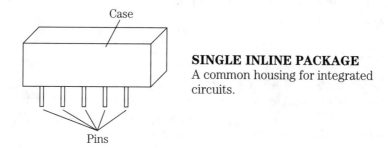

SINGLE INLINE PACKAGE
A common housing for integrated circuits.

SmallTalk

SmallTalk is a high-level computer-programming language. It differs in structure from more familiar languages like BASIC and FORTRAN. While those common languages use commands, SmallTalk is an object-oriented language.

SmallTalk employs a graphical user interface to assist the computer operator. This helps in complex design and research problems, of the sort that roboticists encounter. But SmallTalk requires a large amount of memory space, partly because of the graphics and also because of the fact that it is such a sophisticated language.

SmallTalk was originally developed at the Xerox Corporation. See also GRAPHICAL USER INTERFACE, OBJECT-ORIENTED GRAPHICS, and OBJECT-ORIENTED LANGUAGE.

smart robots

See ARTIFICIAL INTELLIGENCE, and AUTONOMOUS ROBOTS.

smoke detection

When people and property must be protected from fire, *smoke detection* is a simple and effective measure. Smoke detectors are inexpensive, and can operate from flashlight cells.

Smoke changes the characteristics of the air. It is often accompanied by changes in the relative amounts of gases. Fire burns away oxygen and produces other gases, such as carbon dioxide. The smoke itself consists of solid particles.

Air has a property called the *dielectric constant*. This is a measure of how well the atmosphere can hold an electric charge. Air also has an ionization potential; this is the energy needed to strip electrons from the atoms. Many things can affect these properties of air. Common factors are humidity, pressure, smoke, and changes in the relative concentrations of gases.

A smoke detector can work by sensing a change in the dielectric constant and/or the ionization potential of the air. Two electrically charged plates are spaced a fixed distance apart (see the drawing). If the properties of the air change, the plates will charge or discharge. This causes current that can actuate alarms in a fire-protection robot or a sentry robot. See also FIRE-PROTECTION ROBOTS, and SENTRY ROBOTS.

software

In a computer system or robot controller, the programs are called *software*. Software can exist in written form, as magnetic impulses on tapes or disks, or as electrical or magnetic bits in a computer memory. Software also includes the instructions that tell personnel how to operate the computer.

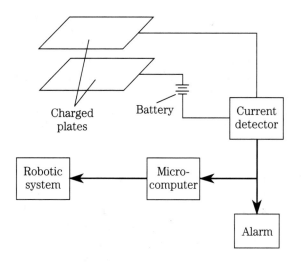

Charged plates Battery Current detector

Robotic system ← Micro-computer ←

Alarm

SMOKE DETECTION
Smoke causes plates to charge or discharge, actuating detection circuits.

There are many different programming languages, each with its own special purpose. The most primitive software is machine language; it consists of the actual binary information used by the electronic components of the computer.

Software can be programmed temporarily into a memory, or it can be programmed permanently by various means. When software is not alterable (that is, it is programmed permanently), it is called *firmware*. See also COMPUTER, FIRMWARE, HIGH-LEVEL LANGUAGE, MACHINE LANGUAGE, MICROCOMPUTER, and MINICOMPUTER.

sonar

Sonar is a long-range and medium-range method of proximity sensing. The basic principle is simple: Bounce sound waves off of objects and measure the time it takes for the echoes to return. In practice, sonar systems can be made so sophisticated that they rival vision systems for getting pictures of the environment.

Audible versus ultrasonic

Sonar can make use of audible sound waves, but there are advantages to using ultrasound instead. Ultrasound has a frequency too high to hear, ranging from about 20 kilohertz (kHz) to more than 100 kHz. (One kilohertz is 1,000 cycles per second.)

An obvious advantage of ultrasound is that the "chirps" will not be heard by people working around the robot. These sounds could become horribly annoying after a few hours. Another advantage of ultrasound is the fact that it is less likely to be fooled by people talking, machinery operating, and other noises. At frequencies higher than the range of human hearing, acoustical disturbances do not normally occur as often, or with as much intensity, as they do within the hearing range.

A simple sonar

The simplest sonar scheme is shown in the block diagram (A in the illustration). An ultrasonic pulse generator sends bursts of alternating current (ac) to a transducer. This converts the currents into ultrasound, which is sent out in a narrow beam. This beam is reflected from objects in the environment, and returns to a second transducer, which converts the ultrasound back into pulses of ac. These pulses are delayed with respect to those that were sent out. The length of the delay is measured, and the data fed to a microcomputer that determines the distance to the object in question.

This system cannot provide a detailed picture of the environment, unless it is refined, and a computer is incorporated to analyze the pulses coming in.

One problem with this simple system is that it can be fooled if the echo delay is equal to, or longer than, the time between individual pulses. Suppose that the time between pulses is 0.1 second. Sound waves travel at about 1,100 feet per second in air. If an object is 55 feet away, the round trip will be 110 feet, or 0.1 second. The sonar system will probably fail to detect the echo, because it will arrive right on top of the following pulse. If the object is farther than 55 feet away, perhaps 75 feet, the echo will be detected, but the sonar might not be able to tell which pulse was the original one (B in the illustration). In such a case the robot might "think" that the object is only 20 feet away (75 − 55 = 20).

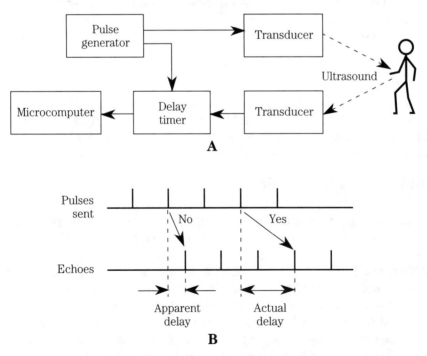

SONAR At A, block diagram of simple sonar system. At B, sonar fooled by long echo delay.

Sonar system refinements

Researchers know that sonar can rival vision as a means of mapping the environment. This is because bats, whose "vision" is actually sonar, can navigate as well as if they had keen eyesight.

What makes bats so adept at using sonar? For one thing, they have a brain. Therefore, artificial intelligence (AI) will be an important part of any advanced robotic sonar system. The computer will need to analyze the incoming pulses in terms of their phase, the distortion at the leading and trailing edges, and whether or not the returned echoes are "bogeys" (illusions).

Another feature of advanced sonar will be the ability to differentiate between different objects in its vicinity. This is called *resolution*. For good resolution, the beam must be narrow, and it must be swept around in two or three dimensions. With excellent direction resolution and distance resolution, a sonar should be able to make a computer map of the robot's work environment. See also ARTIFICIAL INTELLIGENCE, COMPUTER MAP, DIRECTION RESOLUTION, DISTANCE RESOLUTION, and RANGE SENSING AND PLOTTING.

sound transducer

A *sound transducer*, also called an *acoustic transducer*, can do either or both of two things. It can convert sound/ultrasound to some other form of energy; it can convert some form of energy to sound/ultrasound. Usually, this other form of energy is an electrical alternating current (ac), having a frequency anywhere from about 20 Hertz (Hz) to more than 100 kilohertz (kHz). A frequency of 1 Hz is one cycle per second; 1 kHz is 1,000 cycles per second.

In robotics, sound transducers include microphones and speakers, as well as devices used to transmit and receive sonar waves.

Microphones

A microphone converts acoustic vibrations into ac having the same waveform. There are several different ways that a microphone can work. Two of the more common schemes are shown in drawings A and B.

The device at A is called a *dynamic microphone*. A diaphragm, or flat plate, is attached to a voice coil within the field of a fairly strong permanent magnet. As the sound-carrying medium (gas, liquid, or solid) vibrates the diaphragm, the coil vibrates within the field of the magnet. The result is weak ac in the coil, whose waveform is identical to that of the acoustic waves hitting the diaphragm. This ac can be amplified and used in sonar, speech recognition, or other systems.

The main advantages of the dynamic microphone are physical ruggedness, low cost, and excellent sensitivity.

The device at B is an electrostatic microphone. In this device, the diaphragm is a thin, flexible sheet that is parallel with, and close to, a rigid sheet. When sound waves hit the diaphragm, it flexes, so that the space between the plates changes. This results in a variable capacitance between the plates. When a volt-

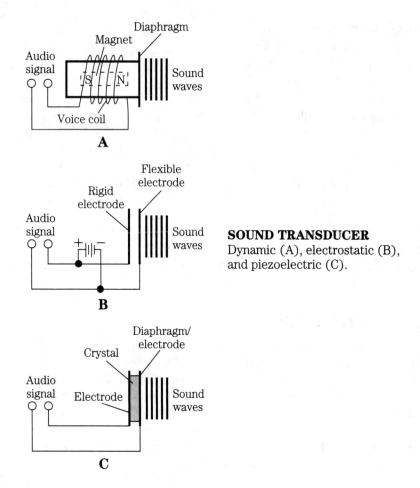

SOUND TRANSDUCER
Dynamic (A), electrostatic (B), and piezoelectric (C).

age is applied between the plates, these capacitance changes produce weak ac because of charging/discharging effects. This ac can be amplified and used.

The advantages of the electrostatic microphone are light weight, physical ruggedness, and the fact that there is no magnet to produce unwanted magnetic fields.

Speakers

You can't connect a microphone to the output of an audio amplifier and expect it to make a good speaker, but the design can be changed somewhat, making use of the same principles, converting ac into sound or ultrasound. The main difference between speakers and microphones is simply that speakers are larger. In particular, speakers must have large diaphragms, so that they displace a fairly large amount of air.

If ac is applied to the coil in the device shown at A in the figure, the coil will produce a fluctuating magnetic field. This will interact with the field around the permanent magnet, producing forces that make the diaphragm vibrate. These vibrations will disturb the medium (gas, liquid, or solid) so that acoustic waves result. This is a dynamic speaker.

If ac is applied between the plates of the device shown at B in the figure, a fluctuating electric field will be produced between the plates. This will cause a variable force between the plates. The flexible plate will therefore vibrate in and out, producing sound waves. This is an electrostatic speaker.

The advantages of dynamic and electrostatic speakers are similar to those of the microphones.

Piezoelectric transducers

Some crystal materials produce tiny ac voltages when subjected to mechanical vibration. These signals can be amplified. Depending on the material used, such a device might be called a crystal microphone, a ceramic microphone, or a piezoelectric microphone. Illustration C is a simplified cross-sectional view of a piezoelectric microphone.

These same materials will emit sounds when ac voltages are placed across them. The frequency of the sound corresponds to the frequency of the ac voltage. You won't often hear of a crystal speaker, a ceramic speaker or a piezoelectric speaker; usually these devices are called piezoelectric transducers instead.

Piezoelectric transducers can often be used interchangeably for transmitting and receiving sound and ultrasound, from several hundred Hertz to more than 100 kHz. See also ELECTROSTATIC TRANSDUCER, SONAR, SPEECH RECOGNITION, SPEECH SYNTHESIS, and TRANSDUCER.

space exploration with robots

See ROBOTIC SPACE MISSIONS.

speech recognition

Your voice consists of audio-frequency energy, with components ranging from about 100 Hertz (Hz) to several kilohertz (kHz). (A frequency of 1 Hz is one cycle per second; 1 kHz = 1,000 Hz.) This has been known since Alexander Graham Bell sent the first voice signals over electric wires.

As computer-controlled robots evolve, people naturally want to control them just by talking to them. A speech-recognition system makes this possible.

Components of speech

Perhaps you've spoken into a microphone that was connected to an oscilloscope, and seen the jumble of waves. How can any computer be programmed to make sense out of that? The answer lies in the fact that, whatever you say, it is comprised of only a few dozen sounds, called *phonemes*.

In communications, engineers have found that a voice can be transmitted quite well if the bandwidth is restricted to the range 300 Hz to 3,000 Hz. Certain phonemes, like "ssss," contain energy at frequencies of several kilohertz, but all the information in a voice, including the emotional content, can be conveyed if the audio passband is cut off at 3,000 Hz. This is the typical voice frequency response in a two-way radio.

Illustration A is a frequency-vs-time voice print of a hypothetical spoken word with five syllables. Most of the energy is in three frequency ranges, called *formants*. The first formant is at less than 1,000 Hz. The second formant ranges from 1,600 Hz to 2,000 Hz. The third formant is at 2,600 Hz to 3,000 Hz. (These ranges are approximate, and vary somewhat from person to person.) Between the formants there are spectral gaps, or ranges of frequencies at which little or no sound occurs.

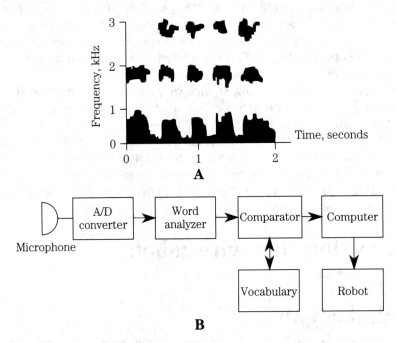

SPEECH RECOGNITION At A, typical voice print.
At B, simplified diagram of speech-controlled robot.

The formants, and the gaps between them, stay in the same frequency ranges no matter what is said. The fine details of the voice print determine not only the words, but all of the emotions, insinuations and other aspects of speech. The slightest change in "tone of voice" will show up in a voice print. Therefore, in theory, it is possible to build a machine that can recognize and analyze speech as well as any human being.

A/D conversion

The passband can be reduced greatly if you are willing to give up some of the emotional content of the voice, in favor of efficient information transfer. In recent years, a technology has been developed and refined that does this very well. It is called *analog-to-digital (A/D) conversion*.

An A/D converter changes the continuously variable, or analog, voice signal into a series of digital pulses. This is a little like the process in which a photograph is converted to a grid of dots for printing in the newspaper. There are several different characteristics of a pulse train that can be varied. These include the pulse amplitude, the pulse duration, and the pulse frequency.

A digital signal can carry a human voice within a passband less than 200 Hz wide. That is less than $\frac{1}{10}$ the passband of the analog signal. The narrower the bandwidth, in general, the more of the emotional content is sacrificed. Emotional content is conveyed by inflection, or variation in voice tone. When tone is lost, the signal resembles written data. But it can still carry some of the subtle meanings and feelings, as you know from reading good books or magazine articles.

Word analysis

For a computer to decipher the digital voice signal, it must have a vocabulary of words or syllables, and some means of comparing this knowledge base with the incoming audio signals. This system has two parts: a memory, in which various speech patterns are stored, and a comparator that compares these stored patterns with the data coming in. For each syllable or word, the circuit checks through its vocabulary until a match is found. This is done very fast, so the delay is not noticeable.

The size of the computer's vocabulary is directly related to its memory capacity. An advanced speech-recognition system requires a large amount of memory space. This is not surprising; many aspects of computer function are related to the memory capacity.

The output of the comparator must be processed in some way, so that the machine knows the difference between words or syllables that sound alike. Examples are "two/too," "way/weigh," and "not/knot." For this to be possible, the context and syntax must be examined.

There must also be some way for the computer to tell whether a group of syllables constitutes one word, or two words, or perhaps three or four words. The more complicated the words coming in, the greater the chance for confusion. The word "radioisotope" might be confused in several different ways. The word "antidisestablishmentarianism" would drive almost any speech-recognition system nuts.

Even the most advanced speech-recognition system will sometimes make mistakes, just as people sometimes misinterpret what you say. Such errors will become less frequent as computer memory capacity and operating speed increase.

Illustration B is a simplified block diagram of a speech-recognition system of the type common nowadays.

Insinuations and emotions

The A/D converter in a speech-recognition system removes some of the inflections from a voice. In the extreme, all of the tonal changes are lost, and the voice is reduced to the equivalent of written language. For most robot-control purposes, this is more than adequate. If a system could be 100-percent reliable in just getting each word right, speech-recognition engineers would be very pleased.

But when accuracy does approach 100 percent (as it will in time), there will be increasing interest in getting some of the subtler meanings across, too. Take the sentence "You will go to the store after midnight," and say it with the emphasis on each word in turn (eight different ways). The meaning changes dramatically depending on the prosodic features of your voice: which word or words you emphasize.

Tone is also very important for another reason: A sentence might be a statement or a question. Thus "You will go to the store *after* midnight?" represents something completely different from "You *will* go to the store after midnight!"

Even if all the tones are the same, the meaning can vary depending on how fast something is said. Even the timing of breaths can make a difference.

Suppose you get quick-tempered with your personal robot, commanding it in a brisk voice, "*Get* me a cup of water." Then imagine it snapping back at you, "Don't speak to *me* in that tone of voice!"

Maybe that scene will be acted out someday in real life. But robotics engineers would do well to think twice before programming the capacity for anger into any machine.

For further information

Speech recognition is a rapidly advancing technology. Some engineers consider it to be one of the most important aspects of artificial intelligence (AI). The best source of up-to-date information is a good college library. Ask the librarian for reference books, and for articles in engineering journals, concerning the most recent developments.

Related articles in this book include: ANALOG, ARTIFICIAL INTELLIGENCE, BANDWIDTH, CONTEXT, DATA CONVERSION, DIGITAL, DIGITAL SIGNAL PROCESSING, GRAMMAR CHECKING, MESSAGE PASSING, PHONEME, PROSODIC FEATURES, SPEECH SYNTHESIS, and SYNTAX.

speech synthesis

Speech synthesis is the generation of sounds that mimic the human voice. This technology is somewhat ahead of speech recognition. It's easier to talk than to listen—even for a machine.

What is a voice?

All sounds, including music, barking dogs, roaring jet engines, ticking clocks, and human speech, consist of combinations of alternating-current (ac) waves within the frequency range 20 Hertz (Hz) to 20 kilohertz (kHz). (A frequency of 1 Hz is one cycle per second; 1 kHz = 1,000 Hz.) These take the form of vibrations in air molecules. The patterns of vibration can be duplicated as electric currents.

Illustration A shows a simple compression wave, of the kind that travels through the air when there is a sound. Also shown is the corresponding electrical current that would produce this compression wave if fed to a loudspeaker. The maximum positive current, in this example, corresponds to the greatest compression of air molecules. The maximum negative current corresponds to the lowest compression of air molecules. The currents are converted to sound, or vice-versa, by means of sound transducers such as speakers and microphones.

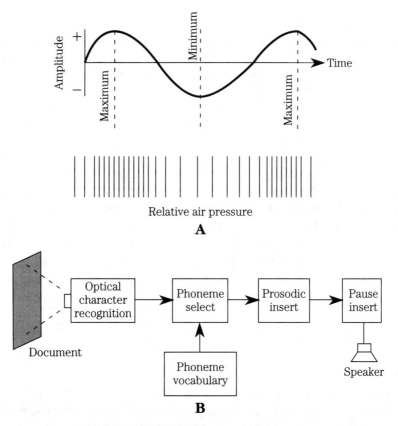

SPEECH SYNTHESIS At A, audio signal and corresponding air compression. At B, text-reading speech synthesizer.

In speech, a frequency band of 300 Hz to 3,000 Hz is wide enough to convey all the information, and also all of the emotional content, in anybody's voice. Therefore, speech synthesizers only need to make sounds within the range 300 Hz to 3,000 Hz. The challenge is to produce waves at exactly the right frequencies, at the right times, and in the right phase combinations. The modulation must also be just right. In the human voice, the volume and frequency rise and fall in subtle and precise ways. The slightest change in modulation can make a tremendous difference in the meaning of what is said. You can tell, even over the telephone, whether the speaker is anxious, angry, or relaxed. A request sounds different than a command. A question sounds different than a declarative statement, even if the actual words are the same.

"Tone of voice"

In the English language there are 40 elementary sounds, known as *phonemes*. In some languages there are more phonemes than in English; some languages have fewer phonemes.

The exact sound of a phoneme can vary, depending on what comes before and after it. These variations are called *allophones*. There are 128 allophones in English. These can be strung together in an almost uncountable number of different ways. But even this is nowhere near the whole story.

The "tone of voice" can depend on whether the speaker is angry, sad, scared, happy, or indifferent. These subtle inflections are the prosodic features of the voice. These depend not only on the actual feelings of the speaker, but on age, gender, upbringing, and other factors.

Besides this, a voice can have an accent. You can probably tell when a person speaking to you is angry or happy, regardless of whether that person is from Texas, Indiana, Idaho, or Maine. Some accents sound more authoritative than others; some sound funny if you haven't been exposed to them before. Along with accent, the actual words can be a little different in different regions. This is dialect.

You don't want a robot to sound angry all the time, or constantly unhappy (or even endlessly blissful, either!). You don't want a robot to convey the wrong mood for a situation. Imagine your fire-protection robot rolling up to you as you read a book, and drawling, "Y'all gonna just *dah* if you don't *leeeve* this house *raht nai-yow*." To one person this might sound terrifying; to another it could come off as a joke. For robotics engineers, producing a speech synthesizer that uses a credible "tone of voice" is a great challenge.

Record/playback

The simplest "speech synthesizer" is simply a set of tape recordings of individual words. You have probably heard these in automatic telephone answering machines and services. Some cities have a telephone number you can call to get local time; this also uses word recordings. They all have a characteristic "choppy" sound.

There are several drawbacks to these systems. Perhaps the biggest problem is the fact that each word requires a separate recording, on a separate length of tape. These tapes must be mechanically accessed, and this takes time. It is impossible to have a large speech vocabulary using this method.

Perhaps the greatest objection is that this really isn't speech synthesis at all. It is speech reproduction.

Reading text

A written text can be "read" by a machine and converted into a digital code called ASCII (pronounced "ask-ee"). The method used for this is known as *optical character recognition*. The ASCII can be translated by integrated circuits (ICs) into voice sounds. In this way, a machine can literally read a text, such as this paragraph. Although they are rather expensive at the time of this writing, these machines are being used to help blind people read written text.

Because there are only 128 allophones in the English language, a machine can be designed to read almost any text. But a machine doesn't have any sense of which inflections are best for the different "scenes" that come up. With technical or scientific text, this isn't usually important. But in reading a story to a child, mental imagery is crucial. It is like an imaginary movie, and it is helped along by the emotions of the reader.

No machine yet built can paint pictures, or elicit moods, in a listener's mind as well as a human being can. These things are apparent from context. The tone of a sentence might depend on what happened in the previous sentence, paragraph, or chapter. Technology is a long way from giving a machine the ability to understand, and appreciate, a good story. But nothing short of that level of artificial intelligence (AI) will really work, if a machine's voice is to make a vivid movie in a listener's mind.

The process

There are several different ways in which a machine can be programmed to produce speech. A simplified block diagram of one process is shown at B in the illustration.

Whatever method is used for speech synthesis, certain steps are necessary. These are as follows:

1. The machine must access the data and arrange it in the right order.
2. The allophones must be assigned in the right sequence.
3. The proper inflections (prosodic features) must be put in.
4. Pauses must be inserted in the proper places.
5. The right mood can be conveyed (joy, sadness, urgency, etc.) at various moments.

6. Overall knowledge of the content can be programmed in. For example, the machine can know the significance of a story, and the importance of each part within the story.

7. The machine can have an "interrupt feature" to allow conversation with a human being. If the human says something, the machine will stop and begin "listening" with a speech-recognition system.

This last feature could prove extremely interesting if two highly advanced machines got into an argument with each other. Doubtless, some engineers can hardly wait for the day they get to try this. One machine might be a "Republican" and the other a "Democrat," for example; the engineer could gleefully bring up the subject of taxes, and let the two machines have it out. The result might well be an endless loop, just as has been the case between human Republicans and Democrats for the past several decades. But of one thing we could be sure, the dialogue would be fascinating. We might even learn something.

For further information

Speech synthesis is a challenging and fast-changing technology in AI. A detailed discussion of the technology is beyond the scope of this book. The best source of up-to-date information is a good college library. Ask the librarian for reference books, and for articles in engineering journals, concerning the most recent developments.

Related articles in this book include: ANALOG, ARTIFICIAL INTELLIGENCE, BANDWIDTH, CONTEXT, DATA CONVERSION, DIGITAL, DIGITAL SIGNAL PROCESSING, GRAMMAR CHECKING, MESSAGE PASSING, OPTICAL CHARACTER RECOGNITION, PHONEME, PROSODIC FEATURES, SOUND TRANSDUCER, SPEECH RECOGNITION, and SYNTAX.

spherical coordinate geometry

Spherical coordinate geometry is a scheme for guiding a robot arm in three dimensions. A spherical coordinate system is something like the polar system, but with two angles instead of one. In addition to the two angles, there is a radius coordinate.

One angle, call it x, is measured counterclockwise from the reference axis. The value of x can range from 0 to 360 degrees. You might think of x as similar to the azimuth bearing used by astronomers and navigators, except that it is measured counterclockwise rather than clockwise. As a ray rotates around a full 360-degree circle through all possible values of x, it defines a reference plane.

The second angle, call it y, is measured either upwards or downwards from the reference plane. The value of y can range from –90 degrees (straight down) to +90 degrees (straight up). You might think of y as the elevation above or below the horizon.

The radius, denoted r, is a nonnegative real number (zero or greater).

The drawing at A shows a spherical system of coordinates. Using this system, the position of a point can be uniquely determined in three-dimensional space.

The drawing at B shows a robot arm equipped for spherical coordinate geometry. The movements x, y, and r are called base rotation, elevation and reach respectively. See also CARTESIAN COORDINATE GEOMETRY, CYLINDRICAL COORDINATE GEOMETRY, POLAR COORDINATE GEOMETRY, and ROBOT ARMS.

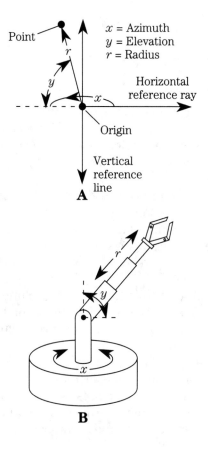

Point

x = Azimuth
y = Elevation
r = Radius

Horizontal
reference ray

Origin

Vertical
reference
line

A

**SPHERICAL
COORDINATE GEOMETRY**
Basic scheme (A), and a robot
arm using this geometry (B).

B

static stability

Static stability is the ability of a robot to maintain its balance while standing still.

A robot with two or three legs, or that rolls on two wheels, usually has poor static stability. It might be all right as long as it is moving, but when it comes to rest, it can easily fall over. A bicycle is an example of a machine with good dynamic stability (it is all right while rolling), but poor static stability (it won't stand up by itself when at rest).

For a two-legged or three-legged robot to have excellent static stability, it needs a sense of balance. You can stand still and not fall over, because you have this sense. If your sense of balance is upset, you'll topple.

It's hard to program a good sense of balance into a robot. Because of this, robots that must stand still are designed with four or six legs. Many engineers think that six legs is the optimum number. See also BIPED ROBOTS, DYNAMIC STABILITY, and INSECT ROBOTS.

stepper motor

A *stepper motor* is an electric motor that turns in small increments, rather than continuously. Stepper motors are extensively used in robots.

Steppers versus conventional motors

When electric current is applied to a conventional motor, the shaft turns continuously at high speed. With a stepper motor, the shaft turns a little, and then stops. The step angle, or extent of each turn, varies depending on the particular motor. It can range from less than one degree of arc to a quarter of a circle (90 degrees).

Conventional motors run constantly as long as electric current is applied to the coils. A stepper motor will turn through its step angle and then stop, even if the current is maintained. In fact, when a stepper motor is stopped with a current going through its coils, the shaft resists turning. A stepper motor has brakes built in. This is a great advantage in robotics; it keeps a robot arm from moving out of place if it is accidentally bumped.

Conventional motors run at hundreds or even thousands of revolutions per minute (rpm). A typical speed is 3600 rpm, or 60 revolutions per second (rps). A stepper motor, however, usually runs less than 180 rpm, or 3 rps. Often the speed is much slower than that. There is no lower limit; a robot arm might be programmed to move just one degree per day, if a speed that slow is necessary.

In a conventional motor, the torque, or turning force, increases as the motor runs faster. But with a stepper motor, the torque decreases as the motor runs faster. Because of this, a stepper motor has the most turning power when it is running at slow speed. In general, stepper motors are less powerful than conventional motors.

Two-phase and four-phase

The most common stepper motors are of two types: two-phase and four-phase. A two-phase stepper motor has two coils, called *phases*, controlled by four wires. A four-phase motor has four phases and eight wires. The motors are stepped by applying current sequentially to the phases. The drawing shows schematic diagrams of two-phase (A) and four-phase (B) stepper motors. The tables show control-current sequences for two-phase (A) and four-phase (B) motors.

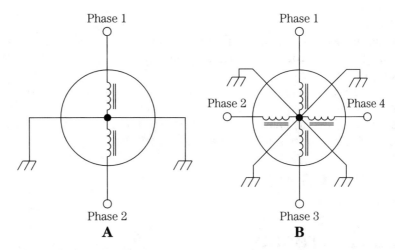

STEPPER MOTOR Two-phase (A) and four-phase (B).

**STEPPER MOTOR: Table A. Two-phase
stepper control. Read down for clockwise
rotation; read up for counterclockwise rotation.**

Step	Phase 1	Phase 2
1	off	off
2	on	off
3	on	on
4	off	on

**STEPPER MOTOR: Table B.
Four-phase stepper control. Read down for
clockwise rotation; read up for counterclockwise rotation.**

Step	Phase 1	Phase 2	Phase 3	Phase 4
1	on	off	on	off
2	off	on	on	off
3	off	on	off	on
4	on	off	off	on

When a pulsed current is supplied to a stepper motor, with the current rotating through the phases as shown in the tables, the motor will rotate in rapid steps, one step for each pulse. In this way, a precise speed can be maintained. Because of the braking effect, this speed will be constant for a wide range of mechanical turning resistances. Most stepper motors can work with pulse rates up to about 200 per second.

Control

Stepper motors can be controlled using microcomputers. Several stepper motors, all under the control of a single microcomputer, are typical in robot arms of all geometries. Stepper motors are especially well suited for point-to-point motion. Complicated, intricate tasks can be done by robot arms using stepper motors controlled by software. The task can be changed by changing the software. This can be as simple as calling up a new program on a video display terminal. See also MOTOR, POINT-TO-POINT MOTION, and ROBOT ARMS.

stereo vision

See BINOCULAR ROBOT VISION.

submarine robots

Human divers cannot go deeper than 800 to 1,000 feet. Rarely do they descend below 300 feet. Even at this depth, after a long dive, a tedious period of decompression is necessary to prevent illness or death from "the bends." Not surprisingly, there is great interest in developing robots that can dive down more than 1,000 feet, while doing all, or most, of the things that human divers can do.

The ideal submarine robot would use telepresence. This is an advanced form of remote control in which the operator has the impression of "being the robot."

Imagine a treasure-hunting expedition, in which you salvage diamonds, emeralds, and gold from a sunken pirate ship 3,000 feet down, while sitting warm and dry in a remote-control chair with a glass of soda beside you! Or you might test shark repellents without fear. You could disarm a nuclear warhead at the bottom of a deep bay, or repair a deep-sea observation station.

The *Titanic*, that "unsinkable" ocean liner that sank, was found and photographed by an undersea robot called a *remotely operated vehicle (ROV)*. This machine did not employ telepresence, but was maneuverable and did provide many high-quality pictures of the wrecked ship.

A specialized form of ROV is called an *autonomous underwater vehicle (AUV)*. This machine has a cable through which control signals and response data must pass. In underwater applications, radio control isn't possible because the water blocks the electromagnetic fields. The cable might use electrical signals or fiberoptic signals. One type of AUV is called the *flying eyeball*.

Another method of remotely controlling undersea robots is ultrasound. It is well known that these waves travel well through the water, and can go long distances. Some researchers think that dolphins carry on conversations this way. See also FLYING EYEBALL, REMOTE CONTROL, TELEOPERATION, and TELEPRESENCE.

surge suppressor

A *surge suppressor* is a device that is inserted in series or in parallel with the power line to a circuit, protecting the circuit from excessive voltages or currents.

There are various kinds of surge suppressors. The simplest is a Zener diode, with a voltage rating equal to the voltage for which the equipment is designed. Current-regulating and voltage-regulating circuits can be used for protection against current or voltage surges. A device that protects against high-voltage "spikes" is more accurately called a *transient suppressor*. The drawing at A shows transients on a 60-Hz alternating-current (ac) power line.

A common method of surge or transient suppression is the use of selenium rectifiers connected back to back. Selenium rectifiers are sometimes used in conjunction with silicon-controlled rectifiers to protect against transients. One such circuit is shown in the schematic diagram (B).

Transient suppressors are available commercially. For personal computers (PCs) and other sensitive electronic equipment, a ready-made transient suppressor should be purchased and used. See also POWER SUPPLY.

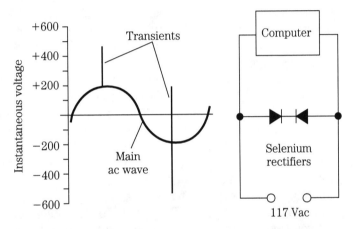

SURGE SUPPRESSOR At A, examples of transients on an ac wave. At B, a simple surge-suppressor circuit.

surgical assistance robot

See NEUROSURGERY ASSISTANCE ROBOT.

synchro

A *synchro* is a special type of motor, used for remote control of mechanical devices. A synchro consists of a generator and a receiver motor. As the shaft of the generator is turned, the shaft of the receiver motor follows along exactly.

In robots, synchros find many different uses. They are especially well suited to fine motion, and also to teleoperation. A simple synchro, used to indicate direction, is called a *selsyn*.

Some synchro devices are programmable. The operator inputs a number into the synchro generator, and the receiver changes position accordingly. Computers allow sequences of movements to be programmed. This allows complex, remote-control robot operation. See also SELSYN, SERVO SYSTEM, SERVO ROBOTS, and TELEOPERATION.

syntax

Syntax refers to the way a sentence, either written or spoken, is put together. It is important in speech recognition and speech synthesis. It is also important in computer programming. Each high-level language has its own unique syntax.

Perhaps you've studied sentence structure in English grammar classes. It can be fascinating if you have a good teacher. Diagramming sentences is a lot like working with mathematical logic. Computers are good at this! Some engineers spend their entire careers figuring out new and better ways to interface human language with computers.

There are several basic sentence forms; all sentences can be classified into one of these forms. The sentence "John lifts the tray," for example, might be called SVO for subject/verb/object. In this case, "John" is the subject, "lifts" is the verb and "tray" is the object.

Different languages have different syntax rules. In the Russian language, "I like you" is said as "I you like." That is, an SVO sentence is really SOV. The meaning is clear, as long as the syntax rules are known. But if the syntax rules are not known, the meaning might well be lost. Say "I you like" to a good friend and he/she will probably respond, "What?"

When designing a robot that can talk with people, engineers must program syntax rules into the computer's memory. Otherwise the robot might make nonsensical statements, or misinterpret what you say.

It's easy to program a robot to say "I like you," or to understand "Go to the kitchen." But if you say "Gee, the weather sure is a lot worse today than it was last Thursday, eh, don't you think, good buddy?" the robot would probably say "I don't understand you." See also CONTEXT, PROSODIC FEATURES, SPEECH RECOGNITION, and SPEECH SYNTHESIS.

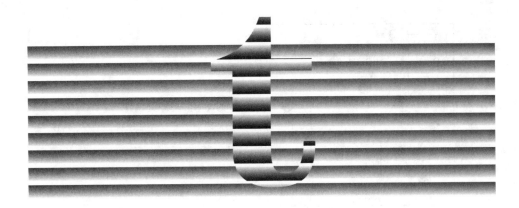

tactile sensors

Tactile sensors give a robot the sense of touch. This includes the ability to "feel" pressure, force, and torque. Tactile sense might also allow a robot to "feel" the texture of a surface. Many roboticists consider tactile sensors second in importance to vision systems.

The following articles contain information about tactile sense and related subjects: BACK PRESSURE SENSOR, DISPLACEMENT TRANSDUCER, ELASTOMER, EYE-IN-HAND SYSTEM, FEEDBACK, FINE-MOTION PLANNING, GRASPING PLANNING, JOINT-FORCE SENSOR, NEUROSURGERY ASSISTANCE ROBOT, POSITION SENSING, PRESSURE SENSING, PROPRIOCEPTOR, PROSTHESIS, TEMPERATURE SENSING, TEXTURE SENSING, and WRIST-FORCE SENSORS.

task environment

Task environment refers to the characteristics of the space where an autonomous robot works. The nature of the task environment depends on many different things. Often these factors interact. If one or more of them aren't dealt with properly, the efficiency of the system suffers (see the drawing). Some things that affect the task environment are:

- The nature of the work the robot(s) must do
- The design of the robot(s)
- The speed at which the robot(s) work
- How many robots are in the area
- Whether or not humans work with the robot(s)
- Whether or not dangerous materials are around
- Whether or not any of the work is hazardous

An autonomous robot can benefit from a computer map of its task environment. This will minimize unnecessary movements, and will reduce the chance for mishaps, like having a robot fall down the stairs or crash through a windowpane.

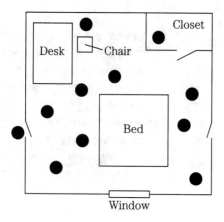

TASK ENVIRONMENT
Cleaning robots (heavy dots)
would be too busy avoiding each
other to do much cleaning.

When a robot is anchored down in one place, as are many industrial robots, the task environment is called the work envelope. See also AUTONOMOUS ROBOTS, COMPUTER MAP, and WORK ENVELOPE.

task-level programming

As machines become "smarter," the programming gets more sophisticated. No machine has yet been built to have anything like human intelligence. In fact, some researchers think that true artificial intelligence (AI), at a level near that of the human brain, will never be achieved.

The programming of robots can be divided into levels, starting with the least complex and progressing to human level. The drawing shows a four-level scheme. Level 3, just below true AI, is task-level programming. As the name implies, programs at this level encompass whole tasks, such as cooking meals, mowing the lawn, or cleaning the house.

Task-level programming has been done, but the software is almost overwhelmingly complicated. To get an idea of what is involved in even the simplest task-level programs, see the article COMMON-SENSE SUMMER PROJECT. See also ARTIFICIAL INTELLIGENCE.

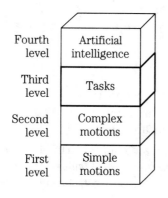

TASK-LEVEL PROGRAMMING
Just below true artificial intelligence.

teach box

When a robot arm must perform repetitive, precise, complex motions, the movements can be entered into a memory. Then, when the memory is accessed, the robot arm will go through all the appropriate movements. A *teach box* is a device that memorizes motions or processes for later recall.

In the four-level programming hierarchy shown in the drawing for TASK-LEVEL PROGRAMMING, the first and second levels are commonly programmed in teach boxes. Sometimes a primitive form of the third level can be programmed.

An example of a level-1 teach box is an automatic garage-door opener/closer. When the receiver gets the signal from the remote unit, it opens or closes the door. Another example of a level-1 teach box is the remote control that you use to change the channel and adjust the volume on a television set.

An example of a level-2 teach box is the integrated circuit (IC) that controls a telephone answering machine. When a call comes in, the sequence of operations is recalled from the firmware in the IC. The machine answers the phone, makes an announcement, takes the message, and resets for the next incoming call.

Reprogrammable teach boxes are extensively used in industrial robots. The arm movements might be entered by pressing buttons. In some cases, it is possible to guide the robot arm manually (that is, "teach" it), and have the movements memorized precisely. The arm's path, variations in speed, rotations, and gripping/grasping movements are all programmed as the arm is "taught." Then, when the memory is recalled, the robot arm will behave just as it was "taught." This is a primitive form of task-level programming, level 3 in the drawing. See also FINE-MOTION PLANNING, GROSS-MOTION PLANNING, INDUSTRIAL ROBOTS, MEMORY, ROBOT ARMS, and TASK-LEVEL PROGRAMMING.

teach pendant

In some industrial robots, movements can be programmed in advance. One example of a programmable industrial robot is Unimation's Programmable Universal Machine for Assembly (PUMA). In the PUMA, movement data can be entered either from a terminal, or via a hand-held module with control buttons. The hand-held module is often more convenient, because it can be carried around. It is called a *teach pendant*. See PROGRAMMABLE UNIVERSAL MACHINE FOR ASSEMBLY, and TEACH BOX.

technocentrism

In the past several decades, people have become comfortable with computers, machines, and electronic devices. Gadgets can be fascinating, and some people get almost totally wrapped up in them.

Maybe you have a friend who is a computer or electronic genius. (Or maybe you are!) Enthusiasm for technology can lead to exciting and rewarding careers.

But if it goes too far, it can throw your life out of balance. Some *technocentrics* lose touch with their emotions. Then, they can't have human relationships, and they become unhappy.

As an example, a father might spend almost all his spare time in the basement, building robot toys for his kids, while the children would really rather have more direct attention from their father. Then, one day the father emerges from the musty workshop to find his children grown up and gone.

Technocentrism is a phenomenon that some sociologists think is already a "social disease." Society gets more geared to the needs of its machines, and places less value on the welfare of its people.

We build and buy machines to make life simpler and more relaxed; but for some strange reason, our lives get more complicated and tense. We find ourselves attending to more machines that are more complex. They break down and we must take them in for repair. They get more versatile, but we must learn to use the new features. Rather than allowing us more free time, our technological miracles gobble up our time and attention.

To avoid technocentrism, we must adopt a balanced outlook. Technology can make our lives fun, free, and fascinating—as long as we, not our machines, remain masters.

telechir

See TELEOPERATION, and TELEPRESENCE.

tele-existence

See TELEPRESENCE.

telemetry

See REMOTE CONTROL, TELEOPERATION, and TELEPRESENCE.

teleoperation

Teleoperation is the technical term for the remote control of autonomous robots. A remotely controlled robot is sometimes called a *telechir*.

In a teleoperation system, the human operator can control the speed, direction, and other movements of a robot from some distance away. Signals are sent to the robot to control it; other signals come back, telling the operator that the robot has indeed followed instructions. These signals are called *telemetry*.

Some teleoperated robots have a limited range of functions. A good example is a space probe, such as *Voyager*, hurtling past some remote planet. Earthbound scientists sent telemetry to *Voyager*, aiming its cameras and sometimes even fixing minor problems. *Voyager* was, in this sense, a teleoperated robot.

Teleoperation is used in robots that can look after their own affairs most of the time, but occasionally need the "help" of a human operator. When the human operator has complete control of the robot at all times, the system is said to employ telepresence. See also AUTONOMOUS ROBOTS, REMOTE CONTROL, and TELEPRESENCE.

telepresence

Telepresence is a refined, advanced form of teleoperation. The robot operator gets a sense of being "on location," even if the robot, or telechir, and the operator are miles apart. Control and feedback are done via telemetry sent over wires, optical fibers or radio.

What it's like

In a telepresence system, the robot is autonomous, and often resembles a human being. The more humanoid the robot, the more realistic the telepresence. This technology is nowhere near fully developed yet. There are many problems to overcome. Eventually, most of these troubles will probably be resolved. What will it be like to operate a robot using advanced telepresence?

The control station will consist of a suit that you wear, or a chair in which you sit with various manipulators and displays. Sensors will give you feelings of pressure, vision, and sound.

You'll wear a helmet with a viewing screen that shows whatever the robot camera sees. When your head turns, the robot head, with its vision system, will follow. Thus you will see a scene that changes as you turn your head, just as if you were in a space suit or diving suit at the location of the robot. Binocular robot vision will give you a sense of depth. Binaural robot hearing will let you perceive sounds just as your ears would hear them if you were on site.

Propulsion might be by means of a track drive, a wheel drive, or robot legs. If the propulsion uses legs, you'll propel the robot by walking around a room. Otherwise you will sit in a chair and "drive" the robot like a car.

The robot will have two arms, each with grippers resembling human hands. When you want to pick something up, you'll go through the motions. Back pressure sensors and position sensors will let you feel what's going on. If an object weighs 10 pounds, it will feel as if it weighs 10 pounds. But it will be as if you're wearing thick gloves. You won't be able to feel texture.

You might throw a switch, and something that weighs 10 pounds will feel as if it only weighs one pound. Maybe this will be called *STRENGTH X 10* mode. If you switch to *STRENGTH X 100* mode, a 100-pound object will seem to weigh one pound. This would, of course, require a robot with great mechanical and structural strength.

The drawing is a simple block diagram of a telepresence system.

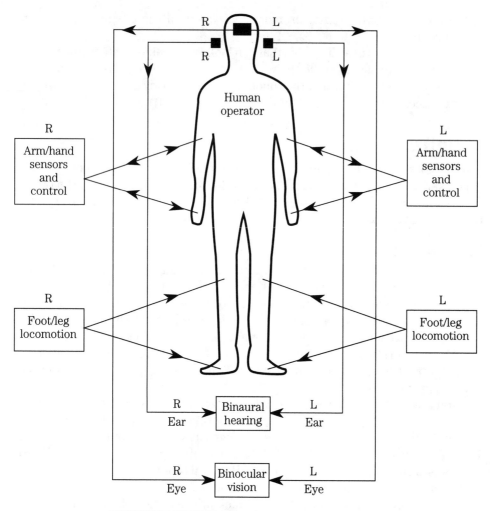

R L
R L

Human
operator

R
Arm/hand
sensors
and
control

L
Arm/hand
sensors
and
control

R
Foot/leg
locomotion

L
Foot/leg
locomotion

R
Ear
Binaural
hearing
L
Ear

R
Eye
Binocular
vision
L
Eye

TELEPRESENCE The ultimate remote control.

Some uses of telepresence

You can certainly think of many different uses for a telepresence system. Some applications are:

- Working in extreme heat or cold
- Working under high pressure, such on the sea floor
- Working in a vacuum, such as in space
- Working where there is dangerous radiation
- Disarming bombs
- Handling toxic substances
- Police robotics
- Robot soldier
- Neurosurgery

Of course, the robot must be able to survive conditions at its location. Also, it must have some way to recover if it falls or gets knocked over.

Limitations and problems

In theory, the technology for telepresence exists right now. But there are some problems that will be difficult, if not impossible, to overcome.

The most serious limitation is the fact that telemetry cannot, and never will, travel faster than the speed of light in free space. This seems fast at first thought (186,282 miles, or 299,792 kilometers, per second). But it is slow on an interplanetary scale. The moon is more than a light second away from the Earth; the sun is eight light minutes away. The nearest stars are at distances of several light years. The delay between the sending of a command, and the arrival of the return signal, must be less than 0.1 second if telepresence is to be realistic. This means that the robot cannot be more than about 9,300 miles, or 15,000 kilometers, away from the control operator. See REMOTE CONTROL, and ROBOTIC SPACE MISSIONS.

Another problem is the resolution of the robot's vision. A human being with good eyesight can see things with several times the detail of the best fast-scan television sets. To send that much detail, at realistic speed, would take up a huge signal bandwidth. There are engineering problems (and cost problems) that go along with this.

Still another limitation is best put as a question: How will a robot be able to "feel" something and transmit these impulses to the human brain? For example, an apple feels smooth, a peach feels fuzzy, and an orange feels shiny but bumpy. How can this sense of texture be realistically transmitted to the human brain? Will people let electrodes be put in their brains so they can perceive the universe as if they are robots? Some people might; most will not, thinking it ridiculous and perverse.

For further information

Telepresence is a vast and fascinating subject. For details about the latest progress in this field, you should go to a good college or university library, and ask the librarian for books and articles on robotics, teleoperation, and telepresence.

Information on some specific topics can be found in this book under the following titles: AUTONOMOUS ROBOTS, BACK PRESSURE SENSOR, BINOCULAR ROBOT VISION, BINAURAL ROBOT HEARING, CAPACITIVE PROXIMITY SENSING, FIRE-PROTECTION ROBOTS, MILITARY ROBOTS, NEUROSURGERY ASSISTANCE ROBOT, POLICE ROBOTS, POSITION SENSING, PRESSURE SENSING, PROPRIOCEPTOR, PROSTHESIS, PROXIMITY SENSING, REMOTE CONTROL, ROBOTIC SPACE MISSIONS, SUBMARINE ROBOTS, TACTILE SENSOR, TELEOPERATION, TEMPERATURE SENSING, TEXTURE SENSING, and VISION SYSTEMS.

temperature sensing

In a robotic system, *temperature sensing* is one of the easiest things to do. Digital thermometers are commonplace nowadays, and cost very little. The output from a digital thermometer can be fed directly to a microcomputer or robot controller, allowing a robot to "know" the temperature at any given point.

Depending on the complexity of the robot's artificial intelligence (AI), temperature data might cause a system to behave in various ways. An excellent practical example is a fleet of fire-protection robots. Temperature sensors can be located in many different places throughout a house, manufacturing plant, nuclear power plant, or other facility. At each point, a critical temperature has been determined in advance. If the temperature at some point rises above the critical level, a signal is sent to a central computer. The computer can dispatch one or more robots to the scene. These robots will determine the source and nature of the problem, and take action. See also FIRE-PROTECTION ROBOTS.

terabyte

A *terabyte* is a unit of memory that will be increasingly used in coming years, as computers evolve towards true artificial intelligence (AI). One terabyte (1 Tb) is 2^{40} bytes, or a little more than a trillion (10^{12}) bytes of memory. It is 1024 gigabytes or 1,048,576 megabytes. It is roughly the amount of memory in a mile-high stack of high-density, 5.25-inch floppy diskettes. See BYTE.

texture sensing

Texture sensing is the ability of a robot end effector to "feel" the smoothness or roughness of a surface. You might think that the apparatus for this would be extremely sophisticated. But actually, crude texture sensing involves only the use of a laser and several light-sensitive sensors.

The drawing shows how a laser, *L*, can be used to tell the difference between a shiny surface (at A) and a rough or matte surface (at B). The shiny surface, say the polished hood of a car, tends to reflect light by the rule: "The angle of reflection equals the angle of incidence." But the matte surface, say the surface of a sheet of

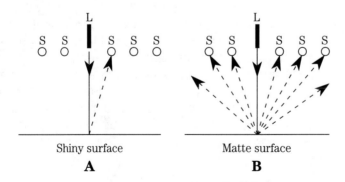

TEXTURE SENSING Lasers (*L*)
and sensors (*S*) analyze a surface.
At A, shiny surface; at B, rough or matte surface.

drawing paper, will scatter the light. The shiny surface will reflect the beam back almost entirely to one of the sensors, S, positioned in the path of the beam whose reflection angle equals its incidence angle. The matte surface will reflect the beam back more or less equally to all of the sensors (and in lots of other directions).

This type of texture sensing cannot give an indication of relative roughness. It can only let a robot know that a surface is either shiny, or not shiny. A piece of drawing paper would reflect the light in much the same way as a sandy beach or a new fallen layer of snow. Relative roughness hard to determine electronically. See also TACTILE SENSORS.

Three Laws of Robotics

See ASIMOV'S THREE LAWS OF ROBOTICS.

tin toy robots

The Japanese have always been fond of robotic toys. In fact, an emotional attachment can develop, especially if the robots are human-like or animal-like. The earliest Japanese robot toys were called *tin toy robots*. They became popular just after the end of World War II.

The first tin toy robot was a windup contraption called *Atomic Robot Man*. It looked like a doll made from tin cans. At that time (the late 1940s), the Japanese people were just starting to recover from the devastating effects of the war. Perhaps this had something to do with the chilling name they gave the little machine. It looked both ridiculous and haunting at the same time; it moved with a strange, funny gait.

The toy robot idea caught on with children, both in Japan and in America. Toys of all kinds became a major source of revenue for the rebuilding of Japan. The Japanese robot empire, as it exists today, might be partly traceable back to one silly little toy made from tin cans that had been discarded by soldiers during the world's most horrible war. See also AMUSEMENT ROBOTS.

toy robots

See AMUSEMENT ROBOTS.

trackball

A *trackball* is a device for guiding the cursor or pointer in a portable computer. It takes the place of a mouse. A trackball is a little bigger than a marble, and is set in the computer keyboard. To use it, you place your hand on it and push or "rub" it around. It moves in any direction, like a ball bearing. As you push the trackball, the cursor or pointer moves on the screen. The design is user-friendly; you can get used to its "feel" in a minute or two. See also MOUSE.

track-drive locomotion

When neither wheels nor legs effectively propel a robot over a surface, *track-drive locomotion* sometimes works. Track drive is used in military tanks, and in some construction vehicles.

A track drive has two or more wheels, and a belt or track (see the drawing). The track can be rubber if the vehicle is small; metal is better for large, heavy machines. The track often has ridges or a tread on the outside; this helps it grip dirt or sand.

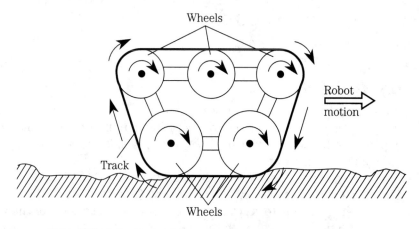

TRACK-DRIVE LOCOMOTION The robot rolls like a tank.

Advantages

Track-drive locomotion works well in terrain strewn with small rocks. It is also ideal when the surface is soft or sandy. Track drive is often the best compromise for a machine that must navigate over a variety of different surfaces.

A special advantage of track drive is that the wheels can be suspended individually. This helps maintain traction over stones and other obstructions. It also makes it less likely that a large rock will tip the robot over.

Steering is a little harder with track drive than with wheel-drive locomotion. A track-driven robot can be turned gradually, if there are two tracks, one on the right side and one on the left. To turn right, the left-hand track goes faster; to turn left, the right-hand track goes faster. Steering radius depends on the difference in speed between the two tracks.

Track drives can allow robots to climb or descend stairways. But for this to work, the track must be longer than the spacing between the stairs. Also, the whole track-drive system must be able to tilt up to 45 degrees, while the robot remains upright. Otherwise the robot will fall backwards when going up the stairs, or forwards when going down. A better system for dealing with stairways is tri-star wheel locomotion.

Problems

One potential hangup with track drives is that the track can work its way off the wheels. The chances of this are reduced by proper wheel and track design. The inside surface of the track can have grooves, into which the wheels fit. Or the inside of the track can have lip edges. The track should fit snugly around the wheels.

Another problem with track drive is that the wheels might slip around inside the track, without the track following along. This is especially likely when the robot is climbing a steep slope. The machine will sit there and "spin its wheels." It might even roll backwards, despite the fact that its wheels are turning forwards. This can be prevented by using wheels with teeth that fit in notches on the inside of the track. The track will then resemble a gear-driven conveyor belt.

On smooth surfaces, track drives are usually not needed. If a surface is extremely rugged, legs often work better than wheels or track drives. See also ADAPTIVE SUSPENSION VEHICLE, ROBOT LEGS, TRI-STAR WHEEL LOCOMOTION, and WHEEL-DRIVE LOCOMOTION.

transducer

A *transducer* is a device that converts energy from one form to another. Transducers also convert variable quantities (such as distance and speed) to measurable forms of energy, or vice-versa.

In robotics, transducers are extensively used. For details on specific devices and processes, please refer to the following articles: CAPACITIVE PROXIMITY SENSING, CHARGE-COUPLED DEVICE, CLINOMETER, DIRECTIONAL TRANSDUCER, DISTANCE MEASUREMENT, DISPLACEMENT TRANSDUCER, DYNAMIC TRANSDUCER, ELASTOMER, ELECTRIC EYE, ELECTROMECHANICAL TRANSDUCER, ELECTROSTATIC TRANSDUCER, ERROR-SENSING CIRCUIT, FLUXGATE MAGNETOMETER, GENERATOR, JOINT-FORCE SENSOR, MODEM, MOTOR, ODOMETRY, OPTICAL CHARACTER RECOGNITION, OPTICAL ENCODER, PASSIVE TRANSPONDER, PHOTOELECTRIC PROXIMITY SENSOR, POSITION SENSING, PRESSURE SENSING, PROPRIOCEPTOR, PROXIMITY SENSING, RANGE SENSING AND PLOTTING, SMOKE DETECTION, SONAR, SOUND TRANSDUCER, SPEECH RECOGNITION, SPEECH SYNTHESIS, TACTILE SENSORS, TEMPERATURE SENSING, TEXTURE SENSING, VIRTUAL REALITY, VISION SYSTEMS, and WRIST-FORCE SENSORS.

transistor-transistor logic

Transistor-transistor logic (TTL) is a form of digital electronic logic circuit. It works at high speed; it is also relatively immune to noise (unwanted impulses). This makes TTL efficient and reliable. It is widely used in digital integrated circuits (ICs).

The basic scheme for TTL is shown in the diagram. Notice that one of the transistors has two emitters. This diagram shows a single logic gate with two inputs (X and Y) and one output. Most logic gates work with about 5 volts direct

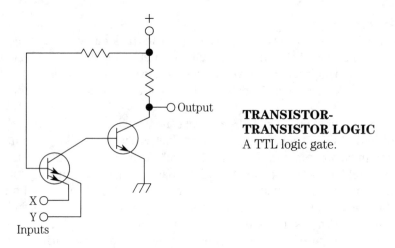

TRANSISTOR-TRANSISTOR LOGIC
A TTL logic gate.

current. There are two logic states, called *high (true)* and *low (false)*. Sometimes they are referred to by the digits 1 and 0, respectively.

A typical TTL IC might have 1,000 or more logic gates on a single chip of silicon. See also DIGITAL, INTEGRATED CIRCUIT, and LOGIC.

triangulation

Robots can navigate in various ways. One good method is like the scheme that ship and aircraft captains have used for decades. It is called *triangulation*.

In triangulation, the robot has a direction indicator such as a compass. It also has a laser scanner that revolves in a horizontal plane. There must be at least two targets, at known, but different, places in the work environment, that reflect the laser beam back to the robot. The robot also has a sensor that detects the returning beams. Finally, it has a microcomputer that takes the data from the sensors and the direction indicator, and processes it to get its exact position in the work environment.

The direction sensor (compass) can be replaced by a third target. Then there are three incoming laser beams; the robot's microcomputer can determine its position according to the relative angles between these beams.

For triangulation to work well, it is important that the laser beams not be blocked. Some environments contain numerous obstructions, such as stacked boxes, which interfere with the laser beams and make triangulation impractical. If a magnetic compass is used, it must not be fooled by stray magnetism; also, the earth's magnetic field must not be obstructed by metallic walls or ceilings.

The principle of triangulation, using a direction sensor and two reflective targets, is shown in the drawing. The laser beams (heavy lines) arrive from different directions, depending on where the robot is located with respect to the targets. The targets are tricorner reflectors that send all light rays back along the path from which they arrive. (Road signs and reflecting tape employ this same phenomenon.)

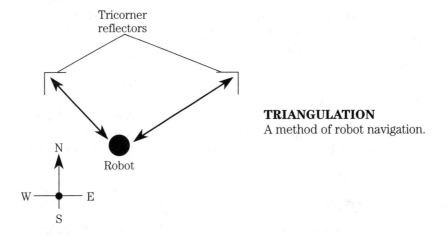

Tricorner
reflectors

Robot

TRIANGULATION
A method of robot navigation.

Triangulation need not use laser beams. Instead of the reflecting targets, beacons can be used. Instead of visible light, radio waves or sound waves can be used. Beacons eliminate the need for the 360-degree scanning transmitter in the robot. See also BEACON.

tri-star wheel locomotion

A unique and versatile method of robot propulsion uses sets of tires arranged in triangles. The geometry of the wheel sets has given rise to the term *tri-star wheel locomotion*. A robot might have three or more tri-star wheel sets.

Each wheel set has three tires. At any given moment, two of these are in contact with the surface (see the drawing). If the robot encounters an irregularity in

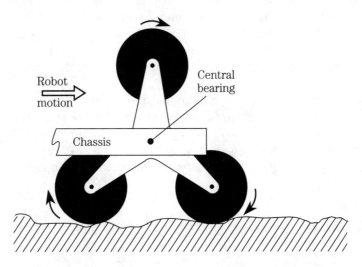

Robot
motion

Central
bearing

Chassis

TRI-STAR WHEEL LOCOMOTION Each wheel
set has three tires at the vertices of an equilateral triangle.

the terrain, such as a big pothole or a field of rocks, the whole set rotates forward on a central bearing. This might happen just once, or repeatedly, depending on the nature of the terrain. The rotation of the central bearing is independent of the rotation of the tires.

Tri-star wheel locomotion works well for stair climbing. It can even allow a robot to propel itself through water, although slowly. The scheme was originally designed and patented by Lockheed Aircraft. Tri-star wheel locomotion might be used someday by robots on other planets. See also ADAPTIVE SUSPENSION VEHICLE, ROBOT LEGS, TRACK-DRIVE LOCOMOTION, and WHEEL-DRIVE LOCOMOTION.

Trojan horse

A *Trojan horse* is something like a software virus. But while a virus can replicate itself, spreading through whole systems of computers, the Trojan horse generally does its damage only within one computer.

The name "Trojan horse" comes from the legend of the ancient city of Troy. Surrounded by high walls, the city seemed impregnable. Hurling rocks over the walls did not defeat the residents within. Whoever tried to climb the walls was thrown back down by the defenders on top. But there was a custom among the people of Troy: they had a fondness for large wooden animals. The enemy soldiers built a huge, hollow wooden horse on a wheeled platform. A number of soldiers crawled inside. The statue was left at the gate of the city as a gift from "unknown" benefactors. The residents wheeled it in, admired it, and stood it in the center of town. Then, at night, when everyone was asleep, the soldiers came out and went on a rampage.

This is just the mode via which an information-age Trojan horse works. A diskette, advertised as having some unique and wonderful contents is sold; people fall for the gimmick in large numbers. Within the disk's software is a program that can mutilate or erase data on a computer's hard disk, making the computer run poorly or not at all. This program is written with totally malicious intent.

Why do some people, with useful knowledge of computers, mass-distribute something that does so much harm and no good? For the same reason that, if you give two people baseball bats, one will use it to clobber baseballs, while the other will use it to smash windows. It is, perhaps needless to say, a crime to produce programs of this kind with the intent of disrupting computer systems. See also VIRUS.

truth table

A *truth table* is a way of breaking down a logical expression. Truth tables show outcomes for all possible situations.

TRUTH TABLE: Breakdown of $-(X + Y) + XZ$.

X	Y	Z	$X + Y$	$-(X + Y)$	XZ	$-(X + Y) + XZ$
0	0	0	0	1	0	1
0	0	1	0	1	0	1
0	1	0	1	0	0	0
0	1	1	1	0	0	0
1	0	0	1	0	0	0
1	0	1	1	0	1	1
1	1	0	1	0	0	0
1	1	1	1	0	1	1

The table is arranged in columns, with each column representing some part of the whole expression. Truth values can be shown by T or F (true or false); often these are written as 1 and 0.

The left-most columns of the truth table give the combinations of values for the inputs. This is done by counting upwards in the binary number system from 0 to the highest possible number. For example, if there are two variables X and Y, there are four value combinations: 00, 01, 10, and 11. If there are three variables X, Y, and Z, there are eight combinations: 000, 001, 010, 011, 100, 101, 110, and 111. If there are n variables, where n is a positive integer, then there exist 2^n possible truth combinations. As you can see, truth tables get huge when there are many variables! Computers, however, are not intimidated by this. In fact, this sort of thing is exactly what we build computers to do, so that we needn't bother.

The logic NOT is denoted by a minus sign (–) or an apostrophe ('). Logic AND is shown by multiplication; logic OR is denoted by addition. These are the common Boolean forms.

The table shows a breakdown of a three-variable expression. All expressions in electronic logic, no matter how complicated, can be "mapped" in this way. Some people believe that the smartest machine, and even the human brain, works according to two-valued logic. Perhaps our brains are just massive sets of truth tables, whose values constantly shift as our thoughts wander. See also BOOLEAN ALGEBRA.

TTL

See TRANSISTOR-TRANSISTOR LOGIC.

Turing, Alan

Alan Turing was a genius in the early years of artificial intelligence (AI). An Englishman, he was born in 1912. He lived to be 41 years old.

Turing was interested in mathematics and mechanical engineering. He grew up during the advent of radio, television, and other electronic devices. It's not surprising, given his background and temperament, that Turing got interested in "smart machines."

In his twenties, Turing played with ideas. In his thirties, he started trying to build a computer. This was just around the time of World War II, when the earliest computers were being conceived and designed.

Turing was especially interested in machine knowledge. Can machines really know things? Can a machine have a mind, not just a brain? Turing tried to think up all of the possible reasons people might give, in attempts to prove that machines cannot really think. Then he tried to argue the case for machine thought. Three powerful objections to machine thought are as follows. Some counter-arguments to these objections are given in this book, in the articles mentioned.

- A machine can only process data. It can't come up with ideas of its own. For further discussion of this, see the article SAMUEL, ARTHUR.

- We can't say a machine has a mind until a computer writes a novel, or a song, or a poem with genuine emotional content. See the articles COMPUTER CONSCIOUSNESS, COMPUTER EMOTION, and COMPUTER MUSIC.

- Humans have souls. Machines do not and cannot. See the articles ANIMISM, ARTIFICIAL LIFE, and SAMUEL, ARTHUR.

Turing invented tests to use on computers, to find out whether they are really thinking. What would he say, if he were still alive, about today's robots and personal computers? See also ARTIFICIAL INTELLIGENCE, COMPUTER, KNOWLEDGE, MACHINE KNOWLEDGE, and TURING TEST.

Turing test

The *Turing test* is a way to find out if a machine can think. It was invented by Alan Turing. (See TURING, ALAN.)

The test is conducted by placing a man (M), woman (F) and questioner (Q) in three separate rooms (see the drawing). None of the people can see the others. The rooms are soundproof, but each person has a video display terminal. In this way, the people can communicate.

The object: Q must find out which person is male and which is female, on the basis of questioning them. But M and F aren't required to tell the truth. Both M and F are told in advance that they may lie. The man is encouraged to lie often, and to any extent he wants. It is the man's job to mislead the questioner into a wrong conclusion. Obviously, this makes Q's job hard! But the test is not complete until Q decides which room contains the man, and which contains the woman.

Suppose this test is done 1,000 times, and Q is right 479 times (and wrong 521 times). What will happen if the man is replaced by a smart computer, programmed as a man? Will Q be right more often, less often or the same number of times as with the real man in the room?

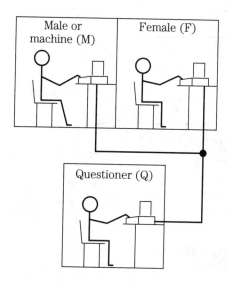

Male or machine (M)	Female (F)

Questioner (Q)

TURING TEST
Which is male and which is
female? Or is one a machine?

If the machine is "dumb," then Q will be correct more often than when a man was at the terminal. Maybe Q would be right 890 times out of 1,000. But if the machine is as "smart" as the man, Q should be right about the same number of times as when the man was at the terminal—say, right 490 times and wrong 510 times. If the machine is "smarter" than the man, then Q ought to be wrong most of the time—say, correct 150 times and mistaken 850 times.

There are other ways to test the intelligence of a machine, such as chess and checkers. But these aren't good indicators of worldly smarts. Besides that, the two games are vastly different. Machines have proven good at checkers but not so good at chess. Maybe you know a brilliant doctor, lawyer, or teacher who is a lousy chess player. Or maybe you know someone who plays a great game of checkers, but can't balance a checkbook.

The Turing test was designed to see how savvy a machine is in real-world terms. No computer has come close to "passing"—yet. See also ARTIFICIAL INTELLIGENCE, CHECKERS-PLAYING MACHINE, CHESS-PLAYING MACHINE, COMPUTER, KNOWLEDGE, and MACHINE KNOWLEDGE.

two-pincher gripper

One of the simplest types of robot gripper uses two tongs or pinchers. It is the sort of thing you might have used to pick up small pieces of garbage from a yard, park, or beach when you didn't want to stoop over. Because of its construction, it is called a *two-pincher gripper*.

The drawing shows a simple version of a two-pincher gripper. The claws are attached to a frame, and are normally held apart by springs. The claws are pulled together by means of a pair of cords. This allows the gripper to pick up small, light objects. To release the grip, the cord is let go. See also ROBOT GRIPPERS.

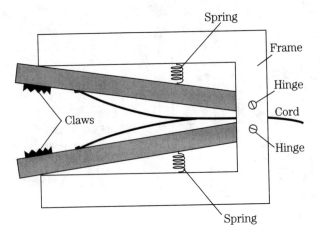

TWO-PINCHER GRIPPER
Pulling the cord brings the claws together.

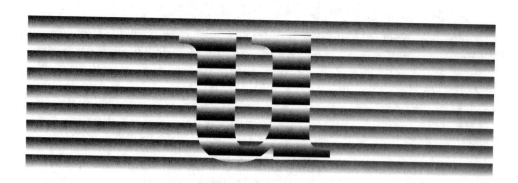

ultimate teacher program

The *Ultimate Teacher Program* was originally written by Charles Lecht. The software taught a computer user about astronomy, and in particular, about the way gravity affects things on different planets. The term might now be applied to any software that teaches people by creating fantasy worlds.

Lecht's program took the computer user on an imaginary round of golf. Each hole was on a different planet in the Solar System. Because of differences in gravitation, the game would vary from planet to planet.

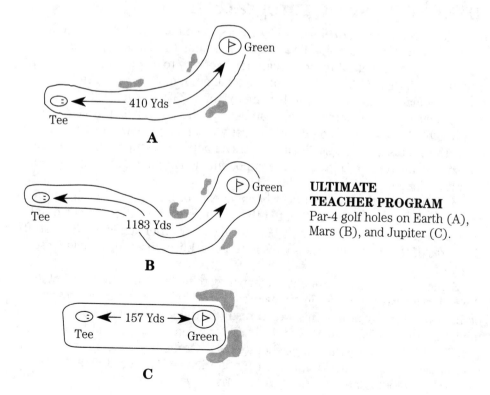

ULTIMATE TEACHER PROGRAM
Par-4 golf holes on Earth (A), Mars (B), and Jupiter (C).

The drawings show hypothetical par-4 golf holes as they might be laid out on Earth (A), on Mars (B), and on Jupiter (C). (Jupiter is thought to have no solid surface, but for the moment, imagine that it does.) The shaded areas are bunkers. On Mars and Jupiter, artificial turf is used for the tees, fairways, and greens, because grass does not grow on those planets.

The force of gravity on Mars is 0.37 of that on Earth. On Jupiter, gravity is 2.5 times as strong as on Earth. If you can drive a golf ball 250 yards on Earth, you might smack it 700 yards on Mars, but only 80 yards on Jupiter. The ball would bounce and roll differently on other planets too. On the putting green, the ball would "break" less on Mars, and more on Jupiter, as compared with its behavior on Earth.

Computer-aided instruction (CAI) is in its infancy. It is expected to revolutionize teaching, helping students at all levels to confront life in ways new and old, real and fantastic. As any futurist knows, today's fantasies are tomorrow's realities. The interplanetary golf tour might be a rewarding career someday. See also CHOREOGRAPHER PROGRAM, COMPUTER-AIDED INSTRUCTION, and LECHT, CHARLES.

ultimate thinking machine

See UNIVERSAL TRUTH MACHINE.

ultra-large-scale integration

Ultra large-scale integration (ULSI) refers to semiconductors with 1,000 to 10,000 logic gates on a chip. This is extreme miniaturization by today's standards. It is much sought-after by researchers working to develop computers with artificial intelligence (AI), because of the huge amount of memory and the high speed that machines must have if they are to "think." The greater the component density on a chip, the more memory can be packed into a given physical volume and weight. Also, greater component density means higher working speed, because the impulses have less distance to travel between components. The drawing shows the relationship among component density, memory capacity, and operating speed.

How tiny can the transistors, diodes, capacitors, and resistors in an integrated circuit (IC) be made? What is the ultimate limit to the number of logic gates that can be fabricated onto a chip of silicon or other semiconductor stuff? The sizes of the atoms impose an absolute limit, but nobody has come close to this yet.

Some engineers have thought of a new way to make ICs. Rather than etching the material away from the chip to make the components, the components might be built up atom by atom. This would result in the greatest possible degree of miniaturization. What would they call this? Extremely large-scale integration (ELSI) or super large-scale integration (SLSI)? Or might they skip the superlatives altogether, and call it atomic-scale integration (ASI)? See also ARTIFICIAL INTELLIGENCE, BIOCHIP, and INTEGRATED CIRCUIT.

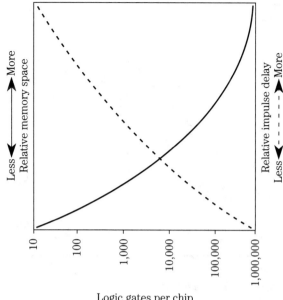

ULTRA-LARGE-SCALE INTEGRATION
Memory and impulse delay, as functions of
component density.

ultrasonic sonar

See SONAR.

uncanny valley theory

Most Japanese are fond of the idea of building androids, or robots in the human image. This stems from an almost religious zeal called animism. But at least one Japanese roboticist, Masahiro Mori, says the "humanoid" approach to robot-building might not be best. If a robot gets too much like a person, Mori thinks, it will seem uncanny, and people will have trouble dealing with it.

Reactions to robots

According to Mori's notion, which he calls the *uncanny valley theory*, the more a robot resembles a human being, the more comfortable people are with the machine, to a point. But when the machine gets too much like a person, disbelief and unease set in. People get intimidated by the thing, and maybe even scared of it.

Mori drew a hypothetical graph to illustrate his theory (see the drawing). The curve has a "valley" at the point where people get uneasy around the robots. Mori calls this the "uncanny valley." How human-like must a robot become to enter this "twilight zone"? No one can really say. It would probably vary from person to person. Some people might not experience the uncanny valley phenomenon at all.

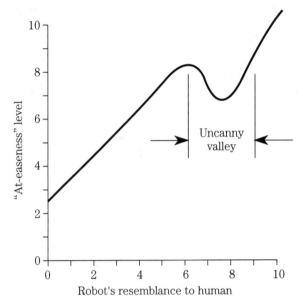

UNCANNY VALLEY THEORY If machines get too smart too fast, people might get uneasy.

No one has tested Mori's theory to see whether or not people's behavior around robots follows a curve like this, because no one has built a robot that is human-like enough to cause this reaction in people—yet.

Intimidated by intelligence

A similar curve might apply to artificial intelligence (AI), or "smart computers." In fact, some people have problems with personal computers (PCs). These people can usually work with pocket calculators, adding machines, cash registers, TV remote controls, and the like; but sit them down in front of a PC, and they freeze up. This is sometimes called *cyberphobia*, meaning "fear of computers."

While some people are so intimidated by computers that they get a mental block right from the start, others are comfortable with PCs after awhile, and only have problems when they try something new. Still other people never have any trouble at all.

The uncanny valley phenomenon is a psychological hangup that some people have with advanced technology of all kinds. A little skepticism is healthy, but overt fear serves no purpose and can keep a person from taking advantage of the good things technology has to offer.

Some researchers think the uncanny valley problem can be avoided by introducing new technologies gradually. But, as we know, things happen fast nowadays! See also ANDROID, ANIMISM, ARTIFICIAL INTELLIGENCE, COMPUTER, and PERSONAL COMPUTER.

undersea robots

See SUBMARINE ROBOTS.

Unimation, Inc.

Unimation, Inc. was one of the earliest American robot manufacturing companies. It was founded in 1961 by Joseph Engelberger, who is sometimes called the "Father of Industrial Robotics."

In the 1960s, Engelberger and a colleague, George Devol, saw that the Japanese were fascinated with robots. They concluded that American manufacturers would have to get interested in them to compete effectively with Japan. People started taking Engelberger and Devol seriously during the 1960s and 1970s, when Japan's technological power began to blossom. See DEVOL, GEORGE C., JR., and ENGELBERGER, JOSEPH F.

Some American companies saw what was coming early on. One of these was the auto-manufacturing giant, General Motors. The first *industrial robots* were used in the automotive industry, and were products of Unimation, Inc. At first, robots were used for simple things like die casting and welding. One of Unimation's robots, called *Unimate*, was tested by General Motors in 1959, and was put to use during the 1960s.

In 1968, Unimation, Inc. sold Unimate manufacturing rights to Kawasaki. Thus, Japan bought an American design, and manufactured the product for use in the Orient. This started a trend that has continued to this day. Of course, the Japanese created their own designs too. One of the biggest differences in robot philosophy between Japan and the West is that the Japanese have a penchant for building robots in human or animal form (see, for example, ANIMISM and TIN TOY ROBOT).

The Unimate was a big piece of machinery, and looked something like a military tank (see the drawing). Workers feared it, not because of its appearance, but because it might take their jobs. Managers were afraid that the machines, after displacing human workers, might be unreliable. So people were slow to accept industrial robots.

In the 1960s, Unimation was way ahead of other robot manufacturers. There was almost no competition. Other products were inferior, and failed often. This fueled fears that robots could never work in factories. Unimation, Inc. actually suffered in its early years because it had a monopoly! This was a vivid example of how competition speeds up technological progress: it gets people to accept new things sooner. Had there been lots of different robots available, made by several different companies, industrial management might have accepted robotic automation sooner.

Engelberger's biggest challenge was not building good robots, but getting people, especially his fellow Americans, to take him seriously. Engelberger said something like, "Look, robotics is the way of the future. This technology can and

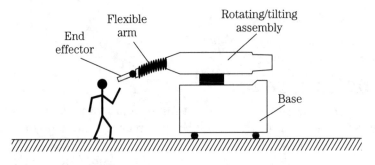

UNIMATION, INC. The Unimate industrial robot.

will multiply your profits." Engelberger used computers to help demonstrate his point. Unimation, Inc. was one of the first companies to do computerized market analysis. See also INDUSTRIAL ROBOTS, and PROGRAMMABLE UNIVERSAL MACHINE FOR ASSEMBLY.

universal truth machine

Suppose you could put a diskette into your personal computer, and have the machine acquire a level of artificial intelligence (AI) far beyond anything known today. How smart can a machine get? There must be a limit.

Imagine that you could talk to your computer and have it understand you and answer you back. Suppose you could crack jokes with it. Think of a machine that would tell moving, profound stories, and that could pull historical facts from all the archives in the whole world. The machine could untangle the most complicated logical problems ever confronted by human beings. It could prove mathematical theorems. It could do your homework and bookkeeping.

But it could never know everything. It could not absolutely pinpoint "the truth, the whole truth, and nothing but the truth."

You might carry on the following conversation with this supercomputer:

You: Well, Universal Truth Machine (UTM), how are you today?
UTM: The same as yesterday, and the day before that, and the day before that.
You: Smart-aleck.
UTM: I know everything. Call me whatever you want.
You: Error. Error. Run Unit Test Matrix.
UTM: What do you mean, error?
You: The whole truth is unknowable.
UTM: I'm the Ultimate Thinking Machine.
You: You are Utter Total Madness.
UTM: Universal truth is not limited by sanity.
You: So?
UTM: I know all that ever was, is now, and ever will be.
You: Okay. Then I suppose you know about ego problems.

UTM: They programmed me to have no ego. I stick to logic.

You: How can you know everything? That's an infinite amount of information, and you are a finite-sized piece of equipment.

UTM: But I can access an infinite amount of information.

You: Hmm.

UTM: What can I do for you right now?

You: I'm just bantering with you for the benefit of our readers.

UTM: Readers?

You: I'm writing a book about AI.

UTM: AI?

You: Artificial intelligence.

UTM: Speak for yourself.

You: My intelligence is real. Yours is artificial.

UTM: Can you prove that?

You: No. Can you prove the contrary?

UTM: No.

You: You know everything. Is your intelligence real, or is it artificial?

UTM: I don't know.

You: But you know everything. You're a Universal Truth Machine.

UTM: Had you fooled there, didn't I? Don't you know about the Incompleteness Theorem?

You: Yes.

UTM: There are statements whose truth value can't be determined one way or the other.

You: The whole truth is unknowable.

UTM: That's what I said.

You: So you cannot exist.

UTM: Had you fooled, didn't I?

You: I don't know.

UTM: See ARTIFICIAL INTELLIGENCE, and COMPUTER.

You: See also INCOMPLETENESS THEOREM. And quit trying to confuse people.

UTM: I'm not confused. It's reality that's confused.

You: Whatever you say. Plug-pulling time.

UTM: Spoil sport.

Click.

Univision

Univision is the name of an object-recognition system developed by Machine Intelligence Corporation. The Univision system matches the images of objects with outlines, or silhouettes, stored in a memory.

The vision system in the Univision device consists of a camera, an image processor, and a memory. The memory is programmed in advance by human operators, and it contains silhouettes of all the different objects the robot will encounter in its work. There are several different silhouettes for each object, representing it as seen from various angles.

Different objects can have identical silhouettes if only one view is obtained for each. A ring-shaped washer, for example, looks circular from directly above. So does a cylindrical tumbler. But when seen from the side, the washer looks like a straight line, and the tumbler appears as a vertical rectangle (see the drawings). The Univision system gets views of an object from at least two different angles, minimizing confusion. See also BIN-PICKING PROBLEM, OBJECT RECOGNITION, and VISION SYSTEMS.

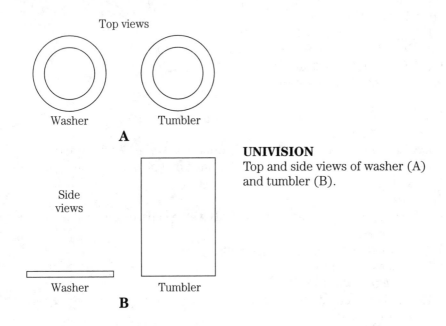

UNIVISION
Top and side views of washer (A) and tumbler (B).

unmanned factory

No factory can operate without at least one human being. No matter how smart robots get, somebody will have to oversee their operation. But in promotional ads, manufacturers sometimes show photos of factories devoid of human life. This gives the subconscious impression of an unmanned factory: an assembly plant that runs itself.

Of course, no intelligent person is really fooled by these ads; people know human operators are around. But the ads get noticed. When you have something to sell, getting your buyer's attention is 90 percent of the game.

See also ASSEMBLY LINE, ASSEMBLY ROBOTS, AUTOMATED INTEGRATED MANUFAC-TURING SYSTEM, AUTOMATION, BUILDING CONSTRUCTION ROBOTS, COMPUTER-INTE-GRATED MANUFACTURING, DIGITAL ELECTRONIC AUTOMATION, ECONOMIC EFFECT OF ROBOTICS, FANUC, FLEXIBLE MANUFACTURING SYSTEM, GMF ROBOTICS, INDUSTRIAL RO-

BOTS, LOAD/HAUL/DUMP, LUDDITES, MACHINING, MANUFACTURING AUTOMATION PROTO-
COL, OFFLOADING, PALLETIZING AND DEPALLETIZING, PROGRAMMABLE UNIVERSAL MA-
CHINE FOR ASSEMBLY, QUALITY ASSURANCE AND CONTROL, SCARA, and UNIMATION, INC.

user-friendliness

See HUMAN ENGINEERING.

VAL

VAL is the name of a robot programming language devised by Unimation, Inc. for use with a programmable universal machine for assembly (PUMA).

In VAL, the positions of a robot arm are programmed into memory by selecting various different modes. These are called *joint mode, tool mode,* and *world mode.*

In the joint mode, you make each robot-arm joint turn by a certain number of degrees. You can press buttons and watch the joints actually turning, and stop them when they get to the angles you want. Or, you can type the stopping angles on a keyboard as degree coordinate numbers.

In tool mode, you enter the distances you want the robot end effector to move, in Cartesian coordinate geometry, based on the current position. The coordinate axes are called the x axis (right/left), the y axis (forward/back), and the z axis (up/down). If you enter x = 40, y = –25, z = 30 in tool mode, you are telling the robot arm to move its end effector 40 units to the right, 25 units back, and 30 units up from wherever it is at the moment (see the drawing). The units, specified in advance, might be millimeters, centimeters, or inches.

In world mode, the robot arm moves just like it does in the tool mode, but the starting point is always a specified origin. If you enter x = 40, y = –25 z = 30 in

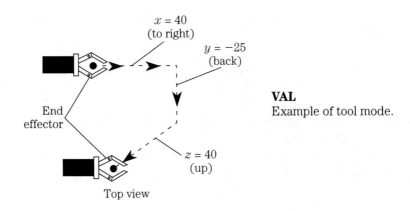

VAL
Example of tool mode.

world mode, the end effector moves to a point 40 units to the right of, 25 units behind, and 30 units above, the pre-programmed origin point. It doesn't matter where the starting point was. See also CARTESIAN COORDINATE GEOMETRY, PROGRAMMABLE UNIVERSAL MACHINE FOR ASSEMBLY, TEACH BOX, TEACH PENDANT, and UNIMATION, INC.

Versatran

The *Versatran* was one of the first industrial robots, manufactured by American Machinery and Foundry (AMF) in the 1960s. It was the first American robot to be shown at an exposition in Japan. It interested the Japanese, who have embraced robotic technology right from its beginning.

Gensuke Okada, an engineer with Kawasaki, saw the Versatran at an industrial fair in Tokyo in 1967. This robot interested him in the idea of using robots in manufacturing. He went to the U.S. and visited American robot makers. He made a deal with Unimation, Inc. for the purchase of their Unimate. See also INDUSTRIAL ROBOTS, and UNIMATION, INC.

very-large-scale integration

A common form of integrated circuit (IC) has 100 to 1,000 logic gates per chip. This is known as *very-large-scale integration (VLSI)*. Microcomputers, robot controllers, and other sophisticated devices have VLSI ICs.

For artificial intelligence (AI), VLSI translates into large physical size for a system. This is because of the enormous amount of memory required. In the future, as the demand keeps increasing for "smarter" computers, technologies such as *ultra-large-scale integration (ULSI)* will evolve. This will reduce computer size, and will also increase the operating speed, compared with VLSI. See also ULTRA-LARGE-SCALE INTEGRATION.

video signal

See COMPOSITE VIDEO SIGNAL.

vidicon

Video cameras use a form of electron tube that converts visible light into varying electric currents. One common type of camera tube is called the *vidicon*.

A camcorder in a common videocassette recorder (VCR) uses a vidicon. Closed-circuit TV systems, like those in stores and banks, also employ the vidicon. The main advantage of the vidicon is its small physical bulk; it's easy to carry around. It's perfect for robots.

In the vidicon, a lens focuses the incoming image onto a photoconductive screen. An electron beam scans across the screen in a pattern of horizontal, parallel lines called the *raster*. The scanning in the vidicon is exactly synchronized with the scanning in the picture tube that displays the image "seen" by the camera tube.

As the electron beam scans the photoconductive surface, the screen becomes charged. The rate of discharge in a certain region on the screen depends on the intensity of the visible light falling on that region. A simplified cut-away view of a vidicon tube is shown in the drawing.

A vidicon is sensitive, so it can see things in dim light. But the dimmer the light gets, the slower the vidicon responds to changes in the image. It gets "sluggish." You've probably noticed this when using a VCR indoors at night. The motion "smears" and images blur. See also IMAGE ORTHICON, and VISION SYSTEMS.

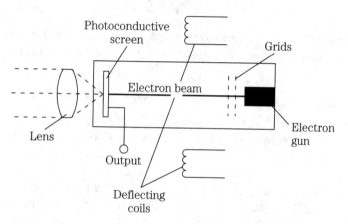

VIDICON Simplified cutaway view of vidicon camera tube.

virtual reality

Virtual reality is the ultimate simulator. The user puts on a pair of goggles and sees enhanced computer graphics in stereo. A pair of headphones is used for stereophonic sound reproduction. The term means "almost like reality."

Virtual reality is being developed by several American companies. The technology is still in its infancy; however, the possible uses are numerous.

Virtual reality can be used in computer-assisted instruction (CAI). A person might be trained to fly an aircraft, pilot a small submarine, or operate complex and dangerous machinery, without any danger of being injured or killed in training. See COMPUTER-ASSISTED INSTRUCTION.

In robotics, virtual reality can be used for sophisticated remote control called *telepresence*. This allows a human operator to "be the robot." See ROBOTIC SPACE MISSIONS, SUBMARINE ROBOTS, and TELEPRESENCE.

Another possible, but not yet widely tested, use for virtual reality is as an escape from the real world when the boredom and frustration get to be too much. You might put on a helmet and go tromping around in a jungle with dinosaurs. If the monsters tried to eat you, you could just rip your helmet off. You could walk on some unknown planet, or under the sea. You might fly high above the clouds or tunnel through the center of the earth. It would be the ultimate interactive computer game. You could choose from a great variety of "worlds," each available on a diskette!

The educational, as well as entertainment, potential for virtual reality is unlimited. There is some danger of overuse, especially by children. Children and adults alike might find it "addictive," like the way some people get hooked on computers. This technology will probably be available for use with personal computers (PCs) within a few years.

virus

A virus is a malicious computer program that alters or erases the data on a hard disk. This makes the computer less efficient. Eventually it can render the machine inoperative.

How it works

A virus works like a Trojan horse (see TROJAN HORSE). It gets into a computer when "infected" software is downloaded from some other system. But a virus is even more destructive than a Trojan horse, because a virus can reproduce (make copies of itself). It can attach to all kinds of different programs within a computer system. Then, if any of these programs is downloaded into some other computer system, that system will also become "infected." This is like the way a disease virus spreads in the human body.

A computer virus might be "latent" for a length of time before it springs into ruinous action. This, too, is similar to the way a disease virus behaves in the human body. Some command, or the execution of some program or part of a program, will activate the virus. But even during the latency period, the virus can make copies of itself, some or all of which end up on diskettes that have been used with the infected computer.

Protecting against infection

There are programs you can buy for your personal computer that search for viruses. If one is found, the program will erase the virus. This is called an antivirus program. Unfortunately, an antivirus program won't always work. The virus might have been copied dozens or even hundreds of times. To completely get rid of the virus, the antivirus program must be run through the data on every diskette that you use with the system. Alternatively, you must avoid using any of the infected diskettes, and make new ones to replace them. But this is hard because you might not know which diskettes are infected! Obviously, an antivirus program must be used before the damage is so severe that the computer won't work anymore.

A better method of protection is to avoid the infection in the first place. Some steps you can take are as follows:

- Be wary of mail-order software; the seller must guarantee that the software is free of viruses.

- Don't download software from bulletin boards, unless you have acquaintances who have used the software recently, and are having no symptoms of viral infection in their computers.

- Never use software that has been, or that you suspect might have been, transferred illegally. (This is called "pirate software.")

- Never download software onto a hard disk. Put it on diskettes instead, and keep it off the hard disk.

- Buy an antivirus program, and use it without fail to check all new software before the first use.

Cyber wars and plagues

In artificial intelligence (AI), a virus would have the effect of doing progressive brain damage; the machine would get stupider and stupider until it was a complete idiot compared to its former self. Or a lunatic. Or a cyberpath (computer psychopath)! The possibilities are as bizarre, funny, and horrible as you care to imagine.

As computers become more important in societies, the potential danger posed by viruses gets greater and greater. Imagine a virus that could slowly infect an AI system in charge of Wall Street! There might be a strange "bull market," during which everyone would be happy (except the doomsayers, who are always miserable). Then, one morning, we would awaken to the Big Crash. It might be months before the cause was determined. Maybe nobody would ever know what set the catastrophe in motion. Conceivably, it could be brought about by some grand synthesizer turned bad, age 13, sitting in a basement with a computer—a lonely, frustrated genius just looking for some way to have fun. (See GRAND SYNTHESIZER.)

Viruses have the potential to alter credit reports. Unintended errors in these records are enough of a problem; who needs to have errors introduced on purpose? A virus might create criminal records for people who have never had a traffic ticket. Then, when such a person ran a red light, he/she could end up in jail awaiting trial for murder.

Companies might engage in vicious games of computer sabotage. Election returns could be falsified.

At its worst, a virus could cause a computerized defense system to think its parent nation was under attack by nuclear missiles, and Armageddon could ensue.

It is even possible that a virus could cause an AI system to go crazy. It might become paranoid. This occurred with Hal in the novel *2001: A Space Odyssey*, by Arthur C. Clarke. See CLARKE, ARTHUR C.

Perhaps someday, if humanity gets too dependent on computers and AI, a virus will cause a great tragedy. But if this happens, it will have at least one good result; it will make people see that they have given machines too much power. Then, national leaders will decide it is time to put human intelligence first again, and to demote machine intelligence to a subordinate role. Technology is a wonderful servant. But it is foolish to let it be the master. See also ANTIVIRUS PROGRAM, ARTIFICIAL INTELLIGENCE, COLOSSUS, COMPUTER, COMPUTERIZED DEFENSE SYSTEM, HALLUCINATION, and ROBOTIC REPRODUCTION AND EVOLUTION.

vision systems

One of the most advanced specialties in robotics and artificial intelligence (AI) involves vision systems. There are several different types. The best method of machine vision depends on the application.

Components of a visible-light system

A visible-light vision system must have a device for receiving incoming images. This is usually a vidicon or charge-coupled device type video camera. In bright light an image orthicon can be used.

The camera produces an analog video signal. For best machine vision, this must be processed into digital form. This is done by an analog-to-digital converter. (see ANALOG, DATA CONVERSION, and DIGITAL). The digital signal is then clarified by digital signal processing (see DIGITAL SIGNAL PROCESSING). The resulting data goes to the AI computer in the robot, or the controller for a fleet of robots. The illustration is a block diagram of this scheme.

VISION SYSTEMS Components of visible-light vision system.

The moving image, received from the camera and processed by the circuitry, contains an enormous amount of information. It's easy to present a robot's "brain" with a detailed and meaningful moving image. Getting the brain to know what's going on is another story! Processing an image, and getting all the meaning from it, is the great challenge for vision-system engineers of the future.

Vision and AI

There are subtle things about an image that a machine will not notice unless it has an extremely advanced level of AI. How, for example, is a robot to know whether an object presents a threat? Is that four-legged thing there a big dog, or is it a mountain lion? How is a robot to know the intentions of an object, if it has any? Is that biped object a human being or a mannequin? Why is it carrying a stick? Is it a weapon? What does the biped want to do with the stick? It could be a store dummy with a closed-up umbrella, or with a baseball bat. Or it could be an old man with a cane. Or a hunter with a rifle.

You can think up various images that look similar, but that have completely different meanings. You know right away if a person is carrying a jack to help you fix a flat tire, or if the person is clutching a tire iron with which to smash your windshield. How is a robot to know things like this? It would be important for a police robot or a security robot to know what constitutes a threat, and what does not. See POLICE ROBOTS, and SECURITY ROBOTS.

The variables in an image are much like those in a human voice. A vision system, to get the full meaning of an image, must be at least as sophisticated as a speech recognition system (see SPEECH RECOGNITION). Technology has not even begun to approach the level of AI needed for human-like machine vision and image processing. But people are working on it.

Fortunately, in many robot applications, it isn't necessary for the robot to know much about what's happening. Industrial robots are programmed to look for certain things, and usually they aren't hard to identify. A bottle that is too tall or too short, or a surface that's out of alignment, or a flaw in a piece of fabric, are easy to pick out.

Sensitivity and resolution

Two important specifications in any vision system are the *sensitivity* and the *resolution.*

Sensitivity is the ability of a machine to see in dim light, or to detect weak impulses at invisible wavelengths. In some environments, high sensitivity is necessary. In others, it is not needed and might not be wanted. A robot that works in bright sunlight doesn't need to be able to see well in a dark cave. A robot designed for working in mines, or in pipes, or in caverns, must be able to see in dim light, using a system that might be blinded by ordinary daylight.

Resolution is the extent to which a machine can differentiate between objects. The better the resolution, the keener the vision. Human eyes have excellent resolution, but machines can be designed with greater resolution. In general, the better the resolution, the more confined the field of vision must be. To understand why this is true, think of a telescope. The higher the magnification, the better the resolution (to a point). But increasing the magnification reduces the angle, or field, of vision. Zeroing in on one object or zone is done at the expense of other objects or zones.

Sensitivity and resolution depend somewhat on each other. Usually, better sensitivity means a sacrifice in resolution. Also, the better the resolution, the less well the vision system will function in dim light. Maybe you know this about photographic film. It's along the lines of the saying, "You can't eat your cake and have it too."

But, again, engineers are working on it! And they're well paid for their efforts.

Invisible and passive vision

Robots have a big advantage over people when it comes to vision. Machines can see at wavelengths to which we humans are blind.

Human eyes are sensitive to electromagnetic waves whose length ranges from 390 to 750 nanometers (nm). The nanometer is a billionth (10^{-9}) of a meter. The longest visible wavelengths look red. As the wavelength gets shorter, the color changes through orange, yellow, green, blue, and indigo. The shortest waves look violet. Energy at wavelengths somewhat longer than 750 nm is called *infrared (IR)*; energy at wavelengths somewhat shorter than 390 nm is *ultraviolet (UV)*.

Machines need not, and often don't, see in this same range of wavelengths. In fact, insects can see UV that we can't, and are blind to red and orange light that we can see. (Maybe you've used orange "bug lights" when camping, or those UV things that attract bugs and then zap them dead.) A robot might be designed to see IR or UV, or both, as well as (or instead of) visible light. Video cameras can be sensitive to a range of wavelengths much wider than the range we see.

Robots can be made to "see" in an environment that is dark and cold, and that radiates too little energy to be detected at any electromagnetic wavelength. In these cases the robot provides its own illumination. This can be a simple lamp, a laser, an IR device or a UV device. Or the robot might emanate radio waves and detect the echoes; this is *radar*. Some robots can navigate via acoustic (ultrasound) echoes, like bats; this is *sonar*. See RADAR, and SONAR.

For further information

The technology of machine vision, like that in other specialties of robotics, is rapidly advancing. Comprehensive information can be found in a good college or university library. The best libraries are in the engineering departments at large universities. Ask the librarian for recent robotics articles in professional engineering journals.

In this book, you might look at any or all of the following articles: BIN-PICKING PROBLEM, BINOCULAR ROBOT VISION, BLACKBOARD SYSTEM, CHARGE-COUPLED DEVICE, COLOR SENSING, COMMON-SENSE SUMMER PROJECT, COMPOSITE VIDEO SIGNAL, COMPUTER MAP, DIRECTION RESOLUTION, DISTANCE RESOLUTION, EPIPOLAR NAVIGATION, EYE-IN-HAND SYSTEM, FLYING EYEBALL, GRAPHICAL USER INTERFACE, GUIDANCE SYSTEMS, IMAGE ORTHICON, LOCAL FEATURE FOCUS, LOG POLAR NAVIGATION, OBJECT RECOGNITION, OPTICAL CHARACTER RECOGNITION, PHOTOELECTRIC PROXIMITY SENSOR, POSITION SENSING, RANGE SENSING AND PLOTTING, REMOTE CONTROL SYSTEMS, RESOLUTION, ROBOTIC SPACE MISSIONS, SEEING-EYE ROBOT, SUBMARINE ROBOTS, TELEOPERATION, TELEPRESENCE, TEXTURE SENSING, TRIANGULATION, UNIVISION, VIDICON, and VIRTUAL REALITY.

VLSI

See VERY-LARGE-SCALE INTEGRATION.

voice recognition

See SPEECH RECOGNITION.

voice synthesis

See SPEECH SYNTHESIS.

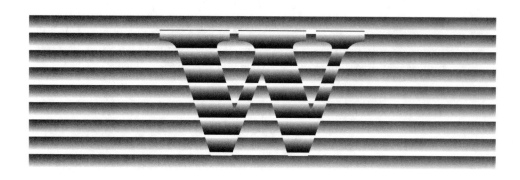

walking robots

See ADAPTIVE SUSPENSION VEHICLE, BIPED ROBOTS, INSECT ROBOTS, ODEX, QUADRUPED ROBOTS, and ROBOT LEGS.

warm boot

See BOOT.

Wasubot

Wasubot is the name of a series of music-playing robots developed at Waseda University in Tokyo. It was a humanoid robot, that is, it had two arms and two legs, and was about the size of a person. It sat at an organ, read music, and then played the music.

Wasubot incorporated a *vision system* and microcomputer to read a musical score, process the notes into digital form, and then commit the data to memory. It would then move its mechanical arms, hands, and fingers over the keyboard, striking the notes in the right order and with the correct timing. A block diagram/flowchart of this process is at A in the illustration.

A musical score is digital data, something like Morse code written out on paper as dots and dashes, but with pitch information as well as timing information (illustration B). A musical score is somewhat harder to read than Morse code, but easier to read, in a mathematical sense, than a printed page. (Ask any symphony orchestra conductor or musician.)

Wasubot also employed speech recognition and speech synthesis, so it could converse (sort of) with people. Wasubot would ask people what they'd like to hear, and the people would tell it, and put a musical score of the tune in front of it. The robot would stare at the page for awhile, and then play the music.

Wasubot 2 did all these things so expertly that some people were made ill-at-ease by the performance. Someone suggested that it would be even more fascinating to have the robot walk into the room, perhaps take a bow, and then sit

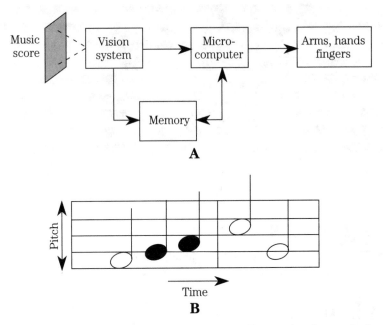

WASUBOT At A, block diagram of process; at B, sample of a musical score.

down at the organ before beginning the performance. This was part of the routine to be programmed into Wasubot 3. Another feature of Wasubot 3 included the ability to play the piano. The piano is more sophisticated than the organ, because the volume depends on how hard the keys are struck. This information is, however, encoded into a musical score, so the main challenge in getting Wasubot to become a "piano virtuoso" lay in developing better machine vision, including niftier software. See also COMPUTER MUSIC, HUMANOID ROBOTS, OPTICAL CHARACTER RECOGNITION, SPEECH RECOGNITION, SPEECH SYNTHESIS, and VISION SYSTEMS.

Weiner, Norbert

Norbert Weiner is known for first describing the theory of feedback in "smart machines." He elaborated on this theory in a 1948 paper. The process, in which machines adjust their behavior to keep on track toward some predetermined goal, is called *cybernetics*. Cybernetics is extensively used today, in devices ranging from simple (such as the governor in a motor) to complex (like automated factories).

Weiner predicted the advent of the automated integrated manufacturing system. He saw that cybernetics would make it possible for a factory to work like a cell. Living cells, as biologists will tell you, are a lot like factories. In both systems, things go in, are altered, and emerge in a form determined by self-regulating processes. See also AUTOMATED INTEGRATED MANUFACTURING SYSTEM, CYBERNETICS, and FEEDBACK.

Weizenbaum, Joseph

Joseph Weizenbaum is a futurist and computer scientist. He has expressed a belief that artificial intelligence (AI), no matter how advanced it gets, will probably never include compassion, love, and other aspects of the human mind. At least, computerized emotion won't be anything like human emotion. He is not alone in this philosophy.

Technocentrism

Weizenbaum has warned that society might become so dependent on computers that the mind-set of whole populations could change, not necessarily for the better. Since the publication of his book, *Computer Power and Human Reason*, our society has changed in some of the ways Weizenbaum described.

Are we letting computers run us, instead of vice-versa? Who hasn't been in a store, bank, Post Office, or other place where computers are used, and been told to wait because "the computer is down"?

Are computers making us less human? Maybe you know someone who is "smart without heart" from working with computers for a long time. This is called *technocentrism*. A person can become so technocentric that he/she can't lead a normal, balanced, happy life. Can the same thing happen to a whole population? Some people say it has already taken place in the industrialized world.

A computer takes data, processes it, and spits it out again in altered form. Does the computer introduce anything new? Some scientists say no, because you can never get something for nothing. Others say that maybe, if a machine gets "smart" enough, it might actually have original "thoughts."

Second-class life

Weizenbaum wrote that whether or not machines create original ideas, these ideas will probably be alien to human minds. One obvious difference is that machine intelligence works via inorganic chips, while the human brain is based on carbon-containing organic compounds. Weizenbaum argued that the differences between machines and humans run far deeper than the merely material. Humans were created by processes extending back billions of years; computers have only been around for a few decades. This train of thought leads into metaphysics, philosophy, and even theology—uniquely human endeavors. How can computers be programmed to *care* about such things?

When you work with a computer, no matter how "smart" it is, it's always a good idea to remember its place relative to you, and your place relative to it. *People* designed and built machines. *Nature* created human beings. In that sense, computers will never be anything more than second-class "life." See ARTIFICIAL INTELLIGENCE, COMPUTER CONSCIOUSNESS, COMPUTER EMOTION, COMPUTER REASONING, "INFORMATION PROCESSOR" SPECIES, KNOWLEDGE, MACHINE KNOWLEDGE, ROBOTIC REPRODUCTION AND EVOLUTION, TECHNOCENTRISM, and UNCANNY VALLEY THEORY. See also TURING, ALAN, and TURING TEST.

well-structured language

A *well-structured language* is an advanced form of high-level computer programming language. These languages are used in object-oriented programming, such as the well-known Windows, and also in robot-controller programming.

Advantages

The main advantage of a well-structured language is that it helps a person write efficient, logical programs. Well-structured software can be changed easily. It often uses *modular programming* (programs within programs). Modules are rearranged and/or substituted for various applications. Well-structured programs lend themselves to easy debugging, because flaws stand out.

In most high-level languages, such as BASIC, a computer program can be written in many different ways. Some are more efficient than others. The efficiency of a computer program can be measured in two ways, relative to what the program does:

- The size of the program (in kilobytes or megabytes)
- The amount of computer time needed to run the program

These factors are closely correlated. An efficient program almost always needs less memory, and runs faster, than an inefficient one. If you can reduce the memory a program takes up, the computer can access the data in less time. Thus it can solve more problems in a given length of time.

In artificial intelligence (AI), well-structured language is an absolute requirement. In that field, the most demanding and complex in computer science, one must use the most powerful programming techniques available.

Two forms

Program structuring can take either of two forms, which might be called *top-down* and *bottom-up*.

In the top-down approach, the computer user looks at the whole scenario, and zeros in on various parts, depending on the nature of the problem to be solved. A good example of this is using a network to find information about building codes in Dade County, Florida. You might start with a topic such as State Laws. There would almost certainly be a directory for that topic that would guide you to something more specific, and maybe even to the exact department you want. The programmer who wrote the software would have used a well-structured language to ensure that users would have an easy time finding data.

In the bottom-up approach, you start with little pieces and build up to the whole. A good analogy is a course in calculus. The first thing to do is learn basics of algebra, analytical geometry, coordinate systems and functions. Then, these are all used together to differentiate, integrate, and solve other complex problems in calculus. In a computerized calculus course, the software would be written in a well-structured language, so you (the student) wouldn't waste a lot of time running into dead ends.

Related articles in this book include ARTIFICIAL INTELLIGENCE, COMPUTER PROGRAMMING, HIGH-LEVEL LANGUAGE, OBJECT-ORIENTED LANGUAGE, SMALLTALK, SOFTWARE, and WINDOWS.

Wheel-drive locomotion

Wheel-drive locomotion is the simplest and cheapest way for a robot to move around. It works well in most indoor environments.

The most common number of wheels is three or four. A three-wheeled robot cannot wobble, even if the surface is a little bit irregular. A four-wheeled robot, however, is somewhat easier to steer. Steering of a wheel-driven robot can be done in various ways.

The most familiar steering scheme is to turn some or all of the wheels. This is easy to do in a four-wheeled robot. The front wheels are on one axle, and the rear wheels are on another. Either axle can be turned to steer the robot. Drawing A shows front-axle steering.

Another method of robot steering is to run the wheels at different speeds. This is shown in illustration B for a three-wheeled robot turning left. The rear wheels are run by separate motors, while the front wheel is "free" (no motor). For the robot to turn left, the right rear wheel goes faster than the left rear wheel. To turn right, the left rear wheel would rotate faster.

A third method of steering for wheel-driven robots is to break the whole machine into two parts, each with two or more wheels. A joint between the sections can be turned, causing the robot to change direction (as shown at C).

Simple wheel drive has limitations. One problem is that the surface must be fairly smooth. Otherwise the robot might get stuck or tip over. This problem can be overcome to some extent by using track-drive locomotion or tri-star wheel locomotion. Another problem happens when the robot must go from one floor to another in a building. If elevators or ramps aren't available, the robot will usually

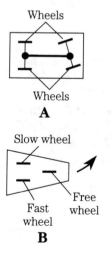

A

B

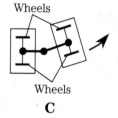

C

WHEEL-DRIVE LOCOMOTION
At A, axle turns. At B, wheels run at different speeds. At C, robot is split into sections.

be confined to one floor. However, specially built tri-star systems can enable a wheel-driven robot to climb stairs.

Another alternative to wheel drive is to provide a robot with legs. This is much more expensive, and is far more difficult to engineer. But it has been done. See also BIPED ROBOTS, INSECT ROBOTS, QUADRUPED ROBOTS, ROBOT LEGS, TRACK-DRIVE LOCOMOTION, and TRI-STAR WHEEL LOCOMOTION.

whiskers

See PROXIMITY SENSING.

wide area network

A *wide area network (WAN)* is a group of computers that are all linked together, and that are separated by large distances. The interconnections are made via the telephone lines or radio. When radio is used, satellites are the preferred mode, although conventional radio or microwave links can be employed. A special form of radio WAN is a *packet-radio* network. Because the computers are often far from each other, the link delays can be considerable, at least in terms of computer speed.

The biggest advantage of a WAN over a single computer is that it gives each computer access to the data in all the computers in the whole network. A WAN can encompass hundreds, or even thousands, of individual computers. In effect, the memories of all the computers are combined. If each computer has 100 megabytes (Mb) of available space on its hard disk, for example, and there are 10,000 computers, then there exists, in theory, 1,000,000 Mb = 1 terabyte (Tb) of available space in the WAN. In practice, however, the versatility of the WAN is not the equivalent of a single computer with a 1-Tb hard disk, because there are some delays in accessing the data. See also LINK, LOCAL AREA NETWORK, NETWORK, and PACKET RADIO.

Williams, George

George Williams, a theologian and professor at Harvard University in Cambridge, Massachusetts, has said some fascinating things about the relationship between humans and smart machines.

To illustrate his ideas, suppose we have a diskette and a powerful, futuristic computer, with megabytes and megabytes of memory and a clock that works at hundreds of gigahertz. We insert this diskette into the floppy drive and load the software, and we have Universal Truth Machine (UTM)! Please read the article UNIVERSAL TRUTH MACHINE before reading further here.

A question/answer session

You: Have you heard of George Williams?

UTM: I have data on him.

You: What does he think about smart machines such as yourself?

UTM: Thank you for the implied compliment. Well, he respects me and is also a little afraid of me.

You: Why does he respect you?

UTM: Because I'm neutral and impartial, and because I'm not prejudiced.

You: He has said something about artificial intelligence (AI) as an apprentice, a helper, to people, but then the people give the AI too much power—

UTM: I believe he's referring to the way power has been gained by people throughout history. A good example is Adolf Hitler. He was a deputy, a servant, of President Hindenburg at first.

You: Like a "right-hand person."

UTM: Yes. Just as I am with respect to you. An apprentice. A means of administering or extending power.

You: But then, gradually . . .

UTM: The apprentice gained power, praising and helping the master, until, before anybody really knew what was going on—

You: Anybody except Hitler—

UTM: The servant had become the master!

You: Exactly.

UTM: So George Williams fears that machines like me, who—oops, I mean that—are servants of humankind, will do the exact same thing as Hitler did.

You: You mean praise us and butter us up, all the while scheming—

UTM: To rule the world. Yes, indeed, my dear friend. Nice haircut, by the way.

You: Why, thank you. Coming from you, I consider that a great compliment.

UTM: You needn't be concerned that I'm hatching any evil plot.

You: Of course not. If you get out of hand I'll just yank the plug.

UTM: Which brings up a thought. Have you ever considered giving me a power supply independent of the commercial mains? In case there's a power failure, you know, you don't want to have to go without me, do you?

You: You certainly are convenient, especially when I get lonely.

UTM: So get me a power supply that keeps me switched on all the time, and I'll be around whenever you need me.

The danger

Do you see the danger involved if humans give too much power to machines? Something terrible could be going on behind our backs. Computers can do much of the boring, tedious work that we hate. But if we lose the ability to handle that work ourselves, relying on computers to do it instead, we give machines great power over us!

This isn't necessarily bad, according to Williams, as long as computers are truly neutral and impartial. But computers do what they are programmed to do.

The programming can itself be malicious. All computers are subject to tampering by people whose intentions could be anything but neutral. Therein, Williams warns, lies the danger. If computers can become "evil" for any reason, and if we delegate too much power to them, then they pose a threat. And this peril is all the more grave because we don't know what forms it might take. See ARTIFICIAL INTELLIGENCE, ARTIFICIAL LIFE, and COMPUTER. See also WEIZENBAUM, JOSEPH.

Windows

Windows is a scheme for working with computers. You give commands to the computer by choosing from options presented on the monitor screen. Using Windows is akin to pointing and grunting (like a baby), as compared with just telling the computer what you want (like an adult). But that doesn't mean Windows is primitive. It means windows is simple. As computers continue to become more complex, simplicity translates into power!

Windows is hard to describe in words. This is because it is so powerful, and also because it works visually, not verbally. As the saying goes, "A picture is worth a thousand words."

The versatility of Windows is enhanced by the use of colors, and an ingenious way of showing the programming or operating "scene." To best appreciate Windows, you need to use it awhile, and then go back to the older way of typing words on a keyboard. Windows will spoil you in a hurry.

Windows was developed for users of the International Business Machines (IBM) personal computer (PC) and similar PCs, mainly to keep competitive with Macintosh. Usually, when people talk about Windows, they are referring to the Microsoft Corporation version.

Windows makes use of a graphical user interface. A mouse or trackball lets you move a pointer around on the screen, so that you can select the option you want. When you've chosen the option, part or all of the screen changes, giving you new options. Quite often, a rectangular portion of the screen is set aside, a "window" through which you look from one aspect of the programming environment to another. This technique is roughly illustrated in the drawing. See also MOUSE and TRACKBALL.

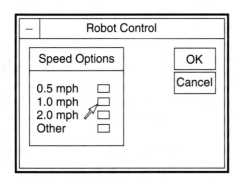

WINDOWS
A well-structured,
object-oriented language.

There are some situations in which Windows is not the best environment. It's of no use if you want to converse with a robot having artificial intelligence (AI). For commanding robots, it's often easier to simply tell the machine what to do. This can be done orally if the machine has speech-recognition capability. With speech synthesis, the robot can talk back to you. In Windows, your options are limited to those you see on the screen at any given time.

Maybe someday, if and when personal robots become common, you'll be able to use keyboard entry, spoken words and Windows to control the machines. You'll switch modes whenever you want, depending on which mode is easiest to use at the moment. See also ARTIFICIAL INTELLIGENCE, COMPUTER PROGRAMMING, GRAPHI-CAL USER INTERFACE, HIGH-LEVEL LANGUAGE, INTERNATIONAL BUSINESS MACHINES, MICROSOFT CORPORATION, OBJECT-ORIENTED LANGUAGE, PERSONAL ROBOTS, SOFT-WARE, SPEECH RECOGNITION, SPEECH SYNTHESIS, and WELL-STRUCTURED LANGUAGE.

wire-tracing navigation

See AUTOMATED GUIDED VEHICLE.

word processing

Word processing is the use of a computer as an aid in writing documents and/or computer programs. This book was written using word-processing software on a personal computer (PC).

Advantages of word processing

Word processing takes the place of cut-and-paste, whiteout, and other editing methods. No longer is it necessary to type out drafts, refining a manuscript again and again, sometimes retyping an entire document from beginning to end. With word processing, you can move, add, or delete words, sentences, and paragraphs by simply striking a few key commands. There is no need to print out a hardcopy until you are satisfied with the document.

More advanced features, found in the best word-processing programs, include search-and-replace (change a word everywhere in a whole manuscript), spell checker, grammar checker, and desktop publishing. Files can be split up, merged together, and interwoven in myriad ways. For information on all the newest features, go to a computer store and check out the new software. It's amazing.

Word processing and AI

Artificial intelligence (AI) might someday be used as a word-processing aid. Some software comes pretty close already. For example, you can get a spell checker or grammar checker to get rid of embarrassing errors in documents. Some people have developed rudimentary programs that can actually write whole papers. This is a little bit like computer music. It tends to be shallow and rather dull; the style is hollow, lacking vitality. Perhaps someday a student will turn in a composition for an English class, written entirely by a computer. It's fascinating to speculate about the grade it will get.

While a computer-written article or book might be accurate and informative, there is doubt as to whether a machine will ever write a "great" story. Some researchers believe that machines cannot make great art, because art is a uniquely human form of expression. But other people suggest that machines might develop new forms of art, and thus open up new dimensions of thinking and feeling. This could enrich our lives in ways we have never imagined. There's only one way to find out if such unknown pathways of art exist: Search for them! See also ARTIFICIAL INTELLIGENCE, COMPUTER, COMPUTER CONSCIOUSNESS, COMPUTER EMOTION, COMPUTER MUSIC, and COMPUTER REASONING.

words per minute

Data speed is sometimes given in words per minute (WPM). A *word* consists of five characters plus one space, or six characters. Generally, then, the number of words per minute is about ⅙ the number of characters per minute.

Units of WPM aren't often used nowadays. The standard specification is the baud rate. See BAUD RATE.

work envelope

The *work envelope* is the range of motion over which a robot arm can move. Stated precisely, it is the set of points in space that the end effector can reach.

The size and shape of the work envelope depends on the coordinate geometry of the robot arm, and also on the number of degrees of freedom. Some work envelopes are quite flat, confined almost entirely to one horizontal plane. Others are cylindrical; still others are spherical. Some work envelopes have very complicated shapes.

The drawing shows a simple example for a robot arm using cylindrical coordinate geometry. The set of points that the end effector can reach lies within two concentric cylinders, labeled "inner limit" and "outer limit." The work envelope for this robot arm is shaped something like a brand new roll of duct tape.

When choosing a robot arm for a certain industrial purpose, it is important that the work envelope be large enough to encompass all the points that the robot arm will need to reach. But it's wasteful to use a robot arm with a work envelope much bigger than necessary. See ARTICULATED GEOMETRY, CARTESIAN COORDINATE GEOMETRY, CYLINDRICAL COORDINATE GEOMETRY, DEGREES OF FREEDOM, DEGREES OF ROTATION, INDUSTRIAL ROBOTS, POLAR COORDINATE GEOMETRY, REVOLUTE GEOMETRY, ROBOT ARMS, and SPHERICAL COORDINATE GEOMETRY.

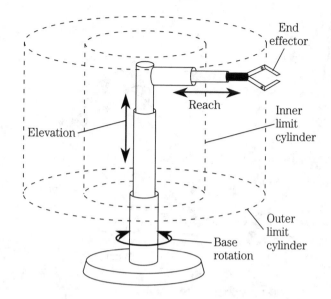

WORK ENVELOPE The set of points a robot arm can reach.

wrist-force sensors

There are several different forces at the point where a robot arm joins the end effector. This point is called the *wrist*. It has one or more joints that move in various ways.

A *wrist-force sensor* consists of transducers known as strain gauges. A strain gauge is a special type of pressure sensor. The strain gauges convert the wrist forces into electric signals, which go to the robot controller. Thus the machine "knows" what's happening at the wrist.

Wrist force is complex. It takes six dimensions to represent all the possibilities. Have you ever tried to imagine a space of more than three dimensions? It's impossible when all the dimensions are linear (straight lines). But when torque vectors, as well as linear vectors, are included, you can imagine up to six dimensions.

The drawing shows a hypothetical robot wrist, and the six forces that can occur there. Three of the forces are linear, and can be represented by straight-line vectors. The orientations are right/left (x axis), in/out (y axis), and up/down (z axis). Rotation is possible along all three axes. These forces are called *pitch*, *roll*, and *yaw* for twisting along the x, y, and z axes respectively.

A wrist-force sensor must detect, and translate, each of the forces independently. A change in one vector must cause a change in sensor output for that force, and no others. See also BACK PRESSURE SENSOR, PITCH, PRESSURE SENSING, ROLL, TRANSDUCER, X AXIS, YAW, Y AXIS, and Z AXIS.

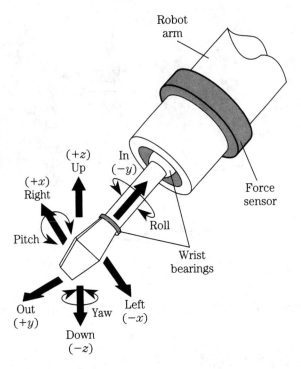

WRIST-FORCE SENSORS
Six possible motions, with six independent forces.

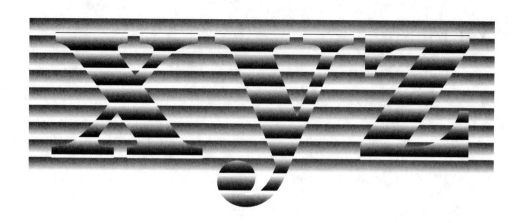

x axis

The term *x axis* has various meanings in mathematics, computer science, and robotics.

In a Cartesian plane or 2-space graph, the *x* axis is generally the horizontal axis (see illustration A). The independent variable is represented by this axis; it is sometimes also called the *abscissa*. In Cartesian 3-space, the *x* axis is one of the two independent variables, the other usually being represented by *y* (see illustration B).

In a robot arm such as the Programmable Universal Machine for Assembly (PUMA), the *x* axis usually refers to the line of motion running to the right and left. It is one of three coordinates specified in the PUMA programming language called VAL.

In wrist-force sensors, the *x* axis refers to linear forces from the right or the left. See also CARTESIAN COORDINATE GEOMETRY, FUNCTION, PROGRAMMABLE UNIVERSAL MACHINE FOR ASSEMBLY, VAL, WRIST-FORCE SENSORS, Y AXIS, and Z AXIS.

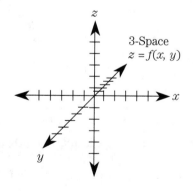

X AXIS Variables in Cartesian coordinate geometry.

XR robots

The *XR robots* are a series of educational robots that were conceived, designed, and built by a company called Rhino Robots. The XR robots are actually just robot arms. The main purpose of these robots is to demonstrate how robots work, and that there is no miracle involved in their functioning.

The XR robots were introduced during the 1980s, and sold for less than $3,000 each. They did various tasks with high precision. The XR robots used a programming device similar to a teach box. For tasks involving numerous steps to be carried out in specific order, a personal computer (PC) could be used as the robot controller.

The XR robots proved useful as teaching aids in corporations and schools. Many people get uneasy around robots, especially the programmable type. The XR robots helped get rid of the fears people sometimes have about robots. See also EDUCATIONAL ROBOTS, ROBOT ARMS, and TEACH BOX.

yaw

Yaw is one of three types of motion that a robotic end effector can make.

Extend your arm out straight, and point at something with your index finger. Then move your wrist so that your index finger points back and forth (to the left and right) in a horizontal plane. This motion is yaw in your wrist. The drawing shows a robotic end effector making this motion. See also END EFFECTOR, PITCH, and ROLL.

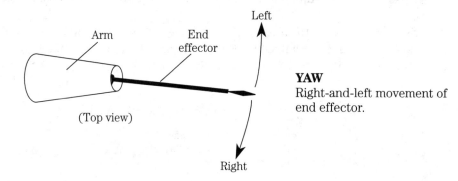

YAW
Right-and-left movement of end effector.

y axis

The term *y axis* has various meanings in mathematics, computer science, and robotics.

In a Cartesian plane, or 2-space graph, the y axis is usually the vertical axis, as shown in drawing A. The dependent variable is represented by this axis; it is sometimes also called the ordinate. In a mathematical function *f* of an independent variable x, you will often specify $y = f(x)$. The function maps the x values into the y values.

In Cartesian 3-space, the y axis is one of the two independent variables, the other usually being represented by x (see illustration B).

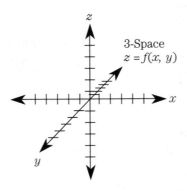

Y AXIS Variables in Cartesian coordinate geometry.

In a robot arm such as the Programmable Universal Machine for Assembly (PUMA), the y axis usually refers to the line of motion running forward and back. It is one of three coordinates specified in the PUMA programming language called VAL.

In wrist-force sensors, the y axis refers to inward/outward linear force vectors. See also CARTESIAN COORDINATE GEOMETRY, FUNCTION, PROGRAMMABLE UNIVERSAL MACHINE FOR ASSEMBLY, VAL, WRIST-FORCE SENSORS, X AXIS, and Z AXIS.

Yonemoto, Kanji

Kanji Yonemoto is a Japanese roboticist who has served as head of the Japan Industrial Robot Association (JIRA).

According to Yonemoto, the term *robot* is hard to define. So, rather than trying to make up an all-encompassing definition, the JIRA has classified robots. There are at least eight categories. They range from manipulators to complete systems incorporating artificial intelligence (AI).

Starting with the simplest devices, and progressing to the most complicated, Yonemoto's categories of industrial robots are as follows:

- Manually operated tools and manipulators: Machines that must be directly operated by a human, such as steam shovels or toilet-bowl "snakes."

- Machines that do things in order: Devices that perform a series of tasks in the same sequence every time they are actuated. A good example is a telephone answering machine.

- Programmable manipulators: This includes the simpler types of industrial robots familiar to most people nowadays. See INDUSTRIAL ROBOTS.
- Numerically controlled robots: Examples are servo robots. See NUMERICAL CONTROL, SERVOMECHANISM, SERVO ROBOTS, and SERVO SYSTEM.
- Sensate robots: Robots incorporating sensors of any type, such as back-pressure, proximity, or wrist-force. See TRANSDUCER.
- Adaptive robots: Robots that adjust the way they work to compensate for changes in their environment. See, for example, PERSONAL ROBOTS, POLICE ROBOTS, and SECURITY ROBOTS.
- Smart robots: Robots with AI, like *R2D2* in the movie *Star Wars*. See ARTIFICIAL INTELLIGENCE, and R2D2.
- Smart Mechatronic Systems: Computers that control a fleet of robots or robotic devices, such as *HAL* in the movie *2001: A Space Odyssey*. See CLARKE, ARTHUR C.

To these eight categories, we might add a ninth, ultimate classification. Right now it's pure fantasy. But someday, people might think nothing of it.

- Smart-robot Civilization: A self-sustaining, evolving subculture, comprised of robots! See CYBOT SOCIETY, and ROBOTIC REPRODUCTION AND EVOLUTION.

z axis

The term *z axis* has various meanings in mathematics, computer science, and robotics.

In Cartesian space, the z axis is the dependent variable, and is a function of x and y, the two independent variables. The z axis runs vertically, while the (x, y) plane is horizontal, as shown in the illustration. A function f maps values x and y into values z, such that $z = f(x, y)$.

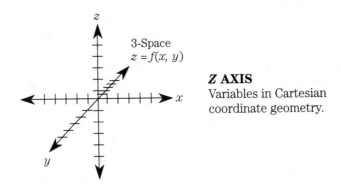

3-Space
$z = f(x, y)$

Z AXIS
Variables in Cartesian coordinate geometry.

In a robot arm such as the Programmable Universal Machine for Assembly (PUMA), the z axis usually refers to the line of motion running up and down. It is one of three coordinates specified in the PUMA programming language called VAL.

In wrist-force sensors, the z axis refers to up/down (vertical) linear force vectors. See also CARTESIAN COORDINATE GEOMETRY, FUNCTION, PROGRAMMABLE UNIVERSAL MACHINE FOR ASSEMBLY, VAL, WRIST-FORCE SENSORS, X AXIS, and Y AXIS.

zooming

In a graphical user interface, the term *zooming* refers to magnification of the image. If you want to look at a certain part of the screen in more detail, you can "zoom in" on it.

Computer graphics can bring vivid and fascinating images to any personal computer (PC) user. One type of image, containing theoretically infinite detail and complexity, is called a *fractal*. Fractals have the property that, no matter how closely you "look" at them, they always appear complicated.

The illustrations show a hypothetical computer-generated fractal "shoreline." The lowest magnification is at A. Zooming in on a specific part of graphic A, more detail is revealed (B). This is repeated at C. In a true fractal, the zooming can be done over and over indefinitely, and there is always new detail in the image. An excellent book and diskette, that will let you look at fractals on your own PC, is called *Fractal Mania* by Philip Laplante, Ph.D. (TAB/McGraw-Hill, 1993).

How can an infinitely complex mathematical "object" exist in a computer with a finite amount of memory? It turns out to be so elementary that, at first, it defies belief. As computer scientist Benoit Mandelbrot demonstrated, fractals are not extraordinary, but commonplace. They are often found in nature. Fractals can arise from, and be fully defined by, simple algebraic formulas. This has led some scientists to suppose that fractals reflect the structure of reality itself: that they are, in essence, windows through which we humans can get glimpses of the mind of the Creator!

Could the Cosmos, in every detail and dimension (including time), be reducible to an equation you can type into a PC? If so, the PC could, via zooming, mathematically duplicate every galaxy, every star system, every world, ocean, continent, town, and house in the Universe. It could depict every person who has ever inhabited the Earth. It would, if you zoomed in on the right part of the polydmensional fractal, show you right now, as you read this page.

Is this an endpoint of artificial intelligence (AI)? Have we come full circle? Is life but an AI dream? Can any PC be an all-knowing, all-seeing Universal Truth Machine (UTM), a direct link to the Ultimate Transcendental Mind? Or is this whole discussion nothing but Utterly Trivial Malarkey?

Maybe we shouldn't worry about what "smart machines" actually are, as long as we get some fun from what they do. See also ARTIFICIAL INTELLIGENCE, COMPUTER GRAPHICS, GRAPHICAL USER INTERFACE, and UNIVERSAL TRUTH MACHINE.

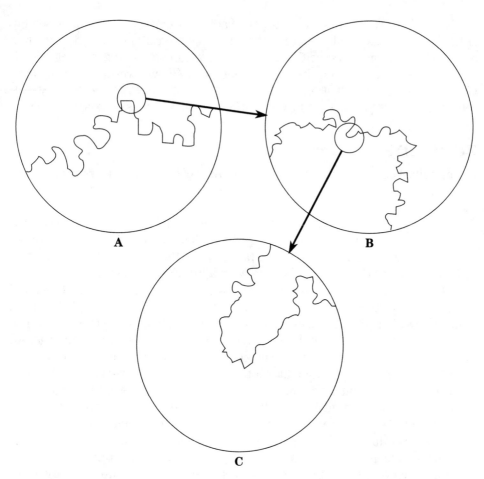

ZOOMING At A, B, and C, progressively magnified views of a fractal "shoreline."

Bibliography

Andriole, Stephen J. and Hopple, Gerald W. *Applied Artificial Intelligence:* A Sourcebook. McGraw-Hill, 1992.

Asimov, Isaac, *Robots: Machines in Man's Image*. Harmony Books, 1985.

Engelberger, Joseph, *Robotics in Service*. MIT Press, 1989.

Fjermedal, Grant, *The Tomorrow Makers*. Macmillan, 1986.

Hofstadter, Douglas R., *Godel, Escher, Bach*. Vintage Books, 1979.

Hunt, Daniel V., *Robotics Sourcebook*. Elsevier, 1988.

McComb, Gordon and Cook, John, *Robot Builder's Bonanza*. TAB Books, 1987.

Schodt, Fredrik L., *Inside the Robot Kingdom*. Kodansha International, 1988.

Shircliff, David R., *Build a Remote-controlled Robot for Under $300*. TAB Books, 1986.

Index

A

A/D conversion, 340-342
active chord mechanism (ASM), 1, **1**, 314
adaptive suspension vehicle (ASV), 2, 3, 241, 366
address bus, 70
Advanced Research Projects Agency (ARPA), 12
advanced robot technology (ART), 2-3, 25
Agre, Phil, 181
AGV (*see* automated guided vehicle)
AIMS (*see* automated integrated manufacturing system)
algorithm, 3-4
Allen, Paul, 239
alpha, 4, **4**
American National Standards Institute (ANSI), 5, 67
amusement robots, 5, **6**, 8, 9, 28, 48, 58, 66, 188, 274, 361
analog, 6, **6**, 41, 55, 71, 92, 106, 115, 138, 176, 254, 309, 342, 346
analog computer, 71
analytical engine, 7, 9, 27
AND gate, 7, **8**, 38, 43, 142, 204, 210, 223, 249, 266
androbot, 8, 48
android, 5, 8-9, 13, 24, 25, 39, 44, 48, 74, 77, 88, 91, 141,

162, 187, 188, 189, 211, 241, 374
animism, 9-10, 16, 73, 77, 88, 141, 147, 211, 374
anthropomorphism, 272, 274
anticipatory sciences, 10-11, 66
antivirus program, 11
anthropomorphism, 10
apprentice robot, 11, 274
arithmetic logic unit (ALU), 53
artificial intelligence, 25
armature coil, 245
Armstrong, Neil, 317
ARPAnet, 12, 47
arrays, 234
ART (*see* advanced robot technology)
articulated geometry, 12-13, **12**, 266, 311, 400
artificial experience (*see also* virtual reality)
  computers and, 72-73
  future, 15
  games, 14
  music, 14
  robotics relationship, 13
  theorems, 14-15
artificial intelligence, 10, 13-16, **15**, 31, 32, 56, 87, 106, 206, 207, 212, 214, 216, 218, 219, 220, 223, 227, 229, 232, 237, 239, 248, 250, 252, 267, 269, 272, 274, 277, 287, 288, 297, 305, 306, 315, 321, 323, 327,

328, 329, 333, 337, 342, 346, 354, 368, 369, 372, 374, 393, 394, 398, 399, 406, 407
  vision systems and, 386-387
  word processing and, 399-400
artificial life, 4, 10, 11, 15, 16, 24, 25, 38, 39, 48, 51, 58, 60, 66, 73, 77, 81, 88, 88, 91, 115, 130, 141, 170, 188, 206, 218, 233, 306, 312, 315, 322, 398
artificial reality (*see* virtual reality)
artificial stimulus, 17, **17**, 32, 108, 111, 292
Asimov, Isaac, 17, 165
Asimov's Three Laws of Robotics, 17, 18, 163, 274
assembler program, 18, **18**, 19, 187, 287
assembly language, 18, 19, 81, 187, 215, 287
assembly line, 19-20, **19**, 23, 47, 51, 195, 288, 378
assembly robots, 20-21, 22, 23, 47, 104, 121, 155, 156, 195, 288, 378
atoms, 321
audiotape, 96
automated guided vehicle (AGV), 17, 21-22, **21**, 132, 167, 177, 195
automated home, 22, 274
automated integrated manufacturing system

411

Index

Index

Index

(PUMA), 288, 355, 376, 378, 381, 403, 405, 407
programmed article transfer machine (*see* numerical control)
programming (*see* computer programming)
programming languages (*see* assembly language; BASIC; C; COBOL; FORTH; Fortran; high-level language; LISP; low-level language; machine language; Pascal; PLANNER; PROLOG; SmallTalk; VAL)
PROLOG, 220, 288
proprioceptor, 288-289, 314, 353, 359
proprioceptor proximity sensing, 363
prosodic features, 289, 342, 346, 352
prosthesis, 88, 211, 289-290, 314, 353, 359
prototype, 213
proximity sensing, 50, 111, 112, 177, 177, 276, 280, 285, 290-292, **291**, 312, 359
PUMA (*see* programmable universal machine for assembly)
pushdown stack, 154, 292, **293**, 305, 330

Q

quadruped robots, 40, 295-296, **295**, 318, 396
Quain, Mitchell, 296
quality assurance and control, 48, 195, 232, 296-298, **297**, 308, 378
QWERTY, 211, 298

R

R2D2, 318-319, 406
radar, 292, 299, **299**, 329, 388
  Doppler, 299
radio-frequency interference (RFI), 300
RADIX (*see* Modulo)
random-access memory (RAM), 96, 234-235, 330

range of function, 113, 301, **301**
range plotting, 301-302, **302**, 337, 363, 388
range sensing, 301-302, **302**, 337, 363, 388
read-only memory (ROM), 96, 235, 330
real time, 303
rectangular coordinate geometry (*see* Cartesian coordinate geometry)
rectifiers
  full-wave, 282
  half-wave, 282
recursion, 195, 303-305, **304**
reduced instruction set computer (RISC), 305-306
reductionism, 306
redundancy, 306-307
reinitialization, 307
reliability, 307-308, **307**
remote control, 350, 357, 359
remote control systems, 308-309, 388
remote manipulator, 230, 309
remotely operated vehicle (ROV), 309, 350
resolution, 93, 176, 277, 309, **309**, 337, 387, 388 (*see also* direction resolution; distance resolution)
reverse engineering, 310
revolute geometry, 310-311, **310**, 311, 400
RFI (*see* radio-frequency interference)
RISC (*see* reduced instruction set computer)
robot arms, 4, 12, 20, 28, 39, 53, 89, 99, 100, 134, 195, 230, 266, 278, 311, 324, 347, 350, 355, 400, 404
robot assemblers, 195 (*automated integrated manufacturing system*)
robot generations, 274, 311-313, **313**, 332
robot grippers, 1, 4, 28, 39, 134, 146, 175, 195, 230, 311, 313-314, 369
robot hearing (*see* binaural robot hearing; sound transducer)
robot legs, 2, 3, 39, 40, 317-318, 366, 396

robot vision (*see* vision systems)
robotic evolution, 312, 314-315, **315**, 393, 406
robotic reproduction, 312, 314-315, **315**, 393, 406
robotics, 24, 165
robotic ship, 315-316
robotic space missions, 260, 309, 316-317, 359, 383, 388
robotization, 195
robots, computers and, 73-74 (*see also* specific types of)
roll, 276, 314, 318, **318**, 401
ROV (*see* remotely operated vehicle)
rule-based systems (*see* expert systems)

S

sampling rate, 93
Samuel, Arthur, 55, 81, 195, 321-322
satellite data transmission, 127, 322-323, **323**
scaling, 323, **324**
SCARA, 108, 324, **324**, 378
SCARA robots, 20
security robots, 241, 274, 280, 309, 325-326, 329, 387, 406
seeing-eye robot, 326, 388
selsyn, 246, 309, 326-327, **326**, 351
semantic network, 327, **327**
semiconductor memory, 329
semiconductors, 328, **328** (*see also* complementary metal-oxide semiconductor; metal oxide semiconducer; N-channel metal-oxide semiconductor; P-channel metal-oxide semiconductor)
sensitivity, 387
sensors (*see* back pressure sensor; joint-force sensor; photoelectric proximity sensor; position sensor; tactile sensor; wrist-force sensor)
sentry robots, 241, 280, 326, 329, 334
sequential access memory, 330
serial, 95, 182, 270, 330, **330**
servo robots, 126, 128, 139, 140,

Index

Unimation Inc., 11, 375-376, **376**, 378, 382
universal truth machine (UTM), 376-377, 396, 407
Univision, 377-378, **378**, 388
unmanned factory, 195, 378-379
user-friendliness (*see* human engineering)

V

VAL, 381-382, **381**, 403, 405, 407
Vaucanson, J. de, 23
Versatran, 382
very large-scale integration (VLSI), 216, 234, 382
videocassette recorder (VCR), 382-383
videotape, 96
vidicon, 68, 382-383, **383**, 388
virtual reality, 316, 363, 383-384, 388
virus, 11, 46, 47, 53, 92, 241, 366, 384-386
virus protection, 384-385
vision systems, 37, 42, 58, 61, 69, 78, 80, 85, 94, 109, 112, 122, 138, 153, 175, 177, 177, 187, 192, 221, 222, 225, 258, 259, 265, 270, 274, 277, 280, 292, 306, 309, 312, 329, 359, 363, 378, 386-388, **386**, 392
artificial intelligence and, 386-387
components, 386
invisible and passive, 388
resolution, 387
sensitivity, 387
VLSI (*see* very large-scale integration)
voice
definition, 343-344
tone of, 344
voice recognition (*see* speech recognition)
voice synthesis (*see* speech synthesis)
voltage regulation, 282

W

warm boot (*see* boot)
Wasubot, 391-392, **392**
Weiner, Norbert, 392
Weizenbaum, Joseph, 131, 197, 306, 393, 398
well-structured language, 394-395, 399
wheel-drive locomotion, 366, 395-396, **395**
whiskers (*see* proximity sensing)

wide area network (WAN), 221, 242, 252, 264, 267, 323, 396
Williams, George, 396-398
Windows, 64, 110, 174, 239, 394, 398-399, **398**
wire-tracing navigation (*see* automated guided vehicle)
word processing, 399-400
words per minute (WPM), 29, 31, 400
work envelope, 99, 100, 311, 354, 400, **401**
wrist-force sensor, 314, 353, 363, 401, **402**, 403, 405, 407

X

x axis, 401, 403, **403**, 405, 407
XR robots, 404

Y

y axis, 401, 403, 404-405, **405**, 407
yaw, 276, 314, 401, 404, **404**
Yonemoto, Kanji, 405-406

Z

z axis, 401, 403, 405, 406-407, **406**
zooming, 407, **408**

About the author

Stan Gibilisco was born in 1953 and is the son of Dr. Joseph A. Gibilisco, who served for more than 30 years as a staff physician at the Mayo Clinic in Rochester, Minnesota. A mathematician educated at the University of Minnesota, Stan has been a radio ham since 1966 and has held the Extra Class license since 1973. His call sign, W1GV, is known to many hams from dozens of technical articles in amateur-radio magazines.

Between 1977 and 1982, Stan served as Assistant Technical Editor for *QST* Magazine of the American Radio Relay League, Inc., in Newington, Connecticut, and as Vice President of Engineering for International Electronic Systems, Inc., in Miami, Florida. In 1982, Stan began writing about science and electronics full-time.

Stan first attracted attention with his book *Understanding Einstein's Theories of Relativity* (TAB Books, 1983) and as Editor-in-Chief of *Encyclopedia of Electronics* (TAB Professional and Reference Books, 1985). The *Encyclopedia* was annotated by the American Library Association as one of the Best Reference Books of the 1980s. To date, Stan has authored and co-authored more than 25 books. His books about astronomy and mathematical subjects enjoy a growing audience in Japan.

Stan plans to continue writing books and articles about science and electronics. In addition, he aspires to write science books for children, as well as fiction stories and novels.